Narrative of the Canadian Red River Exploring Expedition of 1857

And of the Assinniboine and Saskatchewan Exploring Expedition of 1858

VOLUME 1

HENRY YOULE HIND

CAMBRIDGE
UNIVERSITY PRESS

CAMBRIDGE
UNIVERSITY PRESS

University Printing House, Cambridge, CB2 8BS, United Kingdom

Published in the United States of America by Cambridge University Press, New York

Cambridge University Press is part of the University of Cambridge.

It furthers the University's mission by disseminating knowledge in the pursuit of education, learning and research at the highest international levels of excellence.

www.cambridge.org
Information on this title: www.cambridge.org/9781108070881

© in this compilation Cambridge University Press 2014

This edition first published 1860
This digitally printed version 2014

ISBN 978-1-108-07088-1 Paperback

This book reproduces the text of the original edition. The content and language reflect the beliefs, practices and terminology of their time, and have not been updated.

Cambridge University Press wishes to make clear that the book, unless originally published by Cambridge, is not being republished by, in association or collaboration with, or with the endorsement or approval of, the original publisher or its successors in title.

The original edition of this book contains a number of colour plates, which have been reproduced in black and white. Colour versions of these images can be found online at www.cambridge.org/9781108070881

CAMBRIDGE LIBRARY COLLECTION

Books of enduring scholarly value

North American History

This series includes accounts of historical events and movements by eye-witnesses and contemporaries, as well as landmark studies that assembled significant source materials or developed new historiographical methods. The works range from the writings of early U.S. Presidents to journals of poor European settlers, from travellers' descriptions of bustling cities and vast landscapes to critiques of racial inequality and descriptions of Native American culture under threat of annihilation. The commercial, political and social aspirations and rivalries of the 'new world' are reflected in these fascinating eighteenth- and nineteenth-century publications.

Narrative of the Canadian Red River Exploring Expedition of 1857

Born in Nottingham, Henry Youle Hind (1823–1908) moved to Canada in 1846. He joined the newly formed Canadian Institute in 1849 and later taught chemistry and geology at Trinity College in Toronto. In 1857–8, he made a range of observations during two expeditions to investigate underexplored areas of Canada and their agricultural and mineral potential to support future settlement. Illustrated with a number of plates based on photographs, this two-volume work first appeared in 1860. Intended for a broad readership, the narrative is regarded as a classic of nineteenth-century exploration literature, noted especially for its descriptive use of language and eye for detail. Volume 1 covers the entire Red River expedition of 1857 and the first part of the 1858 expedition through parts of the Assiniboine, Saskatchewan and other valleys.

Cambridge University Press has long been a pioneer in the reissuing of out-of-print titles from its own backlist, producing digital reprints of books that are still sought after by scholars and students but could not be reprinted economically using traditional technology. The Cambridge Library Collection extends this activity to a wider range of books which are still of importance to researchers and professionals, either for the source material they contain, or as landmarks in the history of their academic discipline.

Drawing from the world-renowned collections in the Cambridge University Library and other partner libraries, and guided by the advice of experts in each subject area, Cambridge University Press is using state-of-the-art scanning machines in its own Printing House to capture the content of each book selected for inclusion. The files are processed to give a consistently clear, crisp image, and the books finished to the high quality standard for which the Press is recognised around the world. The latest print-on-demand technology ensures that the books will remain available indefinitely, and that orders for single or multiple copies can quickly be supplied.

The Cambridge Library Collection brings back to life books of enduring scholarly value (including out-of-copyright works originally issued by other publishers) across a wide range of disciplines in the humanities and social sciences and in science and technology.

CANADIAN RED RIVER

AND

ASSINNIBOINE AND SASKATCHEWAN

EXPEDITIONS

VOL. I.

LONDON

PRINTED BY SPOTTISWOODE AND CO.

NEW-STREET SQUARE

KA-KA-BEKA FALLS, KAMINISTIQUIA RIVER.

Printed by Spottiswoode and Co.]

[New street Square, London.

NARRATIVE

OF

THE CANADIAN RED RIVER

EXPLORING EXPEDITION OF 1857

AND OF THE

ASSINNIBOINE AND SASKATCHEWAN

EXPLORING EXPEDITION OF 1858

BY

HENRY YOULE HIND, M.A. F.R.G.S.

PROFESSOR OF CHEMISTRY AND GEOLOGY IN THE UNIVERSITY OF TRINITY COLLEGE, TORONTO

In Charge of the Assinniboine and Saskatchewan Expedition

In Two Volumes

VOL. I.

LONDON

LONGMAN, GREEN, LONGMAN, AND ROBERTS

1860

PREFACE.

THE objects for which the Explorations described in these volumes were undertaken, necessarily involved a more minute topographical examination than would be thought necessary in a general survey of a comparatively unknown country.

It was desirable to ascertain the practicability of establishing an emigrant route between Lake Superior and Selkirk Settlement, and to acquire some knowledge of the natural capabilities and resources of the Valley of Red River and the Saskatchewan.

The country between Lake Superior and Red River is therefore minutely delineated with reference to the object of the exploration of 1857, and the first four chapters are mainly devoted to topographical details of less interest to the general reader than the subsequent narrative. The same remark applies, though in a less degree, to the description of the country west of Red River, the object being to show its fitness, or the contrary, for settlement.

The establishment of a new Colony in the Basin of Lake Winnipeg, and the discovery of a FERTILE BELT of

A 3

country extending from the Lake of the Woods to the
Rocky Mountains, give to this part of British America a
more than passing interest. The idea of a route across
the Continent of America lying wholly within British
Territory, is daily becoming more settled and defined.

The trade of China and Japan, now on the point of
being opened to British enterprise, the gold wealth of
British Columbia, and the FERTILE BELT forming the
northern boundary of the great American desert, all give
importance to the Basin of Lake Winnipeg, which increases with our contemplation of its possible and indeed
probable future.

The illimitable wastes of Siberia, extending over eighty
degrees of longitude, are traversed by Russian couriers
in far less time than with all our appliances of steam and
telegraph, we can receive "news" from China. The same
postal system which there prevails can be far more easily
maintained in British America, and with this vast advantage, that from the Lake of the Woods to the Rocky Mountains the route would lie through a tract of country not
only remarkably fertile, but possessing rich stores of
timber for fuel, lignite coal, iron, and salt—the most
important elements of industry and wealth.

The chief difficulty in the way of rapid transit across
the continent lies between Lake Superior and Rainy Lake.
The liberality which has already been manifested by the
Parliament of Canada, in voting supplies to explore and
open this line of communication, will doubtless be perse-

vered in until the route is well established. The Governor of British Columbia sees in "means of communication" the most expeditious way of calling the inert gold wealth of that distant colony into activity, and it remains for the Imperial Government to determine how soon a postal communication shall be established across the Basin of Lake Winnipeg, and the first step taken in establishing a permanent route through British Territory, between the Atlantic and the Pacific.

London, October 1860.

CONTENTS

OF

THE FIRST VOLUME.

THE

CANADIAN RED RIVER EXPLORING EXPEDITION

OF 1857.

CHAPTER I.

TORONTO TO FORT WILLIAM. — LAKE SUPERIOR.

CHAP. II.

THE KAMINISTIQUIA ROUTE.—FORT WILLIAM.—LAKE SUPERIOR TO THE HEIGHT OF LAND.

CHAP. III.

THE HEIGHT OF LAND TO RAINY LAKE.

CHAP. IV.

RAINY LAKE TO THE SOURCE OF THE WINNIPEG RIVER.

CHAP. V.

THE WINNIPEG RIVER.

CHAP. VI.

RED RIVER SETTLEMENTS.

CHAP. VII.

THE WEST AND EAST BANKS OF RED RIVER, BETWEEN FORT GARRY AND THE BOUNDARY LINE.

CHAP. VIII.

BRIEF HISTORY OF THE COLONY.—STATISTICS OF POPULATION.—ADMINISTRATION OF JUSTICE.—TRADE AND OCCUPATIONS.

CHAP. IX.

THE MISSIONS AT RED RIVER.

CHAP. X.

EDUCATION IN THE SETTLEMENT.—AGRICULTURAL INDUSTRY.

CHAP. XI.

SKETCH OF THE COUNTRY WEST OF RED RIVER.

CHAP. XII.

THE JOURNEY TO CANADA VIA ST. PAUL.

THE

ASSINNIBOINE AND SASKATCHEWAN EXPLORING EXPEDITION

OF 1858.

CHAP. XIII.

FORT GARRY TO THE BOUNDARY LINE, VIA THE ASSINNIBOINE AND LITTLE SOURIS.

Members of the Expedition. — Iroquois Indians from Caughnawaga. — Detroit.—Sault Ste. Marie.—Grand Portage.—Fort Frances.—Red River.— Expedition into the Interior.—The Start.—Supplies.—Prairie Ridges.— Pigeon Traps. — Stony Mountain. —-Birds. — Saline Efflorescence. — Character of the Big Ridge.—The Assinniboine. — Grasshoppers. — Ojibway Encampment.—Archdeacon Cochrane.—Prairie Portage.—Cliff Swallow. —Thunder Storms.—Ojibways.—The Bad Woods.—Assinniboine Forest. —River.—Rabbits.—Sandy Hills of the Assinniboine.—Latitude.—Dimensions of Valley. — Variation of Compass. — Sand Dunes. — Aspect of Country. — Hail Storm. — "Smokes."— Balsam Spruce. — Pine Creek.— The Little Souris.—Grasshoppers.—Fish.—Sioux.—Cretaceous Rocks.— Blue Hills. — Pembina River. — Backfat Lakes.—Vast Prairie.—Prairie Fires.—Horizontal Rocks.—Inoceramus.—Guelder Rose.—Lignite.—An-

CHAP. XIV.

FROM THE BOUNDARY LINE TO THE QU'APPELLE LAKES VIÂ FORT ELLICE.

CHAP. XV.

THE QU'APPELLE VALLEY.— FROM THE MISSION TO SAND HILL LAKE.

CHAP. XVI.

SAND HILL LAKE TO THE SOUTH BRANCH OF THE SASKATCHEWAN.

CHAP. XVII.

FROM THE QU'APPELLE MISSION TO FORT ELLICE, DOWN THE QU'APPELLE RIVER.

CHAP. XVIII.

FROM THE ELBOW OF THE SOUTH BRANCH OF THE SASKATCHEWAN TO THE NEPOWEWIN MISSION ON THE MAIN SASKATCHEWAN.

CHAP. XIX.

FROM THE NEPOWEWIN MISSION ACROSS THE COUNTRY TO FORT ELLICE.

CHAP. XX.

THE QU'APPELLE VALLEY.—PORT PELLY TO THE SETTLEMENTS ON RED RIVER.

CHAP. XXI.

FROM FORT À LA CORNE, DOWN THE SASKATCHEWAN, TO THE GRAND RAPID AND LAKE WINNIPEG.

CHAP. XXII.

FROM THE GRAND RAPID OF THE SASKATCHEWAN TO THE RED RIVER SETTLEMENTS, VIÂ THE WEST COAST OF LAKE WINNIPEG.

LIST OF ILLUSTRATIONS

IN

THE FIRST VOLUME.

*** *The following Illustrations are from Photographs taken by Mr.* HUMPHREY
LLOYD HIME, *Photographer to the Assinniboine and Saskatchewan Expedi-*
tion, or from Sketches by Mr. JOHN FLEMING, *Assistant Surveyor and*
Draughtsman.

CHROMOXYLOGRAPHS.

WOODCUTS.

MAPS AND PLANS.

THE

CANADIAN RED RIVER EXPLORING EXPEDITION

OF 1857

CANADIAN RED RIVER EXPLORING EXPEDITION

OF 1857.

INTRODUCTION.

In July, 1857, the Canadian Government organised and despatched an expedition to examine the country between Lake Superior and the Red River of the North, with a view to determine the best route for opening a communication between that lake and the settlements on Red River. The expedition consisted of the following members : —

GEORGE GLADMAN, Director, and one Assistant.
HENRY YOULE HIND, Geologist, and one Assistant.
W. H. E. NAPIER, Engineer, and five Assistants.
S. J. DAWSON, Surveyor, and four Assistants.
One road superintendent.

The *voyageurs* were composed of twelve Iroquois Indians from Caughnawaga, near Lachine, a Scotchman from the Ottawa, a French Canadian from Collingwood, a French Canadian (Lambert) from Fort William, a half-breed, engaged on board the steamer Collingwood, where he was employed in the capacity of cook, and twelve Ojibway Indians from Fort William, thus making the number of the party forty-four persons in all, when the canoe voyage commenced.

The following extracts from Mr. Gladman's instructions

will exhibit the designs of the Canadian Government in despatching this expedition : —

" The primary object of the expedition is to make a thorough examination of the tract of country between Lake Superior and Red River, by which may be determined the best route for opening a facile communication through British territory, from that lake to the Red River Settlements, and ultimately to the great tracts of cultivable land beyond them. With this view the following suggestions are offered for your guidance, so far as you will find them practicable, and supported by the topography.

" In the first place, after being landed at Fort William, to proceed by the present Hudson's Bay canoe route — by the Kaministiquia River, Dog Lake, Lake of the Thousand Islands, &c., to Lac La Croix, and thence by Rainy Lake, Lake of the Woods, Winnipeg River to Lake Winnipeg, and up the Red River to Fort Garry.

" From Rainy Lake to Lake Winnipeg, the route as at present affords a good navigation for boats of considerable size, with the interruption however of some short portages; but from Rainy Lake eastward to Lake Superior, the route is very much interrupted, and rendered laborious, tedious and expensive, by the great number of portages, some of considerable length, which have to be encountered to avoid the falls and rapids in the ravines and creeks which this route follows.

" For the establishment of a suitable communication for the important objects aimed at, it is believed that the construction of a road throughout, from some point on Lake Superior, probably either at Fort William, or at or near the mouth of the Pigeon River to Rainy Lake, must be undertaken. To ascertain, therefore, at present, by general exploration, what the route for this road should be, whether in the vicinity of the Hudson's Bay route, or by the line of country in which lies the chain of waters from Rainy Lake to the mouth of Pigeon River; this question can obviously be only satisfactorily determined by the difficult portions of both being tested instrumentally, but in either case, as the construction of such road would be a matter of time and much expense, it is considered necessary that the portages, &c.,

of either of the routes above described should be improved, so as to be made more available and facile, and to be auxiliary to the works of the road by facilitating the transport of men, supplies, &c.

"To determine, therefore, the portages to be improved, and the best mode of doing so, and whether the present reaches of canoe or boat navigation may not be further extended by the removal of shoals or the erection of dams, will be points to which you will direct the attention of the engineering and surveying branches of your party.

"From Rainy Lake by Lake of the Woods, and Lake Winnipeg to Fort Garry, as before described, is now comparatively a good water communication, but very circuitous; and should the character of Rat River, which rises at no great distance from the Lake of the Woods, and falls into the Red River above Fort Garry, be found susceptible of its being made a boat channel, a saving probably of 150 miles in length might be effected; or on an exploration of the country through which that river flows, it may be found more desirable to construct a road along it from Red River, and should this be so, the nature of the communication between Red River and Lake Superior eventually would be about 100 miles of road from Red River to the Lake of the Woods, thence about 140 miles of water communication to the eastern end of Rainy Lake, and from that point a continuous road to Lake Superior of from 160 to 200 miles in length." *

My duties in connection with this expedition are explained in the following instructions:—

"Secretary's Office,
"Toronto, 22nd July, 1857.

"SIR,—I have the honour to inform you that his Excellency, the Administrator of the Government, has been pleased to nominate you Geologist and Naturalist to the party which is to leave this city immediately for Fort William, for the purpose, in the first instance, of examining the lines and state of the

* Report on the Exploration of the country between Lake Superior and the Red River Settlement. Printed by order of the Legislative Assembly, 1858.—Instructions and Communications, page 5.

communication thence to Fort Garry, on the Red River. It being indispensable to the satisfactory result of the expedition, as well as to the safety of the party, that one individual should be invested with the general control and management of it, Mr. Gladman has been intrusted with this authority and responsibility, for which he is considered eminently qualified, from his long residence in the territory, his acquaintance with the leading lines of communication, with the trading posts, with the tribes of Indians with whom the party will necessarily come in contact, and with the extent and nature of the supplies which can safely be calculated on as procurable in the country during the course of the expedition. By him, therefore, will be regulated and determined the movements of the party, the routes to be taken and explored, and all matters connected with the provisioning and transport of the party, the hiring and payment of the men, and all other matters of detail whatever comprised in the general conduct of the expedition.

" From the nature of your duties, it may be necessary that you should occasionally separate yourself from the party. In such cases you will state so to Mr. Gladman, who will take care that you are provided with the necessary provisions and means of transport, and with all such necessaries as you may require: and he will arrange with you as to the places and times for your re-uniting yourself with the main body.

" As you will require the services of an Assistant, the appointment of an efficient one is left with you, his remuneration not to exceed 20l. per month. That of the Geologist, Engineer, and Surveyor is fixed at thirty shillings per day each.

" The objects to which your attention is requested are of a general character, comprising a description of the main geological features of the country you traverse, and of whatever pertains to its natural history which you may have an opportunity of observing and recording.

" In relation to its geology, you will be guided by the memorandum furnished you by Sir William Logan; giving especial attention, as far as lies in your power, to the following points:

1. The boundaries of formations.
2. The distribution of limestone.

3. The collection of fossils.

4. The occurrence of economic minerals.

5. The exact position of all facts, and the attitude of the rocks.

" The distribution of limestone should be made a constant subject of question with every one you meet.

" With reference to natural history, you will, if at the time convenient, and the object capable of transportation, collect whatever may appear to be new or of interest; and you are requested to record in a daily journal, such facts in connection with this subject as may present themselves to your notice, when not susceptible of representation by specimen or illustration.

" A general description of the whole of the country you traverse, from Fort William westward, is very desirable; and it is advisable to note, as minutely as possible, all leading features of topography, vegetation, and soil, along your line of route.

" You will proceed with the main party to Fort William, and continue with it, or with such party as may be detached from it, as much as is consistent with the efficient prosecution of your own exploration and researches. It may, of course, be occasionally necessary, as already adverted to, that you should separate from the others for a short time, for which course Mr. Gladman will afford you all requisite accommodation; but as that gentleman's instructions require him to explore not only the present canoe route of the Hudson's Bay Company, from Fort William by Dog Lake, Lake of the Thousand Islands, Lac Croix, Lake of the Woods, and Lake Winnipeg, to Fort Garry, but also in returning to examine the former North West Company's route by Pigeon River; and further to survey or examine the line of Rat River, from the Red River to its source, and the intervening country between it and the Lake of the Woods; it is not probable that there will be much necessity for your leaving the party for more than a few days at a time, which is desirable, from its limited number and the late season of the year.

" It is arranged with Mr. Gladman, that he is to send a messenger, some time hence, with despatches to the Government, explanatory of the progress made towards carrying out the objects of the expedition; and by this means you will also have an opportunity of making such *ad interim* report as you may consider desirable. You will determine the return route to be taken by you and your assistant, whether by Lake Superior or by St. Paul, as you may be led to believe will most conduce to the attainment of the object of your branch of the exploration.

" When materials for illustrating the geology and natural history of the country accumulate, so as to render their transportation an inconvenience, you will hand them over in packages, properly made up and directed, to Mr. Gladman, who will take care that they are safely lodged at some of the posts, and arrangements made for their being securely conveyed to this city.

" Your reports and communications upon the various subjects to which your attention is directed will be addressed to the Hon. Provincial Secretary; and it is presumed to be unnecessary to impress upon you the propriety and expediency of taking care that the subject of such reports, and the results of your labour, shall be only so communicated.

" I have the honour, &c.

(Signed,)   "T. L. TERRILL,

"Provincial Secretary.

" H. Y. Hind, Esq.
      "Professor, &c., Trinity College."

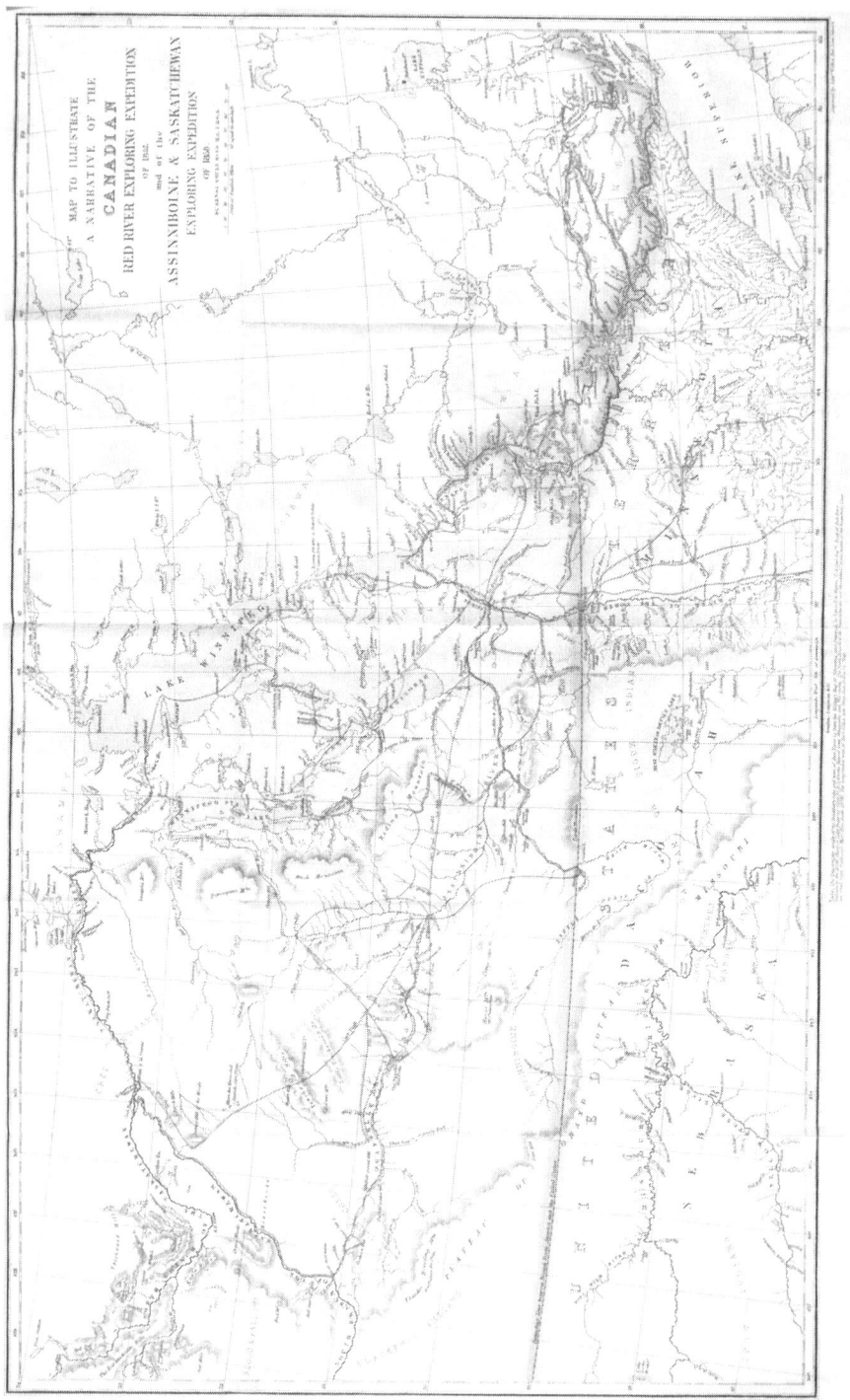

The material originally positioned here is too large for reproduction in this reissue. A PDF can be downloaded from the web address given on page iv of this book, by clicking on 'Resources Available'.

CHAPTER I.

On the 23rd of July, 1857, the Red River Expedition left
Toronto for Collingwood, Lake Huron, and during the
afternoon of the next day embarked on board the steamer
Collingwood, bound to Fort William, Lake Superior. We
passed through the magnificent locks of the Sault Ste.
Marie Canal at 3 P.M. on the 27th, and when entering
Lake Superior were met by an imposing but threatening
spectacle, which instantly arrested and fixed the attention
of all. A huge cloud, dense and black at its base, seemed
to lie with one extremity resting on the Gros Cap de
Superior, the other on Point Iroquois, the distance between those elevations being about six miles. The form
of the cloud was that of a double cone, with the bases
joined together, and the apices resting upon the opposite
heights. A little attention showed that the cloud was in
rapid motion towards us, and as it approached masses

seemed to detach themselves from the main body, and be whirled or driven in its van. For a space of five degrees, the sky beneath the cloud, and between it and the horizon of Lake Superior, was clear and blue, and as the great mass of vapour slowly rose from the lake to the height of about eight degrees, its lower edge became brilliantly tinted by the rays of the sun, which it had hitherto completely obscured; below it a shadow of the deepest purple, sharply bounded by a greenish white line, added extreme beauty and singularity to the spectacle. Its form changed rapidly, and a white line of crested waves beneath it gave warning of an approaching squall, which soon came down with great force, and compelled us to seek shelter in Whisky Bay.

As soon as the morning dawned, preparations were made for steaming out of our harbour of refuge. Fogs, so common in Lake Superior, began to appear about 9 A.M., and continued throughout the day. Fog-bows, several degrees broad, very low, and of little variety of colour, were visible whenever the sun's light succeeded in penetrating our misty screen. On looking over the side of the vessel a double halo of very brilliant colours might be seen encircling the shadow of the observer's head projected on the dark coloured waters. Every man saw "his own halo," but not that of his neighbour. Towards evening a sharp look-out was kept for land on either hand. Caribou Island was supposed to be lying to the south, and Michipicoten Island to the north of our course. The day was very cold; thermometer at 3 P.M., 42°; at 4, 40°; and at 4.30, 39°.5, which was the lowest point it reached. The waters of the lake showed also a temperature of 39°.5. Mr. Wilson, a fellow-passenger, who has resided two summers on Michipicoten Island, says that the Lake Superior summer fogs begin about 9 A.M., and disappear generally at 10 or 11 P.M., but sometimes they last for a week.   They are low, and from the mountain on Michipicoten Island, at an elevation of 800 feet above the lake, they may be seen resting on its waters as far as the eye can reach.

In consequence of the variation of the compass being reported to be much affected by local attractions in this part of Lake Superior, two of our Indians were placed in the bow to look out for land at the approach of night, and in addition to the usual watch, the captain, mate, and some of the passengers were walking the deck until past 11 o'clock.   An evident feeling of anxiety was common to both passengers and crew; several of the former went to their berths without taking off their clothes.   The night was extremely foggy; it was impossible to see more than a few yards beyond the bow of the vessel.   The lead was cast several times, with no bottom at 288 feet.   At a quarter to 12 P.M. no soundings were obtained with twenty fathoms; a few minutes afterwards the lead showed forty-five feet of water; the signal was given to stop her, and then to " back water," but it was too late, a harsh grating noise, a sudden uplifting of the bow of the steamer, and a very decided shock quivering through the vessel, told that she had struck.   The alarm and anxiety inseparable from such an incident followed, and it was several minutes before a rapid inspection by torch light of the ledge of rock on which we had struck showed that there was no immediate danger to be apprehended. Anchors, chains, and fuel were moved aft, but all efforts to get the vessel off were without the least effect.   Her bow was five feet out of the water, her stern in thirty-six feet water; the ledge on which she struck dipped gradually to the south-east, while on either hand, and not removed from the vessel more than fifteen or twenty feet, were huge masses of rock a few feet below the surface of the water.

When morning dawned, and the mists had p rtially cleared away, the steamer was found to be firmly lodged upon a low rocky island of about two acres in area, lying a mile south of Michipicoten Island, and about two miles from Michipicoten harbour. A boat was despatched to Michipicoten Island to procure timber for derricks, with which it was hoped that her bows might be raised from the ledge and the vessel slipped off into deep water, a result which was fortunately attained during the afternoon by the aid of derricks, steam, and a continued rolling from side to side by the united efforts of the passengers running with measured step from one side of the vessel to the other.

Steamer Collingwood on a rock near Michipicoten Island,
Lake Superior.

It was soon ascertained that the sheeting was quite sound, and neither hull or machinery had sustained any material injury from the shock of the preceding night, and the efforts made to move the vessel. Nevertheless the captain thought it would be judicious to go into Michipi-coten harbour and examine her more narrowly, as well as to shift the cargo and coal into their proper places. We reached the entrance of Michipicoten harbour in safety, but had scarcely advanced more than a few hundred yards when we again grounded on a shoal, and remained firmly fixed in a new position. In the evening the Agate Islands in Michipicoten harbour were visited, and very beautiful agates found in great abundance in the trap; but it was difficult to procure good specimens, on account of the hardness of the matrix.

During the afternoon the most singular effects of mirage were continually changing the outline of a few low rocks which projected above the level of the water, some two or three miles from us; and occasionally the steep wooded hills of Michipicoten Island seemed to be filled with brilliant ever-changing little lakes, which, if they preserved their apparent form for more than a minute, were most curiously delusive. All the phenomenon which sunshine and fog are capable of producing seem to be of constant occurrence near the middle of Lake Superior; not an hour passed during the daytime without our witnessing fog-bow, halo, or mirage of very singular beauty. At 4 P.M. on the 30th we steamed out of Michipicoten harbour, and pursued a straight course to Thunder Bay.

Early on the following morning we sighted the Paps and Isle Royale. The magnificent scenery of Thunder Bay and the adjacent shores of Lake Superior were gradually revealed as the mist slowly disappeared with the advancing day. Passed Thunder Cape at 2 P.M., and anchored off Fort William about half-past 4. The scenery of Thunder Bay is of the most imposing description. Pie Island, with its round eminence Le Pâté, 850 feet above the Lake, and Thunder Cape rising boldly 1350 feet, stand on either hand as you enter the deep inlet. Mackay's mountain uplifts a broad front to the height of 1000 feet on the mainland, in the direction of Fort William. The waters of Thunder Bay are coloured by the Kaministiquia for a considerable distance from the three mouths of that river.

Early on the 1st August the expedition, with baggage and stores, were landed at Fort William. Mr. McIntyre, the officer in charge of the Fort, received us with much courtesy and hospitality, kindly placing some of his appartments at our disposal for the night. The Iroquois made themselves comfortable under the canoes we had brought with us from Collingwood, which in the first instance were procured from Lachine, and transported by rail and steamboat to near the western extremity of Lake Superior.

The present position of Lake Superior and its tributaries in relation to Montreal or the Atlantic seaboard, is wholly changed since the period when the old North-West Company, established in 1783, and amalgamated with the Hudson Bay Company in 1821, maintained large establishments at Fort William and at Fort Charlotte, on the Pigeon River, some thirty-five miles in a south-westerly direction from the mouths of the Kaministiquia. In those days of canoe transportation, merchandise was conveyed up the Ottawa, across the height of land to Lake Huron, thence by the north shore of Lake Superior to Fort William, the starting point of the long journey into the great interior valleys of Red River, the Saskatchawan, and the Mackenzie. In these days ships can sail from European or Atlantic ports, and without breaking bulk, land their cargoes at Fort William for less than one-fiftieth part of the cost involved during the period when the North-West

Company became a powerful, wealthy, and influential body.

The completion of the Sault Ste. Marie Canal*, in May, 1855, established an uninterrupted water communication for sea-going vessels between Lake Superior and the ocean. The first ship which sailed from Chicago to Liverpool was the Dean Richmond, in 1856 ; this craft measured 379 tons American measurement, or 266 tons according to the English method of determining the ton·nage of a vessel. Since that period the number of sea-going vessels from the Upper Lake ports has been increasing with great regularity. The trade of Lake Superior is also becoming of unexpected importance. In 1859, between the 1st day of June and the 1st November, the value of the different articles which passed through the St. Mary's Canal amounted to 5,703,433 dollars, and the number of passengers to 11,622. Fifteen years since three schooners constituted the entire fleet engaged in the Lake Superior trade. The number of vessels which passed through the St. Mary's Canal in the seasons of 1858 and 1859 were respectively 443 and 847, with a tonnage 149,307 and 304,860.†

The heights and distances enumerated in the subjoined Table, show a profile of this ship route between Anticosti, in the Gulf of St. Lawrence, and Fort William, at the mouth of the Kaministiquia River, Lake Superior.‡

* The Sault Ste. Marie Canal is one mile and an eighth in length, 70 feet wide at bottom, and 100 at water-line, depth 12 feet. The average lift of the locks is 17 feet 6 inches.

† Detroit Advertiser. From official returns.

‡ See a Map of the Province of Canada, showing the connection by steam navigation of the region of the great lakes with Europe, by the route of the St. Lawrence and the great lakes, prepared for the Canadian Commissioners of the Paris Exhibition, by Thomas Keefer, C. E., Montreal, 1855.

Names.	Distance from Anticosti in miles.	Elevation above the Sea level.	Number of Locks.	Length of Locks in feet.	Breadth of Locks in feet.	Total Lockage in feet.
Anticosti						
Quebec - - - -	410					
Montreal - - - -	590	14				
Lachine Canal - - -	598½	14–58	5	200	45	44¾
Beauharnois do. - - -	614	58·5–141·3	9	200	45	82¼
Cornwall do. - - -	662¼	142·6–185·6	7	200	45	43
Farren's Point do. - -	673½	190·5–195	1	200	45	4
Rapid Plat do. - -	688	195·3–207	2	-	-	12
Pt. Iroquois Canal - -	699½	207–213	1	-	-	6
Galops do. - - -	714½	213–225	2	-	-	8
Lake Ontario - - -	766	234				
Welland Canal - - -	1016	234–564	27	150	26½	330
Lake Erie - - - -	1041	564				
Detroit River - - -	1280	564				
Lake St. Clair						
River St. Clair						
Lake Huron - - -	1355	573				
River Ste. Marie - -	1580	573–582·5				
Sault Ste. Marie Canal. -	1650	582·5–600	2	550	75	17½
Lake Superior - - -	1650	600				
Fort William - - -	1910					
Superior City - - -	2030					

With the single exception of the Sault Ste. Marie Canal, all the great public works which have been contrived and executed for the purpose of reducing the obstacles to uninterrupted navigation between the great lakes and the ocean, lie within Canadian territory, and are under the control of the Canadian Government.*

According to the results of a recent survey of the Ottawa, made with a view to connect the St. Lawrence with Lake Huron by that river, the distance from the

* The cost of the construction of these remarkable links in the chain of unbroken communication which now penetrates a distance exceeding 2000 miles into the interior of the North American continent, approaches 15,000,000 dollars, and the annual revenue has risen from 131,000 dollars, in 1850, to 369,110 dollars in 1858.

mouth of the French River to Montreal by the route surveyed, is 430·76 miles, of which 35·18 miles is already a good navigation, requiring no improvement. Of the other 78·95 miles, 29·32 will require to be canal navigation, and 49·63 miles improved, so as to connect the whole into a first class navigation for vessels drawing 12 feet of water.

The cost, exclusive of deepening the Lachine Canal and Lake St. Louis, and apart from land damages and expenses, is estimated at 12,026,351 dollars.

This route would effect a saving of distance between Chicago and Montreal, over the existing one by the Welland Canal, of 343 miles; but with an increased lockage of 15 locks, and an additional rise and fall of 169·60 feet. The lake navigation by the existing route is 1145 miles in extent, and the inland or river 134; but by the Ottawa, the former is 575 miles, and the latter 401.*

The elevation of Lake Superior above the ocean has been variously estimated by different observers. Captain Bayfield considered it to be 627 feet above the level of the sea, which altitude is adopted by the narrators of Agassiz's tour in that region, and by Messrs. Foster and Whitney, in their report on the geology of the Lake Superior Land District. Sir William Logan, in his Geological Report for 1846–7, states that its surface is 597 feet above the ocean; in Professor Hall's Geology of the 4th District, N. Y., 596 feet is its assigned elevation. Sir John Richardson assumed its level to be 641 feet above the ocean.

The altitude deduced by Mr. Keefer for the map prepared for the Canadian Commissioners at the Paris Exhi-

* Report of the Commissioner of Public Works, 1859, based on the report of T. C. Clarke, Esq., C. E. Engineer Ottawa Survey.

bition in 1855, with the advantages and information derived from the levels obtained in the construction of various railways and canals from the ocean to Lake Superior, established a difference of only three feet in excess of that obtained by Sir William Logan in 1847.

The occasional fluctuations in the level of the waters of Lake Superior certainly exceed three feet, so that an elevation of 600 feet is probably a correct estimate of the mean height of the waters of this Kitchi-gum-mi*, or "Great Lake" of the Ojibways above the ocean.

In the region about Lake Superior the years 1845–6 were unusually dry, and in 1847 the lake had reached a very low stage of water. The years 1849–50 were wet, and the level of the lake in 1851 was from three to three and a half feet above the level of 1847.†

The variations in the levels of the Great Canadian Lakes are phenomena of the utmost importance to commercial interests. ‡ The supply of water to the Erie and Welland Canals is dependent upon the relative height of the waters of Lake Erie. Periods of great anxiety have occurred among mercantile men at Buffalo respecting the supply of water to the great artery which unites Lake Erie with the Hudson River. If Lake Erie should subside to the zero of comparison adopted by Dr. Houghton, the depth of

* Spelt by Longfellow Gitche-Gumee, Big Sea Water (Hiawatha).

† Report on the Geology of the Lake Superior Land District, by J. W. Foster and J. D. Whitney, U.S., Geologists.

‡ The commerce of the Lakes is increasing with marvellous rapidity. Three thousand and sixty-five steamers passed up from Lake Erie to Lakes Huron and Superior, by Detroit, in 1859, and three thousand one hundred and twenty-one passed down. The greatest number up in a single day was eighty-five—down seventy-three. Detroit statistics show that five steamers, seven propellers, four barques, seven brigs, and eighty-five schooners were more or less engaged in the Lake Superior trade during the same year. Forty vessels left during the season for European and outward ports, some of which have returned, and one has taken her second departure.

water on the mitre sill at the Black Rock Guardlock would be less than five feet, through which all the water for the supply of a canal 150 miles long would have to flow. This contingency formed the subject of a memorial to the Legislature of the State of New York in 1854.

The following tables and memoranda are abbreviated from a paper by Major Lachlan, communicated to the Canadian Institute in July 1854.*

VARIATIONS IN THE LEVEL OF LAKE ERIE.

Date.	Comparative Level.	Authorities.
1790	1st *maximum*, being 5 ft. 6 in. above lowest level.	Hall, Higgins, Whittlesey, Mather, &c.
1795	1st *minimum*.	Weld, Whittlesey.
1801	2nd *maximum*.	Higgins, Houghton, Whiting.
1810	2nd *minimum*. Reported 6 feet below 1838.	Whittlesey.
1815	3rd *maximum*, 2 feet less than 1838.	Houghton, Higgins, &c.
1820	3rd *minim*. or ZERO OF COMPARISON.	„ „ Whittlesey.
1827–30	4th *maximum*.	Houghton, Higgins, Whiting.
1832	4th *minimum*.	
1838	5th *maximum*. 5 ft. 3 in. above zero.	American Journal of Science, Dewey, &c. &c.
1846	5th *minimum*. 2 feet above zero.	
1853	6th *maximum*.	
1859	April, 5 feet 6 inches above zero of comparison.	

Years of maximum Level of Lake Erie.	Years of minimum Level of Lake Erie.
1790	1795
1801	1810
1815	1820 zero of comparison.
1827	1832
1838	1841
1853	1846
1859 5 ft. 6 in. above zero.	

---

* See Canadian Journal, 1st Series. On the Periodical Rise and Fall of the Lakes, by Major Lachlan.

Changes in lake levels are important physical pheno-mena in the basin of Lake Winnipeg. The rise or fall of a few feet in Lakes Maintobah and Winnipego-sis deter-mines the character of an immense area of country on the low shores of those great inland lakes. Many hundred square miles of splendid pasturage are accessible at low lake levels, which are converted into marshes or swamps during periods of high water. The shores of the Great Canadian Lakes are generally high, and are not affected by a rise or fall of a few feet. The following additional notices of extraordinary changes in lake levels, which have occurred in the St. Lawrence valley, may tend to throw some light upon similar phenomena in the valley of Lake Winnipeg during recent periods.

From 1788 to 1790, the lakes generally, and Lake Erie in particular, are stated to have been as high as in 1838 (5 ft. 3 in. above zero). Professor Hall mentions evidence of a higher level than in 1838, as recorded by ridges and submerged trees.

In 1819 and 1820, the central and lower lakes are described as being unusually low. Lake Superior in 1827 and 1828 was lower than ever before known. In 1838 Lake Erie was 5 ft. 9 in. higher during the month of August than in 1819. Much land was overflowed, and trees of 100 years' growth destroyed.

In October, 1849, the water level of Lake Ontario was at a minimum; in June, 1853, it was 4 feet 5 inches above the minimum. In the winter of 1855 it again sank to the minimum; and during the summer of 1858 the rise amounted to 4 feet 3 inches. No less than 40 inches, or 3 feet 4 inches, of this rise in the mean level was attained during the summer of 1858.

As the result of observations extended over twelve years (1846 to 1857) in the variations of the level of Lake Ontario, the following facts have been established : —

1. The mean minimum level is attained in January or February.

2. The mean maximum level in June.

3. The mean annual variation is 25 inches.

4. The maximum variation in twelve years is 4 feet 6 inches.

5. There is no periodicity observable in the fluctuation of the lakes, and recent observations tend to show that there is no flux and reflux dependent upon lunar influence.*

The Bishop of Montreal states in his journal that it is only during an extraordinary concurrence of circumstances.that the whole of Lake Superior can freeze over.† He was assured that this remarkable event happened in the winter of 1843, after a calm of four days, and during intensely cold weather. No other instance is said to be on record.

The greatest supposed depth of Lake Superior is 1200 feet. Its area is about 32,000 square miles, its coast line about 1,500 miles, and it contains probably 4,000 cubic miles of water.

The barrier which opposes further progress by steam or boat navigation westward of Lake Superior follows the general direction of the north-western and western coast of that lake. Near Fond du Lac, in the territory of the United States, the dividing ridge separating the valley of Lake Superior from that of the Mississippi, is distant from the St. Louis River about 18 miles in a southerly direction, and here the elevation of the ridge is 475 feet above the waters of the lake.

---

* Chas. Whittlesey and C. Dewey. American Journal of Science and Arts, May 1859.

† Journal of the Bishop of Montreal during a visit to the Church Missionary Society's North West American Missions.

The dividing ridge between the Embarras River, a tributary of the St. Louis River, and Vermilion River, which flows into the valley of Rainy Lake, is about 48 miles in an air line from the north-west coast of Lake Superior. On the Pigeon River, which forms the boundary between the United States and Canada, the dividing ridge separating the St. Lawrence from the Winnipeg basin is only 28 miles in an air line from the north-west coast of the same great water level, but by the canoe route on Pigeon River, the height of land is 53 miles from the coast.

Within the territory of the United States, the country between Lake Superior and the valley of the Mississippi presents no difficulties for the construction of a railroad. The valley of the Mississippi is in direct communication with that of the Red River of the north by traveled roads, so that the approach to the valley of Lake Winnipeg from the head of Lake Superior is only a question of time, and will not involve any considerable outlay when the necessities of the country, or of commerce, render the opening of this line of communication desirable.

Kettle River, flowing into the St. Croix, a tributary of the Mississippi, issues from a small lake not 20 miles from Lake Superior, and the distance of the navigable portion of the Mississippi, adjoining Sandy Lake, is scarcely 45 miles from Fond du Lac. The Mississippi is said to be navigable for steamers of light draught from Crow Wing to beyond this point, and Crow Wing is 130 miles from St. Paul by the traveled road, and less than 120 miles in an air line from Superior City.

St. Paul and Crow Wing will soon be connected by a railway. A large portion of the heavy work on this line is completed, and if no unforeseen events occur, the connection will have been established before the publication of this narrative. This important chain of communication will be referred to in succeeding pages.

The construction of a plank road between Superior City and Crow Wing, is already in contemplation, and the route is even now occasionally traveled. It will no doubt become of great commercial importance to the region of the Upper Mississippi and its numerous tributaries; and it is not improbable that its influence may rapidly extend to other water-sheds, viz. those of Rainy River, Red River, and the Saskatchewan.

In Canadian territory there are two established routes by which access is gained from the valley of Lake Superior to that of Rainy River. The most southerly of these is the old North-west Company's frontier route by Pigeon River, the second by the Kaministiquia River, which forms the subject of the first part of this narrative.

CHAP. II.

THE KAMINISTIQUIA ROUTE. — FORT WILLIAM. — LAKE SUPERIOR TO THE HEIGHT OF LAND.

Thunder Bay.—Fort William.—M'Kay's Mountain.—The Mission of the Immaculate Conception.—The Rev. Jean Pierre Choné.—Indian Treaty. —Mass.—Current River.—Garden at Fort William.—Remains of former Industry.—The first Brigade.—Iroquois and Ojibways.—A Dance.—The River.—Scenery of Kakabeka Falls.—Valley of the Kaministiquia.— Little Dog Lake.—The Great Dog Portage.— Little Dog River.—The Great Falls on Little Dog River.—Their Beauty.—Winter Road to Dog Lake.—Summer Road to Dog Lake.—Area of Dog Lake.—Description of.—Dog River.—Character of the Country.—Prairie River.—Upper Dog River.—Prairie Portage.—Viscous Lakes.—Description of Prairie Portage. —Atmospheric Phenomena.—Scarcity of Animal Life.

THUNDER BAY, which receives the waters of the Kaministiquia*, forms a portion of the north-west expansion of Lake Superior. It is the most southerly of three large and deep land-locked bays, which characterize that part of the coast; and it is situated between the parallels 48° 15′ and 48° 35′ north latitude, and in longitude 89°, and 89° 25′ west of Greenwich. Its greatest length in a north-easterly direction is 32 miles, and its breadth from Thunder Cape to the mouth of the Kaministiquia, upon which Fort William is situated, about 14 miles.

The main entrance to the bay is between the imposing headlands of Thunder Cape, 1,350 feet above the lake level, and Pie Island, 5 miles south-west of the Cape, with

---

* Spelt Kaministikwŏya by Sir Jno. Richardson, " the river that runs far about."

an altitude of 850 feet. The depth of water in this broad entrance exceeds 180 feet, and a measure of 60 to 120 feet is maintained in many parts of the bay.

Seven miles south-east of Thunder Cape the lake is 630 feet deep, with a muddy bottom.

Immediately opposite, and east of the three mouths of the Kaministiquia, the Welcome Islands are distant about two miles, and inside of these islands from 30 to 60 feet of water is shown on Bayfield's chart. Within half a mile of the river's mouth the water shoals rapidly, and the bar has a variable depth of $3\frac{1}{2}$ to $5\frac{1}{2}$ feet water upon it; but within 1,000 yards of the north, or main channel, 12 to 14 feet water is maintained. Land is forming fast near the mouths of the river, and large areas in advance of the increasing delta, sustain a thick growth of rushes.

Fort William, looking up the river.

At a distance of about half a mile from the exit of the northern or main channel, Fort William is situated, upon the left or north bank. Opposite to it is a large island formed by the middle channel of the Kaministiquia,

which branches off from the main stream about one and
a half mile from the bay.   In the time of the North-west
Company, this island was denuded of the trees it sustained, which consisted mainly of tamarack, for fuel and
other purposes, and the greater portion is now covered
with second growth.   A large area south of the fort still
remains denuded of wood, and forms the site of an Ojibway village, besides serving as an excellent open pasture
ground for a herd of cows belonging to the Hudson's Bay
Company, which swim across the river every morning, a
distance of 400 feet, and return at an early hour in the
afternoon to the farm yard in the vicinity of the fort.

The banks of the river here are low and flat, not exceeding ten feet in altitude   In the rear of the fort,
tamarack of small but dense growth prevails.   The soil is
a light sandy loam reposing on yellowish clay.

Two miles above the fort, and in a direction nearly
south from it, the third or southern outlet separates from
the main channel.   The banks of the river continue to
rise above the level of its waters until they attain at the
Mission of the Immaculate Conception, an altitude of 18
or 20 feet.   Near the Mission the Indian Reserve of about
25 square miles begins; it embraces the best and largest
area of cultivable land in the valley of the Kaministiquia,
and much of it being situated on the flanks of McKay's
mountain range, some portions possess many advantages
which do not belong to the available tracts near the
shores of Thunder Bay.

The general course of the river above the Mission for a
distance of nine miles is towards the south-west, by very
tortuous windings.   Five miles from Fort William it approaches the base of the elevated but broken table land
to which McKay's Mountain forms an imposing and
abrupt termination.   McKay's Mountain has an elevation

of 1000 feet above the lake, and is the north-eastern boundary of an irregular but extended trap range, whose south-eastern flank follows the trend of the coast as far as Pigeon River.

It is worthy of remark, that the flanks of McKay's Mountain support a heavy growth of hardwood timber (maple, &c.), and through various sources I was informed that this heavily-timbered land stretches far to the south-west, on the side and borders of the trap range. The rock formations which comprise the country between the Kaministiquia and Pigeon Rivers indicate the presence of a fertile soil on the flank of the irregular table land; the trap with which the slates are associated giving rise upon disintegration to a soil of superior character. At the Mission, a light reddish loam constitutes the soil, having a depth of six feet, and resting upon a bluish grey clay, which extends to the water's edge.

The Mission of the Immaculate Conception is under the charge of the Rev. Jean Pierre Choné, who has resided on the banks of the Kaministiquia for nine years. From that gentleman, who kindly afforded me much information respecting this valley, I obtained numerous facts of interest in relation to its adaptation for settlement. At the Mission there are already congregated from thirty to thirty-five houses, substantially built of wood; in their general arrangement and construction they are far superior to the log houses of Canadian pioneers in the forest. Many of them had gardens attached to them, a few of which were in a good state of cultivation; some small fields fenced with post and rail were in the rear of the most thriving. The river here is from 60 to 70 yards wide, its waters are very turbid, with a current not exceeding two miles an hour.

M. Choné's room, into which we were admitted, gave

us a clue to the prosperity, cleanliness, and appearance of industry which distinguished the mission. A
young tame partridge was hopping about the floor when
we entered. A number of books occupied a small table
in one corner, the other was taken up by a turning lathe,
and various articles manufactured by the curé were lying
about the room. A low bed covered with a buffalo robe
filled another corner, and while we were conversing an
old chief, dressed in scarlet cloth, quietly entered and
placed himself on a chair by the side of a small carpenter's
bench, which filled the remaining angle.

Among many interesting facts with which we were
furnished by the kindness of M. Choné, we learned various
particulars respecting the condition of the Indians and
their relation to the Government of Canada, which an
inspection of the treaty confirmed. In 1850 a treaty was
concluded by the Hon. W. B. Robinson on behalf of Her
Majesty and the Government of the Province with the
Chiefs of the Ojibway Indians, inhabiting the northern
shore of Lake Superior from Batchewanaung Bay to
Pigeon River, and inland to the height of land between
Canada and the territories in the occupation of the Hudson's Bay Company. For the sum of £2000 currency, and
an annual payment of £200, to be paid at Fort William
and Michipicoten, the chiefs surrendered all their right
and title to the above territory, with the exception of the
following reserves made over to them for the purposes
of residence and cultivation, allowance being given under
certain reasonable restrictions that they shall still hunt
over the territory and fish in the waters as heretofore.
The number of Indians included in this treaty was 1240.
The reservations made for their benefit were as follow :—

First. For Joseph Peau de Chat and his tribe ; the reserve to commence about two miles from Fort William on the right bank of the river Kaministiquia, thence westerly six miles parallel to the shores of the Lake, thence northerly five miles, thence easterly to the right bank of the said river so as not to interfere with any acquired rights of the Honourable Hudson's Bay Company.

Second. Four miles square at Gros Cap for Po-to-mi-nai and tribe ; and

Third. Four miles square on Gull River, near Lake Superior, on both sides of the river, for the chief Mish-i-muck-qua.

Our Iroquois being desirous of going to mass at the Mission on Sunday, August 2nd, several of the party accompanied them, and witnessed the rather rare spectacle of a numerous and most attentive Indian congregation engaged in Christian worship. The chapel is a very spacious and well-constructed building of wood, with a semi-circular ceiling painted light blue. The walls were panelled to the height of about four feet, and altogether the interior arrangements and decorations exceeded our anticipations, and everywhere showed the industrious hand or intelligent direction of the Rev. M. Choné. The Indians forming the regular congregation were arranged in the most orderly manner ; the left side of the chapel being appropriated to the men and boys, the right to the women and girls. The boys and girls were placed in front of their seniors. The men were provided with forms, the women sat upon the floor. The utmost decorum prevailed throughout the service, and the chanting of both men and women was excellent, that of the squaws being remarkably low and sweet. Few of the male portion of the congregation took their eyes from the priest or their books during the service. The squaws drew their shawls or blankets over the head and showed the utmost attention. The Curé delivered a long sermon in

the Ojibway language with much energy, and seemingly
with the greatest fluency.   After the ordinary service of
the day was over, being before requested by one of our
party, he delivered an admirable sermon in French.   His
style, language, and manner, were of a very superior order,
and the drift of his words seemed to go far in shadowing
forth the philanthropic impulses which sustained him in
his solitary work of love, so remote from society, comfort,
and civilisation.

In the afternoon I visited the mouth of Current
River, six miles from Fort William.   The river reaches
the Lake by a succession of sloping falls over an argillaceous rock, which in the aggregate exceed forty feet in
height within half a mile from the Lake.   The common
chive was found occupying in abundance the cracks and
fissures of the shale on the banks of the river.

I visited during the day the garden of the fort; its area
is about $1\frac{3}{4}$ acres.   The shallots were small, but the potatoes looked well, being at the time in flower, and Mr.
McIntyre thinks that varieties may be found which will
ripen well near the fort.   Tomatoes do not ripen here ;
turnips and cabbages are very liable to be destroyed by
the cut-worm or grub ; currant bushes procured from the
forest flourish admirably, and produce a very large berry;
the red currant was just beginning to ripen.   This part
of the country appears to abound in currants, raspberries,
strawberries and gooseberries ; they were seen growing
in the woods in every direction, where direct light
penetrated.   A patch of oats in the garden showed a most
remarkable development of stalk and leaf, and the ears
were beginning to show themselves.   The soil of the
garden was brought from the foot of the Ka-ka-beka falls
in the time of the North West Company's glory.

The average period when the Kaministiquia freezes, is

from the third to the fifteenth of November, and it becomes free from ice between the twentieth and twenty-third April. The year 1857 proved an exception in many respects; the ice did not pass out of the river until the thirteenth of May, and on the first of August, the day of my visit, the waters of the river were higher than they had ever been known before at that season of the year.

Indian corn will not succeed in this settlement, early and late frosts cutting it off. Frost occurs here under the influence of the cold expanse of Lake Superior, until the end of June, and begins again towards the end of August. A few miles further up the river, west of McKay's Mountain, the late and early frosts are of rare occurrence, and it was stated that Indian corn would ripen on the flanks of McKay's Mountain.

All kinds of small grain succeed well at the Mission, and the reason why they have not been more largely cultivated is owing to the want of a mill for the purpose of converting them into flour or meal. Near the lake, at Fort William for instance, oats do not always ripen; the cold air from the lake, whose surface, thirty and fifty miles from land, showed a temperature of 39° 5', at the close of the hottest month of the year, is sufficient to prevent many kinds of vegetables from acquiring maturity, which succeed admirably four or five miles up the river.

Fragments of limestone have been procured in the neighbourhood, but the locality could not be pointed out by any of its inhabitants. The ruins of a lime kiln, used by the North West Company, have been discovered, and it is very probable that the limestone was obtained from crystalline layers, the existence of which has been established over wide areas in Thunder Bay, by Sir William Logan, and are noticed by him as being of a " reddish white colour, and very compact, some of which would yield good material for burning." These beds of impure limestone are mentioned by Mr. Murray (Geological Survey of Canada, for 1846-7) as occurring in the lower portions of the formation occupying this valley.

It is worthy of notice that substantial records of far more extensive settlements than now exist, showing a much higher degree of civilisation and improvement, are found at or near the various posts along this route, and particularly at Fort William.

Most of these remains of former industry and art, date from the time when the North West Company occupied the country, and there is reason to believe that much valuable knowledge respecting the resources of particular localities has been forgotten, or is hidden in the memories of those who may not have the opportunity to make it known.

Mr. Keating* mentions the ruins of the old Fort de Meuron, erected by Lord Selkirk. He was also shown the remains of a winter road opened by that enterprising nobleman, from the Kaministiquia to the Grand Portage on the Pigeon River, about thirty-six miles distant. The remains of a road to White Fish Lake is also still to be seen, and, indeed, it forms a winter route for half-breeds and Indians at the present day between the lakes on the Pigeon River, and the valley of the Kaministiquia. The Canadian government have recently laid out the valley of the Kaministiquia below the Ka-ka-beka Falls into two townships, named respectively PAI-POONGE and NEE-BING.

On the 3rd August we prepared for our immediate departure, and were all ready, with the exception of the

---

* Narrative of an Expedition to the Source of St. Peter's River.

Iroquois Indians, by 10 A.M. The delay with them arose from an indisposition to separate and be associated in different canoes with the Ojibways we were obliged to hire; by noon, however, an arrangement was made, it being determined that one brigade of three canoes should proceed at once, the other follow on the morrow. Just before starting a large body of heathen Indians, from the camp on the opposite side of the river, came over in a number of small canoes and commenced a dance outside of the pickets of the fort. They were painted and feathered in various ways, and furnished an admirable subject for our artists. Having danced on the outside of the fort for some minutes, they entered and arranged themselves in a semicircle in the quadrangle. The medicine-man and his assistant, gaudily painted and decked with eagles' feathers, sat on the ground beating a drum, and near to them squatted some half dozen squaws, with a few children. About sixty men and boys, headed by the chief, painted and feathered similar to the medicine man, danced or jumped round the ring. Our party being collected in front of the chief, he made a short speech, which was interpreted by a half-breed attached to the expedition to the following effect :—" They were happy to see us on the soil, they were hungry and required food, and trusted to our generosity and the plenty by which we were surrounded." The pipe of peace was then lit, and handed in turns for each to take a whiff. The picture of a hand across the mouth and cheek was admirably drawn in black on the faces of the chief and medicine-man. The Ka-ki-whe-on, or insignia, consisted of eagle's feathers stuck in a strip of red cloth about four feet long, and attached to a cedar pole. The whole scene was highly ridiculous, and many of the performers were wretched looking crea-

tures, being dreadfully affected with scrofula. Some of
the men, however, possessed splendid looking figures, but
the progress of civilisation will soon close the history of
these wretched Indians of the Kaministiquia.

Our first brigade, consisting of two large five fathom,
and one middle size canoe, containing twenty-six men
in all, started from Fort William at 5 P. M., and arrived
opposite McKay's Mountain at about half-past six. Half
a mile above the mission we noticed a very neat house
in a clearing of about ten acres in extent, the last
effort of civilisation to be seen, with the exception
of an occasional post of the Hudson's Bay Company,
for many hundred miles. The first camp was pitched
about three-quarters of a mile beyond McKay's Mountain.

Opposite this magnificent exposure of trap, the clay
banks of the river are about 14 feet high, and continue
to rise on one side or the other until they attain an
elevation of nearly 60 feet, often, however, retiring from
the present bed of the river, and giving place to an alluvial terrace, some 8 or 10 feet in altitude, and clothed
with the richest profusion of grasses and twining flowering
plants. The current begins to be rapid about nine miles
above Fort William soon after passing Point de Meuron,
the site of the fort established by Lord Selkirk before referred to, and continues so, in the ascending course of the
stream, to the foot of the first demi-portage, called the
Décharges des Paresseux, where an exposure of shale
creates the rapids which occasion the portage. The fall
here is 5 feet 1 inch, in a space of 924 feet. The distance of this portage from the lake, by the windings of
the river, is about 22$\frac{1}{4}$ miles, and the total rise probably
reaches 35 feet.

The current continues rapid to the foot of the Grand
Falls, and high rock exposures commence on the precipitous banks three miles below them.   These gradually
assume the form of mural cliffs, capped with drift, increasing in altitude until they attain at the foot of the
Grand Falls, the height of about 160 feet on the left bank,
while on the opposite side of the river the mountain

Décharges des Paresseux.

portage path winds round the steep hill side of a bold
projecting escarpment, 91 feet in altitude, and nearly half
a mile from the falls.

   At our camp, seven miles below the Grand or Ka-ka-
beka Falls (cleft rock) as they are termed, the level of the
river was estimated to be 40 feet above Lake Superior,
and the foot of the falls 16 feet higher.   The Grand Falls
themselves were found by leveling to have an altitude of
119·05 feet, and involved a portage of 62 chains or three
quarters of a mile.   They are distant from the mouth of
the river by its windings about 30 miles, and in an air
line 17 miles.

As the altitude of these falls has attracted the attention of several observers, the different results obtained may not be without interest: —

	Feet.
Altitude ascertained by leveling (Mr. Dawson, August, 1857) .	119·05
Capt. (now Col.) Lefroy, barometrical measurement . . .	115·00
Mr. Murray, of the Canadian Geological Survey	119·00
Major Delafield	125·00
Sir John Richardson, barometrical measurement	127·00
Lieuts. Scott and Denny*	130·00

Assuming the height of Ka-ka-beka to be 119 feet, the summit will be 175 feet above Lake Superior. This result includes the rapids at the foot of the falls. The levels were taken along the portage path, and, if the rapids be deducted, the true height of Ka-ka-beka probably does not exceed 105 feet.

The scenery of the Grand Falls is extremely beautiful. The river precipitates its yellowish-brown waters over a sharp ledge into a narrow and profound gorge. The plateau above the portage cliff, and nearly on a level with the summit of the falls, is covered with a profusion of blueberries, strawberries, raspberries, pigeon cherry, and various flowering plants, among which the bluebell was most conspicuous. On the left side of the falls a loose talus is covered with wild mint and grasses which grow luxuriantly under the spray. Beautiful rainbows of very intense colour are continually projected on this talus, when the position of the sun and the clearness of the sky are favourable. Numerous small springs trickle down a perpendicular cliff of about 12 feet in altitude at the base of the talus, whose coolness and clearness, compared with the warm, coloured waters of the river, make them a delicious beverage, the difference

* Sir Jno. Richardson's Arctic Searching Expedition.

between the temperature of the springs and river being about 20°. The right side of the cliff at the falls is perpendicular for a height of more than 100 feet, and exposes the stratification with perfect fidelity. The peculiar rounded forms into which the rock divides itself, noticed by Mr. Murray, were well marked.

The alluvial valley of the river from about three miles below the mountain portage to Fort William varies in breadth from a few hundred yards to one mile; the breadth occupied by land of a quality which might fit it for agricultural purposes extends to near the summit of the flank of a low table land, which marks the true limit of the river valley, and the average breadth of this may be double that of the strictly alluvial portion.

The low table land is thinly wooded with small pine, and the soil is poor and dry; the alluvial valley sustains elm, aspen, balsam, poplar, ash, butternut, and a very luxuriant profusion of grasses, vetches, and climbing plants; among which the wild hop, honeysuckle, and convolvulus, are the most conspicuous. The rear portion of the valley, with an admixture of the trees just named, contains birch, balsam-spruce, white and black spruce, and some heavy aspens. The underbrush embraces hazel-nut, cherries of two varieties, &c.

Occasionally the flanks of the low table land approach the river, contract the valley, and give an unfavourable aspect to the country. This occurs near the Décharges des Paresseux, and at most of the heavier rapids. The area available for agricultural purposes below the Grand Falls, probably exceeds twenty thousand acres, but if the flanks of McKay's Mountain be included in the estimate, a large addition may with propriety be assumed.

The Grand Falls mark the limits of a tract of country differing in many important physical aspects from the

valley of the river lower down.   From black argillaceous
slates of Huronian (Cambrian) age we pass to a region in
which granite, gneiss, and chloritic schist prevail, and
where the vegetation is often scanty and poor.

The course of the river is now almost due north to
Little Dog Lake, and its flow is much broken by falls
and rapids, which occasion in a distance of nineteen
miles, six portages and five discharges.   The names, altitudes and distances from Fort William of the falls and
portages are given in a table at the end of the second
volume.

In the forests which lined the banks at the different
discharges, the canoe birch was frequently seen eighteen

Second Falls, Kaministiquia River.

inches in diameter; the underbrush consisted chiefly of
hazel nut.   Whenever the gneissoid and syenitic rocks
prevailed, the valley of the river was much contracted, the
timber light, and the soil shallow and full of boulders or
detached masses of rock.   The volume of water in the
river appeared to be very small, considering its unusual
height at this season of the year: an approximate measurement at one of the rapids gave a breadth of seventy
with an average depth of two feet.

FALL AT THIRD PORTAGE ABOVE KA-KA-BEKA.

Printed by Spottiswoode and Co.]

[New street Square, London.

Extensive areas covered with burnt forest trees, consisting chiefly of pine, occur in the valley of the river as far as Little Dog Lake, when the formidable barrier of the Great Dog Mountain, sustaining a heavy growth of timber, comes into view. Occasionally aspens of large dimensions may be seen from the canoe, but it is not until the plateau of the Great Dog Mountain is attained that they acquire a diameter varying from eighteen to twenty-four inches, five feet from the ground. Trees of this species, and of the above dimensions, are found in abundance on the elevated barrier which separates the region of Great

Entrance to Little Dog Lake, from the Kaministiquia River.

Dog Lake from the valley of the Kaministiquia, 350 feet below.

The Great Dog Portage rises 490 feet above the level of Little Dog Lake, and at the point of greatest elevation the ridge cannot be less than 500 feet over the same lake. The difference between the levels of Little and Great Dog Lakes is 347·81 feet, and the length of the portage between them one mile and fifty-three chains. The view from the summit of the Great Dog is very striking; Little Dog Lake lies at our feet, an unbroken forest of pines dotted with groves of aspen and birch, and in the swamp portions with tamarack, stretches in all directions from east to west, being bounded in the view by the distant undulating outline of the wooded hills which limit the valley of the Kaministiquia. A portion of the abrupt escarpment of the elevated table land in the neighbourhood of McKay's Mountain is distinctly visible in clear weather.

The base of the Great Dog Mountain consists of a gneissoid rock supporting numerous boulders and fragments of the same materials. Micaceous rock was observed in position by Mr. Keating on the east side of the portage.* A level plateau of clay then occurs for about a quarter of a mile, from which rises, at a very acute angle and to an altitude of 283 feet above Little Dog Lake, an immense bank or ridge of stratified sand, holding small water worn pebbles. The bank of sand continues to the summit of the portage, or 185 feet above the clay plateau. The portage path does not pass over the highest part of the sand ridge; east of the path it is probable that its summit is 500 feet, as before stated, above Little Dog Lake.

In an endeavour to reach the head of Little Dog River†, before it begins to make in its short course of about four or five miles a precipitous descent of 347 feet, I found that much of the soil on the flanks of the Great Dog Mountain was far superior to the average quality in the valley of the Kaministiquia. It consisted of a clay loam, with a gravelly subsoil, containing numerous pebbles and

* Expedition to the Sources of the St. Peter's River.

† Little Dog River is a continuation of the Kaministiquia, but, in accordance with the Indian custom, it is named from the lake into which it flows.

Printed by Spottiswoode and Co.

BEGINNING OF GREAT DOG PORTAGE.

[New-street Square, London.

water-worn fragments of rock; this was particularly noticed on the flanks and surface of the lower plateau.

The upturned roots of trees in the track of a tornado, which must have occurred here some years since, afforded an excellent opportunity for examining the soil and sub-soil of the lowest plateau, and the flank of the upper one. The upturned roots of large aspens, birch, and pine showed everywhere a gravelly loam containing pebbles from one to six inches in diameter. On approaching the source of Little Dog River, a black spruce swamp was found to occupy an extensive area but little above the level of the river. The clay soil in this swamp was covered to the depth of two feet with moss, which was again largely overgrown with the Labrador tea plant. Small holes in the moss, filled with clear, cool, limpid fluid, afforded a striking contrast to the heated waters of the rivers and lakes; the temperature of these shallow wells did not exceed 42°, while the water of Great Dog Lake, tested a few hours afterwards (half-past 5 P. M.), was 69°, a difference of 27°.

The Great Dog Mountain derives its name from a murderous conflict between the Sioux and Ojibways, which occurred some centuries since on or near this eminence. The figure of a dog, in commemoration of this event, is carved on the side of the mountain. It was nearly obliterated when Major Long passed through the country in 1823, and we could not discover it in 1857. The Sioux and Ojibways were at war when the French traders and missionaries first visited the head of Lake Superior, which event may be placed as early as the year 1620.*

The great falls of Little Dog River are surprisingly

---

* Schoolcraft. History of the Indian Tribes of the United States. Part VI.

beautiful. The difference in level between Little and Great Dog Lake is 347 feet, which is descended by the foaming torrent in six successive leaps. The course of the canoe route lies some distance to the right of the falls, hence the reason why they have not been described by former travellers in these regions. In picturesque beauty they far surpass Ka-ka-beka, and would probably take rank with the most charming and attractive falls on the continent. They have not the grandeur of the Silver Falls on the Winnipeg, nor do they approach Niagara in magnificence or sublimity, but their extraordinary height, and the broken surface they present, impart to them singular and beautiful peculiarities. The strange aspect they must possess in winter, when fringed with masses of frozen spray, would probably be unrivalled ; and in spring, when the feeding lake is from three to four feet higher than during the summer months, their augmented volume would give them an appearance of magnitude which is lost when the waters are low, in consequence of the succession of ledges of rock over which they leap being partially screened by the foliage of overhanging trees.

The shoals, rapids, and falls on the Kaministiquia will always prevent that river from being used as a mean of communication with the interior for commercial purposes. The first large area of open water on this route is Dog Lake, and with a view to reach that elevated sheet of water a road from the shores of Lake Superior in as direct a line as possible will be required.

About six miles in a north-east by east direction from Fort William the waters of Current River are seen to fall over a precipitous ledge of black argillaceous slate, of Huronian age, within a few yards of their exit into Thunder Bay.

GREAT FALLS ON LITTLE DOG RIVER.

Printed by Spottiswoode and Co.

[New-street Square, London.

The valley of this river forms the winter route of the Indians from Thunder Bay to Great Dog Lake, and while the Great Dog Portage, by the circuitous route of the Kaministiquia, is not less than 43 miles from Fort William, Great Dog Lake is reached by the Valley of Current River, in an eighteen or twenty miles march from Thunder Bay.

In making their winter journey to Great Dog Lake, the Indians generally proceed from the Mission in the neighbourhood of Fort William to the mouth of Current River, and ascend its open and unencumbered course, reaching Dog Lake in one day from Fort William. A cursory inspection of the map will show that the direct line of route from Fort William, or rather from Point Meuron, through the forest, if a track were cleared, would save several miles.*

The height of Great Dog Lake above Lake Superior is 710 feet, and to reach it in canoes or boats by the route of the Kaministiquia involves portages, which in the aggregate amount to 325 chains, or four miles in length.

As a mean of communication between Thunder Bay and Great Dog Lake, the Indian trail up the valley of Current River, appears to be of sufficient importance to require this special notice; a bird's-eye view of the country, from the summit of the Great Dog Portage, showed no mountainous range between that point and Lake Superior apparently equal in altitude to the great barrier of Dog Lake, which exceeds 850 feet above Lake Superior. It acquires additional importance from the fact that a travelled Indian canoe route and winter road exists

---

* In Current River speckled trout are numerous, and its valley abounds with red and black currants, raspberries, strawberries, and gooseberries, wherever sufficient light and air for their growth obtains admittance into the forest which covers the country.

between Dog Lake and Thousand Lakes, on the west side of the height of land.*

The area of Great Dog Lake, according to Mr. Murray†, whose opportunities of examining it were considerably greater than those of the members of the Exploring Expedition, probably exceeds 200 square miles; the country surrounding it is hilly, and covered with forests in which white spruce prevails, interspersed with groves cf aspens, and occasionally dotted with the Weymouth and Banksian pines; white and yellow birch are abundant, and some of them of large dimensions. The lake is bounded by bold primary rocks, and studded with innumerable islands.

The traverse of the canoe route, from the head of the Great Dog Portage to the mouth of Dog River, is about 11 miles in length, and the lake is seen to stretch far to the north of the last named point; the canoe route follows closely the direction of its longest diameter, which is nearly due north and south. The depth of water, as ascertained by occasional soundings along the line of traverse, is very considerable; in one instance, 72 feet was recorded about 200 yards from a low rocky shore.

* During the year 1858, Mr. S. Dawson was employed in examining the country between Lake Superior and Red River with a view to establish a line of communication. In his report of the operations carried on during that year he speaks favourably of a line of road between Thunder Bay and Great Dog Lake. The subjoined extract is from Mr. Dawson's report for 1858 :—"To commence at Lake Superior, a land road would be required from Thunder Bay to Dog Lake, as the navigation of the Kaministiquia is utterly impracticable, except for canoes, and could only be rendered otherwise at an enormous outlay. Dog Lake is distant from Lake Superior 22½ miles, and at a higher elevation by 718 feet, a difference of level which renders a canal out of the question, notwithstanding that the supply of water in the Kaministiquia would be ample. The only way of reaching it, therefore, is by land, and the surveys have progressed so far as to show that a good line may be obtained in a distance of 28 miles."

† Report of Progress for the year 1846–7. (Geological Survey of Canada.)

and another sounding showed 90 feet half a mile from land. The waters of Dog Lake at the time of our visit (Aug. 8th) appeared to be teeming with countless millions of animalculæ visible to the unassisted eye. The water marks showed an elevation slightly exceeding three feet above the level at the time of observation.

In making the traverse I timed the voyageurs at their paddles, and found they made one stroke a second, or sixty a minute, with remarkable accuracy. This would give for a day's work, from 5 A.M. to 7 P.M., with two hours stoppage, 12 hours, at 3600 strokes per hour, or 43,200 strokes a day.

We soon began to find that no feeling of sympathy existed between our Iroquois and Ojibway voyageurs; nor was any effort made by individuals of either nation to assist or enliven those of the other. As an instance of their utter indifference and selfishness, the following trait may be mentioned. One of the Iroquois from Caughnawaga was very ill at Fort William, and not only incapable of working, but unable to walk without assistance. He suffered much from chills and cramps, and was thought to be in a very dangerous condition; with care, medicine, and a good constitution, he grew a little better, and was able to eat, but the salt pork and salt beef we had with us were not very well adapted for a man so much reduced in flesh and slowly recovering from a severe illness. I shot a pigeon, and in the hearing of the Ojibways stated that it was to be cooked for the sick man. The following day I found that the Ojibway who picked up the pigeon had not only eaten it in the presence of the invalid Iroquois, but he and his companions had devoured two partridges which had also been reserved for the Iroquois. Another day, passing near the fire of the Ojibways, I found them roasting a pike which was intended for the dinner of the invalid. In future we committed the pigeons into the sick man's charge, and they were cooked by his friends, but the Ojibways laughed loud and long at the excellent practical joke they had enjoyed, and for many days after they reminded Lambert, the interpreter, about the sick man and his pigeons and partridges. "Tell him," said one waggish fellow, pointing to me, "shoot pigeon for Iroquois, Ojib-way eat it, do Iroquois much good." A joke lasts an Indian a long time, and is continually repeated, both in canoe and in camp; it never appears to lose interest or grow stale.

The former extension of Dog Lake in a westerly direction for fourteen or fifteen miles, up the valley of the river of the same name, is shown by numerous sand ridges which intersect it nearly at right angles to its course, as well as by the probable former extension of a prolongation of the sand ridge which has been described as occurring at the Great Dog Portage, across the valley of the Little Dog River.

Great Dog Lake appears to be a centre of communication to which some degree of speculative interest may be attached; from one of the deep westerly bays, our guides pointed out the direction through which a communication with Thousand Lakes, on the other side of the water shed, has long been known to exist. No doubt the country through which this communication passes embraces extensive marshes, yet, if it avoids the objectionable ascent of Prairie River and Portage, it may be worthy of attention. Thousand Lakes, or Milles Lacs, as it is more commonly called, is eight hundred and thirty-two feet above Lake Superior, consequently one hundred and twenty-two feet above Dog Lake.

This route has long been known to the voyageurs and

Indians about Fort William, and the same may be re-marked of many other routes of which the Indian guides speak and attempt to describe. Thirty-three years ago it was an old "path," and may have been one for cen-turies to the Indians of this region. Communications superior to those now travelled may yet be found, but it seems clear that until the water-shed of Rainy Lake is reached, no connection possessing sufficient water to form a boat route exists, or can be made without numerous dams.

Mr. Keating, so far back as 1823, relates that his party were shown an arm of the lake which extends to the south-west, and which they were informed connects Great Dog Lake by an uninterrupted water communica-tion with the Thousand Lakes. The route is shorter than that by Prairie Portage, but broken by rapids. The same authority says that there is a communication between the Kaministiquia and Thousand Lakes passing more to the south than that from Dog Lake.* This is doubtless the Matawan River which joins the Kaministiquia at Couteau Portage, and rises within five miles of Milles Lacs.

So sluggish is the flow of Dog River, that a rise of ten feet in the level of the lake would push back its waters to a distance of thirty-five miles up the tortuous course of that stream, and the voyageurs relate that in the spring of the year they are accustomed to paddle their canoes over the tops of the willows which fringe its banks below the first rapids, fourteen miles in an air line from the mouth of the river; the greater portion of the intervening valley being then under water.

The banks of Dog River are altogether alluvial for some distance up the valley, with the occasional excep-

* Narrative of an Expedition to the source of the St. Peter's River, &c., &c., by Wm. H. Keating, A.M.S., 1824.

tion of abrupt sand-cliffs already noticed, which come upon
the river and seem to form the termination of ridges,
traversing the valley at nearly right angles to the course
of the stream.

These ridges of sand are probably of very ancient date,
and point to the period when the waters of Dog Lake
were many feet higher than at present.   The wearing
away of the barrier at the mouth of Little Dog River
would be sufficient to account for the former higher
elevation.   Recent water marks showed a rise of 5 feet
within three miles of the mouth of the river.   Further up the stream a rise of 6 feet was indicated.   Its
average breadth is about 80 feet in ordinary seasons;
its general depth at this period of the year cannot
be above 2 or 3 feet, indeed we were informed by
our steersman, that he has often known canoes to be
constantly impeded by shallows and sand bars when the
level was probably much lower than during the present
extraordinary season.   The banks showed alder bushes,
willows, dogwood, and tamarack.

The average height of the banks rises from 4 feet, a
short distance from the mouth of the river, to 10 feet,
14 miles further up.   At nearly every turn, newly formed
oval and elongated ridges of sand protrude and show a
general elevation of 5 feet above the present level.   Low
hills of granite begin to narrow the valley after passing a
small stream coming from the north, and said to lead to
a communication with the Nipigon.

From the summit of a low granite hill, perhaps 200
feet above the river bed, the surrounding country was
distinctly mapped at our feet.   The valley of the river
appeared to have a breadth of a mile at our point of
view, widening out in the direction of Dog Lake, and
contracting towards the Height of Land between low

ranges of hills, which did not seem anywhere to exceed 200—280 feet in altitude.

Some of the hills consisted of bare rock, others were covered with a young forest growth, which appeared to consist chiefly of Banksian pine and aspen. In the distance the tops of a few hills showed clumps of red pine standing erect and tall above the surrounding forest. They may be the remnants of an ancient growth which probably once covered a large portion of this region, having been destroyed by fire at different epochs; wide areas were still strewed with the blackened trunks of trees, and in the young forest, which seems fresh and green at a distance, the ground was found to sustain the charred remains of what had once been a far more vigorous vegetation.

The low ranges of hills bear a great outward resemblance to those which surround Dog Lake. No precipitous escarpments are visible, but most of them have a rounded dome-like aspect, and close inspection of some of them gave indications of the abrading action of ice. Large quantities of Labrador tea (*Ledum palustre*), were seen wherever we landed. The flow of the river for a distance of twenty-five miles from Dog Lake, varies from half a mile to one mile an hour.

The general character of this valley is very uniform, and the idea presented to the mind, in endeavouring to picture its aspect when covered with water in the spring, was that a general rise of 20 or 25 feet would give it an appearance very similar to Great Dog Lake, with analogous deep bays formed by the valleys of its tributaries, and having on its shores hills of the same altitude and similar formation as are found bordering the lake below; in fact, a high (25 feet) dam, as has already been hinted, at the source of Little Dog River, might perhaps convert Dog Lake into a magnificent sheet of water, having in a westerly direction a farther extension of at least 15 miles. It remains, however, to be ascertained whether Dog Lake has not other outlets than the one which leads through Little Dog River. It is not at all improbable that this may be the case.

At our camp on the 9th of August, near the head of a fall of 3½ feet named Barrière Portage, about 3 miles below the mouth of Prairie River, blue berries, not yet ripe, were very abundant, showing a marked difference in the climate of this spot and the Grand Falls, where some days before we had found them perfectly ripe and in the greatest profusion. The difference in elevation is about 542 feet. A quarter of a mile from the camp, in our course up the river, we came upon a bare granite hill, about 250 feet high, rising from the water's edge at an angle of nearly 45°. Its surface consisted of smooth rounded ridges, and 15 feet above the river a collection of water-worn boulders, from 6 inches to 2 feet in diameter, were deposited upon a ledge, leading to the inference that they had been left there by ice during spring freshets, and so far showing some confirmation of the statements of the Indians respecting the remarkable rise of water in the valley during the spring months.

The last portage on Dog River, following the canoe route to Fort Francis, is the Jourdain Portage, four miles in an air line from the height of land, and thirty-seven miles from Dog Lake by the windings of the river, according to Mr. Gaudet, who measured the distance in 1858. It involves an ascent of 8·60 feet by a portage 6½ chains long. A very short distance above it, the mouth and windings of Prairie River are seen with difficulty through the tall rushes which seek to conceal its course for a distance of 200 or 300 yards. Up this little streamlet, scarcely ten feet broad, the canoe route lies, while Dog River, still measuring a breadth of forty feet, can be traced far to the north by a succession of small lakes and ponds which mark its course.

Mr. Murray, of the Canadian Geological Survey, ascended Dog River up to its feeding marsh in 1847, and describes its course after receiving Prairie River, through which our route lay, "as turning off nearly due north, and widening out into a long narrow lake for about two or three miles, after which there follows in the same line a chain of twelve small lakes, or ponds, connected by short rapid streams, comprised within the distance of ten to twelve miles. The uppermost pond appeared at its northern extremity to terminate in a great marsh, which was supposed to be the ultimate source of the river, and to extend far and wide along the height of land, probably joining the Great Marsh of the Savannah Portage on the Red River route."*

Prairie River for a few hundred yards is so thickly fringed with rushes that two canoes cannot proceed side by side, or even pass one another with facility. The distance to Cold Water Lake is about 1¾ mile in an air line, and perhaps nearly double that distance by the windings of Prairie River, whose general course is a few degrees to the south of west. Much of the route towards the high barrier of land at Cold Water Lake, which now comes into view, lies through small marshy lakes or ponds, three in number, very shallow, and much encumbered with aquatic plants. The third or last lake, called Muddy Lake, is about 200 yards long and 100 yards wide. The voyageurs all complained of the great difficulty they experienced in paddling through this small

* Report of Progress, 1846-7.

and shallow sheet of water. It has long been celebrated as the Viscous Lake. Mr. Keating notices this supposed property of the Viscous Lake, but remarks at the same time that no such character was observed when he passed through it. A lake with a similar reputation occurs at the height of land on the Pigeon River route. During our voyage through it in 1858, the voyageurs then persisted in the statement that they experienced great difficulty in urging the canoes forward; one of the gentlemen attached to the expedition, after practically testing the resistance, expressed a strong opinion in unison with that of the voyageurs. The lake was only about three feet deep, and a paddle could be thrust into the soft slime as far as it would reach.

The barrier behind Cold Water Lake, stretching far to the north and south, may rise 220 feet, the western extremity of the portage path, according to measurement, being 157 feet above the lake. It constitutes the great and formidable prairie, or height of land Portage, two miles and five-eighths of a mile long. Cold Water Lake is well named on account of its temperature. Careful observation made it 41·5°, and the large spring or source which feeds it, and gives rise to the Prairie River, one of the sources of the great St. Lawrence, gushes out of the rocky side of the barrier, about 50 feet above the lake, with a temperature of 39·5°. Prairie Portage passes over the height of land, but not the highest land on the route, and its course lies first south-west up a steep wooded hill, without rock exposure, but composed of drift clays, sand, and numerous boulders; it then enters a narrow valley, which terminates in a small lake, about five acres in area and 20 feet deep, occupying a hollow among the hills on the height of land. The portage path continues on in the same direction until the Height of

Land Lake is reached, a small sheet of water, about a square mile in area, and 157 feet above Cold Water Lake. The utmost elevation attained on the Prairie Portage is probably 190 feet above Cold Water Lake, or nearly 900 feet above Lake Superior. No hill within view appeared to possess an elevation exceeding 20 or 30 feet above this limit.

Prairie Portage sustains some spruce and pine of fair dimensions; one *Pinus Banksiana* measured 5 feet 9 inches in circumference four feet from the ground, and many of equal dimensions were seen in the neighbourhood. A considerable portion of the timber is burnt, and the underbrush everywhere shows a profusion of hazel nut, with small shrubs and plants, such as raspberries, blue berries, gooseberries, and strawberries, all of which were here gathered ripe. The Labrador tea (*Ledum palustre*) grew in great profusion in particular spots, and at the termination of the portage, near the Height of Land Lake, the fragrant Indian tea plant (*Ledum latifolium*) abounded in the moss bordering this elevated sheet of water, which is 885 feet above Lake Superior, or 1,485 above the sea.

The following estimates of the heights of Prairie Portage above the sea, are taken from Sir John Richardson's " Arctic Searching Expedition."*

	Feet.
Dog Lake, above Lake Superior	657
Ascent of Dog River	14
Portage to Cold Water Lake	2
West end of Prairie Portage, and Middle Portage . .	161
Lake Superior above the sea	641
Height of Prairie or Middle Portage above the sea . .	1475

" In 1849, the height of the upper end of Dog Portage

* Arctic Searching Expedition; a Journal of a Boat Voyage through Rupert's Land and the Arctic Sea, in search of the Discovery Ships under Sir J. Franklin, by Sir John Richardson, C. B.

was ascertained by me with Delcro's barometer. In the previous season the aneroid barometer gave 328 feet as the height, which was a greater degree of accordance between the instruments than I generally found. Major Long estimates the watershed between Lakes Winnipeg and Superior, at 1200 feet above the tide. Major Delafield calculates the height of Cold Water Lake at 505, to which if 161 be added for the Prairie Portage, and 641 for Lake Superior, we have 1307 feet for the height of Prairie Portage over the sea. Captain Lefroy, by barometrical measurements, made in connection with the Observatory at Toronto, makes the west end of Prairie Portage 1361 feet above the sea ; but the distance between the two places of observation renders the result liable to some error."

At our camp on the Height of Land (Aug. 12th) an atmospheric phenomenon of singular beauty occurred. The night was very beautiful and calm. The moon shone with great clearness and brilliancy, and numerous meteors darted through the sky in the south and west. Early in the morning, before daylight, I noticed a distinct arch of what at first sight I mistook for an aurora, but, observing its position to be nearly due west, referred it to very elevated clouds illumined by the sun's light. Its appearance was like that of a dim auroral arch, well defined, and forming the complete segment of a circle to the height of 45°, its form being persistent as long as observed. The remaining portion of the sky was clear, the moon and planets shining at the time with a very brilliant lustre. It occurred to me that it might be the forerunner of a storm, an idea which the rising sun, lighting up the tops of the trees beneath a perfectly cloudless sky about an hour afterwards, banished for a few hours. Towards noon the sky became overcast from the south-west. About half-past 3 thunder was heard in the distance, and at 4, scud from the south-east began to traverse the sky. At 5 P. M. the clouds in the south-west presented a very magnificent spectacle; they seemed like gigantic waves setting towards the north east. This wave-like appearance occurred in different parts of the heavens, and almost every variety of cloud passed in review. A few minutes before 5 P. M. a very long and vivid flash of lightning shot across the sky in a direction from south to north, succeeded by a distinct snap like that produced by an electrifying machine. About ten seconds afterwards the loud rolling thunder recorded the flash, and at 5 P. M. the rain commenced; the lightning was intensely vivid, and the thunder unusually loud.*

The scarcity of animal life at this season of the year on the canoe route has several times been remarked by travelers. It is probable that the noise inseparable from the passage of several canoes through the lakes and rivers would drive away the game into the interior, but their tracks would be seen if they existed in large numbers. On the Ka-ministiquia the following animals or their fresh tracks were seen: Of quadrupeds: Cariboos, bears, foxes, hares, minks, otters, squirrels, muskrats, and fieldmice. Of birds: eagles, hawks, ducks, pigeons, plover (two varieties), sandpipers, cherry birds, loons, partridges (two varieties), jays, magpies, blackcaps, nighthawks, Canadian nightingales, swallows, humming birds, kingfishers, and owls. There were shot by different members of the expedition, the ruffled grouse or hardwood partridge, spruce or cedar partridge, pigeons, plover (two varieties), squirrels, and one jay. Of fish we

---

* A thunder storm occurred at Toronto this day from 8.30 to 11.30 P.M. See remarks in the Toronto Meteorological Register for August. Canadian Journal, November 1857.

caught one pike. There is no doubt that we might have procured an abundance of pike, but a fear of retarding progress by the drag of the trolling line and its interference with the steersman in the tortuous course of the rivers, as well as in the lakes, led to the discontinuance of its use, until necessity should advise us to adopt it as a means of procuring food.

Mr. Keating states, in 1823, that from Rainy Lake to Lake Superior they did not meet with a single quadruped. The only animals they saw were about thirty or forty birds, chiefly ducks   Among the birds observed were the Canadian jay (*Garrulus Canadensis*); blue jay (*Garrulus cristatus*); hairy woodpecker, Indian hen, golden plover, and woodcock. Partridges (*Tetrao umbellus*) were killed; a whip-poor-will was heard, and a rail seen.

The Bishop of Montreal in 1844 saw one wolf, some of the smaller quadrupeds, innumerable ducks, many loons, some other aquatic birds and a few of the heron tribe.

CHAP. III.

THE HEIGHT OF LAND TO RAINY LAKE.

THE marshy lake which stretches along the narrow level plateau forming the Height of Land in this region, is about one-third of a mile broad, but its length from the north-west to the south-east could not be determined on account of the vast expanse of rushes, with islands of tamarack, which seemed to blend it with an extensive marsh stretching far in both directions. Its elevation above the sea is 1485 feet. A portage about half a mile in length, letting us down $16\frac{1}{3}$ feet, brings Savanne Lake into view. The shores of this small but reedy expanse of water are fringed with Labrador and Indian tea, and here, for the first time, the beautiful Indian Cup or Pitcher Plant (*Sarracenia purpurea*), once so common at the Grenadier's Pond near Humber Bay, Lake Ontario, was seen in great profusion. From the topmost branches of a pine tree which I ascended, a slight depression to the north and north-east of the dividing ridge was observed in the generally level outline of the horizon; by this depression it is not improbable that the waters of the Height of Land Lake and its connecting swamps drain into Dog River. With this exception the horizon appeared to be perfectly uniform, the slight difference in the height of the tamaracks and spruces, which seemed most to abound, furnishing the only deviation from a level expanse in all other directions.

The Savanne Lake with its feeding swamps may therefore be considered to be the source of the waters which, in this latitude, send tributaries to Hudson's Bay; although the Indians say that there exists a connection between the Height of Land Lake and Savanne Lake; the portage between them is named Portage de Milieu; it passes over a low sandy ridge supporting small pine, with tamarack and spruce at its foot. The connections, indeed, which exist between different water-sheds by means of swamps at the Height of Land, impassable even to a small canoe, are by no means of rare occurrence. In the present case we have the Height of Land Lake sending water both to the St. Lawrence and to Hudson's Bay; but if we go a little farther south, we find that in the territory of the United States these interlockages are numerous and complex.* The St. Croix Lake, connecting the Mississippi with Lake Superior; the west fork of Bad River and the Nemakagon at Long Lake, establishing the same connection; and the Big Fork, which flows into Rainy River, thence into Hudson's Bay, is connected with the Ondodawanoan River, a tributary of Lake Winibigoshish, through which the Mississippi flows.

Savanne Lake is about one mile broad; at its southwesterly termination begins the Great Savanne Portage, near the mouth of a small stream, flowing into Savanne

* See Dr. Norwood on this subject, in the Geological Survey of Iowa, Wisconsin, &c. &c.

River, and much encumbered with fallen trees; by this small stream canoes pass when the water is high, and thus avoid the troubles of the Great Savanne Portage.

This common dread of the voyageurs is one mile and forty-one chains in length; it descends $31\frac{1}{2}$ feet to Savanne River, and consists of a wet tamarack swamp, in which moss grows everywhere to the depth of one foot or eighteen inches; the moss is supported by a retentive buff clay, which is exposed at the western extremity of the portage. The remains of an old road formed of the split trunks of trees, probably constructed in the time of the North-West Company, passes through it; it is now in a thorough condition of decay. The same may be said of all the swampy portages along this line of route. In the time of the North-West Company, this portage was doubtless one of the best, considering its length and general character, but now a false step from a rotten or half floating log, precipitates the voyageur into eighteen inches of moss, mud, and water. No physical impediment appears to exist which would prevent this portage from being drained at a very small cost, and converted into one of the best on the whole line of route.

Savanne River, to which it leads, is very rapid a little above the landing place ; but on wading up the stream for about a quarter of a mile, the occurrence of dead water without froth or bubbles, showed that the feeding swamp or lake was near at hand. Savanne River is about twenty-five feet broad here, and it continues a very meandering and crooked westerly course of about eighteen miles to Milles Lacs, or Lake of the Thousand Islands, as it is sometimes termed.

Mr. Gaudet, one of Mr. S. Dawson's assistants, ran a line between Jourdain's Rapids on Dog River and the Savanne River in 1858. The first two miles of the line

were swampy, but not considered impassable even in their
present condition, the depth of black mould over clay
being from six to eighteen inches. From the second mile
post to the seventh the country is well adapted for a road,
consisting of a sandy ridge clothed with Banksian pine
(*Cyprès*). The remaining distance, about one mile and
three-quarters, is a gradual descent to the Savanne River.
The banks of this river are altogether alluvial, and diminish gradually from ten feet in altitude, near its source,
to the level of Milles Lacs, at its entrance into that extensive and beautiful sheet of water. The immediate banks
of Savanne River are clothed with alder, willow, and dogwood ; behind these are seen tamarack, pine, spruce, and
aspen. Near its mouth much marshy land prevails, and
at its confluence with Milles Lacs it is characterised
by a large expanse of rushes and other water plants common in such situations.

While descending this tortuous stream we were surprised and delighted at hearing the exclamation, "canoes!
canoes! " from the lips of our keen-eyed voyageurs, and
soon, sweeping round a distant bend, we observed a north
canoe rapidly approaching. It contained Mr. Bell, an
officer in the service of the Hudson's Bay Company, who
had started from his post on the Mackenzie River in May
of this year, and was on his journey to Montreal, where
he proposed to spend two years of furlough. Mr. Bell
was accompanied by his daughter, a child of about twelve
years, who had journeyed for three months with her
father through the trackless wilderness separating Mackenzie's River from Savanne River. Such early experience of life in the wilderness it is the lot of few to suffer or
enjoy.

Milles Lacs was described by the Indians as extending
in a direction due west much farther than was visible from

the canoe route, on account of the numerous islands with which it is everywhere studded. In the lower portion of the Savanne River many large ponds and reedy lakes, connected together by small water courses, join with the main river, and indicate the great extension which Milles Lacs assumes in an easterly direction during spring freshets. It appears very probable that a length of twenty-five miles, with an average breadth of four miles may be taken as a fair representation of this remote sheet of water; the canoe route through it is twenty-one miles in length, from the mouth of the Savanne to Keg or Baril Portage. Granitic dome-shaped islands are very numerous, and occasional exposures of clay and sand banks come into view on the points and islands along the line of route.

Milles Lacs is drained by the River Seine, which empties into Rainy Lake. The Seine was examined in 1858 by Mr. S. Dawson, and the following notice of its general features is abbreviated from his report published in Toronto in May, 1859. Where the Seine issues from Milles Lacs it is more than one hundred feet wide. It pursues a winding course in a westerly direction through a narrow valley ("flat") thickly wooded with Banksian pine and poplar of large size. On either side low hills rise gradually; they are covered for the most part with a dense growth of poplar, interspersed here and there with tall pines, which rise singly or in groves above the surrounding forest.

"At times the valley contracts, and where it does so the river presents cascades past which a portage has to be made, or little rapids which can be run with a canoe; but between these there is generally a considerable extent of navigable water. This description will apply to the country for about forty miles below Lac des Milles Lacs. The lower part of the valley presents a succession of lakes, varying from a mile to fifteen miles in length, until near

Rainy Lake, into which the river, much increased in volume, discharges itself in a series of cascades, making a plunge of over 112 feet in the distance of five miles and a half. The lakes just referred to are bounded, for the most part, by low hills, generally wooded, but in some cases rocky, with an occasional valley between them presenting a less barren appearance."

The Seine River enters Rainy Lake at Seine Bay, one of the deep north-easterly expansions of that irregular body of water. Seine Bay and a part of the river are shown on Thompson's map of the Boundary Survey executed in 1826. A beautiful reduced copy of the geographical outlines of part of this map is published in the Quarterly Journal of the Geological Society for May, 1854, accompanying a paper on the Geology of Rainy Lake, by Dr. J. J. Bigsby. Seine Bay and the mouth of Seine River are both shown on this reduced copy.

The Seine River receives an affluent called Fire Steel River, which rises in the wide spreading marshes at the height of land, from which also Dog River, flowing into Lake Superior, issues.

Soon after leaving Milles Lacs, the Seine falls by a series of rapids, seven in number, a depth of thirty-six feet in a distance of nine miles; its waters are then precipitated twenty-four feet in two steps at Little Falls, and before reaching Rainy Lake, a distance of sixty-seven miles in a direct line, it falls 350 feet by twenty-nine steps varying in altitude from three to thirty-six feet.

The hills surrounding Milles Lacs here and there bear pine of fair dimensions, while in the narrow and shallow valleys between them there is every indication of hardwood over large areas. Exposures of white quartz are repeatedly seen on the islands and main land at the western extremity of the lake; and not unfrequently are they taken by travellers during their first voyage for the sails of distant boats. The name "sail rock," given to them by the voyageurs, is derived from this erroneous impression. Where the lake narrows on approaching Baril Portage, gneissoid hills and islands about 100 feet high show a well defined stratification dipping north, at an angle of about 15°, and on that side smooth, and sometimes roughly polished; on the south side they are precipitous and abrupt. The same character was noticed at the Baril Portage, which has a length of sixteen chains eighty-five links, with an altitude of $72\frac{1}{2}$ feet, and an *ascent* of 1·86 feet. The north-eastern exposure of the rocks here was smooth, the southern rugged and often precipitous.

Baril Lake is seven and a half miles long, and is the counterpart of the western extremity of Milles Lacs, but it belongs to a different water system, being 1 ft. 10 in. higher than Milles Lacs. It is terminated by the Brulé or Side Hill Path Portage twenty-one chains long, leading to Brulé Lake forty-seven feet below Baril Lake. At Brulé Portage I ascended a steep hill bordering a small rapid stream called Brulé River, and from an altitude of fully 200 feet, had a fine view of the surrounding country. The vegetation upon the hill side and summit was truly astonishing, and the term Brulé Portage received an unexpected interpretation on finding hidden by a rich profusion of brushwood, the dead trunks of many noble pines. Throughout the day the tall trunks of white pine, branchless and dead, rising in clumps or in single loneliness far above the forest, had attracted attention, and on the side of the Brulé Hill we observed many prostrate half-burnt trees of the largest size. One dead trunk was measured, and found to be twelve feet in circumference, five feet from the ground. A living tree, tall, clean, and apparently quite sound, measured nearly ten feet in circumference, and many of the prostrate pines were of equal dimensions.

There can be little doubt that these were the remains of a magnificent white pine forest, which formerly extended over a vast area in this region, since from the summit of the hill the forms of scattered living trees, or tall, branchless and scathed trunks, met the eye in every direction. The young second growth indicated a soil not incapable of sustaining pine trees of the largest proportions; black cherry, birch (both the white and black), alder, small clumps of sugar maple, and a thick undergrowth of hazel nut now occupies the domain of the ancient forest. The south west side of this hill formed a precipitous escarpment 150 feet above the waters of a long clear lake. All around the eye rested upon low dome-shaped hills dipping towards the north-east, and covered with a rich profusion of second growth. The vast wilderness of green was studded with black islands of burnt pine, and a few isolated living trees, serving by their surprising dimensions to tell of the splendid forest which must have once covered the country.

The soil wherever examined consisted of a red sandy loam, covered with a thin coating of vegetable mould. Occasionally bare rock exposures protruded, and granitic boulders were numerous. The uniform size of the second growth timber on the Brulé Hill, seemed to prove that the great fire which devastated this region may have occurred about thirty years since. The hill round which the portage path winds is considerably higher than any observed range on the height of land, and its summit, from which a view of the surrounding country was obtained, is probably about 100 feet above the Height of Land Lake, or 1,585 feet above the ocean; McKay's mountain having an elevation of 1,600 feet above the same level. Brulé Lake bears another name of terrible import. It is called Win-de-go or Cannibal Lake, a term applied to it in commemoration of an unnatural deed committed here by a band of Ojibways in 1811. Although not less, it is stated, than forty in number, yet they were unable to procure sufficient food to preserve them from famine. Many perished with hunger and the survivors sustained existence by feeding upon the dead bodies of their companions. The whole of the band with one exception perished, the survivor, a woman, preserved existence by murder; she, however, was not long permitted to live. Meeting with another party of Indians, who drew the dreadful secret from her, she was put to death, under the impression that those who have once fed on human flesh always retain a desire for it, which they are not unscrupulous in gratifying when opportunity offers. Several instances of cannibalism were mentioned to us by the voyageurs as having occurred on this route; and in the following summer noted spots in the basin of Lake Winnipeg were pointed out, which preserve a similar dreadful reputation. Both voyageurs and Indians always spoke of these horrible deeds in subdued tones and with an expression of anxiety and alarm.

The impression produced by a survey of the solitudes about the western extremity of Milles Lacs and of Baril Lake was rather of a favourable character. If in the course of time mineral wealth should be found to exist in profitable distribution about Milles Lacs, there would be no scarcity of arable soil between the low hill ranges of that beautiful but desolate lake to supply the wants of a mining population; or, in the event of a line of communication between Thunder Bay and Rainy Lake being established, its western shores and those of Baril and Brulé Lakes offer suitable localities for village depôts.

From Brulé Lake to French Portage, a distance of four miles, the canoe route lies through a series of lovely lakelets and short rapid streams fringed with cedar and spruce; behind these are fair-sized red pine, birch, aspen, and large spruce. French Portage bearing due west, is $1\frac{3}{4}$ miles long, and lets us down $99\frac{3}{4}$ feet into French Portage or Pickerel Lake. The timber on this portage consists of aspen, red pine, and spruce. On the shores of the lake low hills appear, wooded with an extensive forest of red pine, varied with patches of spruce, aspen, and birch. Ice formed on the upturned canoes during the night of August 16th.

Pickerel Lake, through which in a direction nearly due south-west the canoe route continues, is a fine sheet of water, thirteen miles long by two to four broad; its shores consist of low hills covered with a thick forest of pine, with spruce, aspen, and birch in the valleys. On the east side of the lake, the remains of an ancient pine forest are often visible in the forms of noble, isolated trees. These occur about six miles from its head, and further on there may be observed small groups of the same trees rising far above the comparatively young growth which now surrounds them. The half-burned standing trunks of huge dimensions, show the extent and character of the earlier forest, and the cause which destroyed their companions. White pines in large numbers still remain at the foot of the lake, and were seen at the portage, which is called Portage du Pin, also Portage des Morts. The first name is evidently derived from the prevalence of large red and white pine; the second has a melancholy reference. It commemorates the death of a voyageur, who being over anxious to cross the portage while supporting the bow of a north canoe, lost his footing, and was so much injured by the heavy burden crushing him as he sank to the ground, that he died after the lapse of a few hours. A north canoe often weighs between three and four hundred pounds when soaked by long immersion in water; this unwieldy burden is borne by two voyageurs, one at the bow, the other at the stern, when crossing the portages; and bruises, sprains, or ruptures are the frequent consequences of over-exertion, rendered necessary, however, by the present condition of the portage paths.

Portage des Morts is twenty-six chains long, and it overcomes a descent of seven feet in the small stream connecting Pickerel Lake with Doré Lake, a sheet of water about a mile across, but extending much further in a north-westerly direction.

Among the trees observed here remarkable for their size, cedar, ash, white and red pine, with birch of two kinds, may be enumerated. The cedar is far superior to any before seen. A clay sub-soil is found in the valley of a small river running near the portage path, and the upturned roots of trees on the hill-side showed fine washed white sand upon which a sandy loam rested. The foot of Doré Lake brings us to the Portage des Deux Rivières, which lets us down 117·21 feet into Sturgeon Lake, in a distance of 32 chains.

The whole country seems to sink with the French and the Deux Rivières Portage. The hills about Sturgeon Lake at its upper end are not above 100 feet high, and if the valleys and lakes were filled up between the tract of country south-west of French Portage, it would be nearly a level plain, with a slight south-westerly descent. In Sturgeon River, leading to the lake of that name, we met with the first marshy place since leaving the mouth of the

Savanne River. The canoes here were forced through a profusion of aquatic plants, among which the beautiful white water lily, with its golden-hued companion, frequently occurred. Willows, small aspen and alder, grew on the banks, but no hill or elevated tableland was visible from the shallow but tortuous river. Once on the open lake, hills about 200 feet high rose into view at some distance on the eastern side. The bushy tops of what appeared to be a grove of elms, were seen near the head of this large and beautiful sheet of water; again wide tracts of burnt land attracted attention, with a few white pines, remains of a forest long since destroyed. The northeastern termini of hill ranges slope to the water's edge, and, when bare, are found to be evenly smoothed and ground down. Everywhere on the shores of the first large expansion of the lake, remains of an ancient forest lay black and branchless, or still flourished green and erect amidst a vigorous undergrowth of spruce and aspen.

Sturgeon Lake and River, or rather a succession of lakes and rivers bearing the above names, extend for thirty-six miles from the Portage des Deux Rivières to Island Portage, which leads into Pine Lake, a small sheet of water connected by means of a broad river about three and a half miles long, with the great Nequauquon Lake or Lac la Croix.

Nine miles from its head, Sturgeon Lake was found to have forty-five feet depth of water, with a muddy bottom. The temperature of the lake was 68° at six P.M.; the pines and balsams growing near the shore were seen to be scraped or barked for about a foot near the ground by Indians, for the purpose of procuring gum or resin.

No lake yet seen on the route can bear comparison for picturesque scenery with Sturgeon Lake. The numerous deep bays, backed by high-wooded hills or rocks, rugged or smooth, according to their aspects, its sudden contraction into a river breadth for a few yards between large islands and the equally abrupt breaking out into open stretches of water, offered a constant and most pleasing variety of scene. The high jutting points of granite rock, which here and there confine the channel, offer rare opportunities for beholding on one side an intricate maze of island scenery, and on the other an open expanse of lake, with deep and gloomy bays, stretching seemingly into the dark forest as far as the eye can reach.

Here we met several Ojibways in their elegant birch bark canoes. They were very friendly, and apparently delighted with a small present of tobacco and tea. One young hunter with his squaw hurried to the shore as we approached, but soon returned gaudily painted with patches of vermilion on his cheeks and in bars across his forehead.

The fourth large expanse of Sturgeon Lake is bounded by low, densely-wooded shores, with high hill ranges in the far distance. The first cascades, with a fall of four and a half feet, occur at the foot of this last expansion; these are quickly followed by the second falls of six and a quarter feet descent, then occurs a narrow reach of river for three miles, which is terminated by the third rapids of two and a half feet fall, leading to another expanse with a general direction nearly due west; the fourth and fifth rapids then occur within four miles of one another, and are followed by Island Portage two miles further on.

Island Portage lets us down ten feet, and involves a portage of fifty yards. Crossing the small Pine Lake, the river now assumes a course nearly due west, and within a distance of four miles, brings us to a north-

eastern arm of Lac la Croix. The canoe route passes
near the north shore of this extensive and beautiful lake.
High precipitous rock exposures begin to show themselves, often clothed with dense groves of pine rising
above the mass of light green aspen foliage which prevails.
Although Lac la Croix is fourteen or fifteen miles long,
yet our traverse did not exceed eight, for we entered the
Nameaukan River, which issues from the north-western
coast, and takes a circuitous north-westerly direction,
bringing us to Rattlesnake Portage, where the river descends by a beautiful cascade 12·14 feet, involving a portage of 110 yards.

We camped at the edge of the cascade, the portage
path offering the only even spot where our blankets could
be spread. The guide pointed significantly to the surging
waters at the foot of the falls, and with a quiet smile said,
" better not walk much in night." Three steps from my
resting-place would have precipitated me into the rapid,
and as a somnambulist happened to be one of the party,
he was carefully warned not to indulge in midnight explorations. The noise of the cascade effectually drove
sleep from my eyes, and although the night was really
short, it seemed an interminable age. Generally my
sleep was excellent, however hard the bed or stormy the
night, yet if rain did not penetrate the canvas tent, I slept
soundly and well, invariably awakening with the first
streak of day.

The dawn of morning and the early start in this rocky
wilderness possess some characteristics peculiar to the
country and the strange companions with whom necessity
compels you to associate. Rising from a bed on the hard
rock, which you have softened by a couple of rugs or a
north blanket, and if time and opportunity permitted by
fresh spruce or pine boughs, the aspect of the sky first

claims and almost invariably receives attention. The morning is probably calm, the stars are slightly paling, cold yellow light begins to show itself in the east; on the river or lake rests a screen of dense fog, landwards a wall of forest impenetrable to the eye. Walking a step or two from the camp a sudden rush through the underbrush tells of a fox, mink, or marten prowling close by, probably attracted by the remains of last night's meal. From the dying camp fires a thin column of smoke rises high above the trees, or spreads lakewards to join the damp misty veil which hides the quiet waters from view. Around the fires are silent forms like shrouded corpses stretched at full length on the bare rock or on spruce branches carefully arranged. These are the Indians, they have completely enveloped themselves in their blankets, and lie motionless on their backs. Beneath upturned canoes, or lying like the Indians, with their feet to the fire, the French voyageurs are found scattered about the camp; generally the servant attached to each tent stretches himself before the canvas door. No sound at this season of the year disturbs the silence of the early dawn if the night has been cold and calm. The dull music of a distant waterfall is sometimes heard, or its unceasing roar when camped close to it as on the Rattlesnake Portage, but these are exceptional cases, in general all nature seems sunk in perfect repose, and the silence is almost oppressive. As the dawn advances an Indian awakes, uncovers his face, sits on his haunches and looks around from beneath the folds of his blanket which he has drawn over his head. After a few minutes have thus passed, not observing his companions show any sign of waking or disposition to rise, he utters a low " waugh "; slowly other forms unroll themselves, sit on their haunches and look around in silence. Three or four minutes are allowed to pass away when one of them rises and arranges the fire, adding fresh wood and blowing the embers into a flame. He calls a French voyageur by name, who leaps from his couch, and in a low voice utters "lève, lève." Two or three of his companions quickly rise, remain for a few minutes on their knees in prayer, and then shout lustily "lève, messieurs, lève." In another minute all is life, the motionless forms under the canoes, by the camp fires, under trees, or stretched before the tent doors, spring to their feet. The canvas is shaken and ten minutes given to dress, the tent pins are then unloosened and the half dressed laggard rushes into the open air to escape the damp folds of the tent now threatening to envelope him. Meanwhile the canoes are launched and the baggage stowed away. The voyageurs and travellers take their seats, a hasty look is thrown around to see that no stray frying pan or hatchet is left behind, and the start is made. An effort to be cheerful and sprightly is soon damped by the mist into which we plunge, and no sound but the measured stroke of the paddle greets the ear. The sun begins to glimmer above the horizon, the fog clears slowly away, a loon or a flock of ducks fly wildly across the bow of the first canoe, the Indians and voyageurs shout at the frightened birds or imitate their cry with admirable accuracy, the guide stops, pipes are lit, and a cheerful day is begun.

After leaving Rattlesnake Portage, rapids and falls follow one another in quick succession. The most important are Crow Portage, with 9·88 feet fall; the Grand Falls Portage 16 feet, and the great and dangerous Nameaukan Rapids, letting the river down in steps between fifteen and sixteen feet. In descending the Grand Rapids, my canoe had a narrow escape. Lambert acted as steersman, and

Printed by Spottiswoode and Co.

GRAND FALLS OF THE NAMEAUKAN RIVER.

[New-street Square, London.

Charley, an Ojibway Indian, as bow-man Lambert was not strong enough to give the proper direction to the canoe in order to avoid a rock jutting out at the head of the rapid. Just as we made the leap, the stern, borne swiftly round by the current, grazed the rock and tore the bark, without, however, doing serious damage. The moment Charley felt the graze, he turned round, brandished his paddle, and shook it at the unfortunate Lambert; we shot down the rapid with great velocity, and embraced the opportunity afforded by the first safe eddy to examine the bark of the canoe. We were deeply laden, and the bottom of the canoe was so covered with our baggage, that no part was visible. "Put your fingers to the bottom of the canoe, monsieur," said Lambert to me; "how much water?" "Two inches," I replied. "That will do, we shall not make more water now we are out of the rapid, it is only a crack, and the bark is tough." We made, however, three inches of water in a short time, and as the baggage was in danger of being wetted, it was deemed advisable to gum the leak without unnecessary delay.

The shores of Nameaukan River are fringed with the Banksian pine, and where an alluvial soil has accumulated, the aspen grows to a large size. Where the river debouches into Nameaukan Lake, there is a fine grove of ancient elms, and underneath their wide-spreading branches we found a large encampment of Indians.

The traverse across Nameaukan Lake is six and a half miles in length, the lake itself extending for more than double that distance in a direction due west. At the extremity of the traverse is the Nu Portage, where the descent is eight and a half feet, leading us into a narrow circuitous river, without perceptible current, which meanders through a reedy expanse, fringed with low willows' for about three miles. The canoe route then takes a winding course, whose general direction is nearly due north for a distance of two and a half miles, when turning westward we suddenly arrive at the open and beautiful, but indescribably barren and desolate region of Rainy Lake.

The canoe route followed by the North-West Company, commonly called the Pigeon River route, joins the chain of communication which has just been described in Nequauquon Lake; from this point they both pursue the same course by the Winnipeg to Red River. The difficulties of the Grand Portage induced the North-West Company to establish their chief depôt at Fort Charlotte, nine miles west of Grand Portage Bay. Fort Charlotte was connected with Point des Meurons by a traveled road in the time of the North-West Company, but owing to neglect and the diversion of traffic, it has long been choked up with young trees, and now only serves as an Indian path. There is also an old cart road between Point des Meurons and White Fish Lake, which is close to Arrow Lake.

In 1858 I took the expedition under my charge by the Pigeon River route, with a view to compare its facilities with those of the more northern communication by the Kaministiquia. The following brief notes of this route will complete the description of the available water communications between Lake Superior and Rainy Lake within British territory.

The Grand Portage made to overcome the falls of Pigeon River, 120 feet high, has been often cited as the chief obstruction to the Pigeon River route. Its length is eight miles fifteen chains. The road is dry, and in comparison with some of the portages on the Kaministiquia route, in good condition. It is passable for an ox team, which is employed by the people in charge of the American trading post in forwarding their supplies.

I endeavoured to procure a waggon and team from the American traders at Grand Portage Bay to transfer the heavy baggage from the east to the west end of the portage, but although the vehicle was available the team was not; one ox having died during the winter, and the other was in such a miserable condition that he could scarcely draw the empty waggon.

The passage of the Grand Portage consequently occupied five days instead of two, and in making a comparison between the two canoe routes to Lake Winnipeg, these facts must be borne in mind. In 1857 the Red River expedition landed at Fort William on the 31st of July, and reached the Settlements on the 4th of September, having been thirty-four days on the road, or forty from Toronto. The expedition of 1858 reached Grand Portage on the 5th of May, and arrived at the Stone Fort on the 2nd of June, a period of twenty-eight days, or thirty-four from Toronto. The Grand Portage lying within the territory of the United States loses all interest as the terminus of a Canadian route; but that part of the water communication which forms the boundary line, and the country between Arrow Lake, White Fish Lake, and Fort William, seems to acquire importance in proportion to the extension of our knowledge respecting its capabilities and resources.

The waters on the rivers and lakes on the east side of the Height of Land, the Lake Superior water-shed, were high in 1858, while those on the west side, or the tributaries to Lake Winnipeg, were unprecedently low. In many of the lakes recent water-marks, four and five feet above the present level, were frequently observed. This remarkable lowness of the water was attributed by the half-breeds and

Indians to the very small quantity of snow which fell on the western slope during the winter of 1857.

It is important to bear in mind that the voyage of the expedition of 1858 was made under the great disadvantages inseparable from unusually low water, and whatever superiority the route appears to possess over that of the Kaministiquia by Fort William, will be much more apparent in ordinary seasons, when the lake and river levels are from two to five feet above their present altitude.

On our arrival at Moose Lake, May 12th, a glistening sheet of solid ice overspread its surface, and seemed to threaten a long delay; but by noon on the following day, under the influence of a hot sun and a gentle breeze, lanes of water opened, through which we succeeded in passing the canoes, and on the evening of the same day a high wind, accompanied by rain, completely broke up the ice in the higher lakes, and opened the communication.

The part of the Pigeon River route to which this notice refers, commences at Arrow Lake, a fine expanse of water close to White Fish Lake, lying in a north-easterly direction, and within thirty miles of the Kaministiquia.

From Arrow Lake, a short portage brings us into Rose Lake on the course of the old North-West Company's route, following the boundary line.

The portages between Rose Lake and the Height of Land are short and low, while the Height of Land Portage is not 500 yards long, and does not rise above fifty feet. The passage from the St. Lawrence water-shed to that of Lake Winnipeg is short, easy, and dry, incomparably superior to the Prairie Portage, and the Great Savanne on the Kaministiquia route. In consequence of the very low stage of the water that year, numerous small rapids were formed in the rivers connecting Gun Flint

Lake with Lake Seiganagah ; during ordinary seasons, these rapids are passed without difficulty, but in 1858 they involved the portage of a portion of the baggage, and the letting of the canoes down them by rope.

From Lake Seiganagah * an Indian route passes into Little Seiganagah Lake, which is connected with Sturgeon Lake on the northern route. The Little Seiganagah is a favourite wintering place of numerous families of Indians ; it abounds in fish, and near its shores the winter road to Fort William runs.

Between Knife Lake and Birch Lake there are two routes, one coinciding with the boundary line, the other, which we followed, passing in a north-westerly direction, causing us to make two portages instead of one, but escaping some rapids.

From Nequauquon Lake one route passes into the Nameaukan River, and another, turning south, follows the boundary line through Loon's Narrows, and then north into Nameaukan Lake. Our guide preferred going by Loon's Narrows, fearing that the always dangerous Nameaukan Rapids would be almost impassable for heavily laden canoes, on account of the low stage of the water.

In Loon's Narrows we found a shallow river with a strong current and many boulders, and in making the north-westerly turn, instead of the broad channel shown on the map of the Boundary Commissioners, a very tortuous, sluggish and shallow stream, led us into the south arm of Sand Point Lake.

Sand Point Lake is connected with the Nameaukan Lake by a broad channel, and it is at this point that the route through Loon's Narrows coincides with the more

---

* Seiganagah or "Full of Islands."

northern route, and follows the boundary line through Rainy Lake to Fort Frances.

At the close of the 2nd volume there is a table of the portages, décharges, rapids, lakes, lake straits, and navigable channels on the Pigeon River route, from Lake Superior to Rainy Lake, showing their lengths and distances from Lake Superior.

CHAP. IV.

RAINY LAKE TO THE SOURCE OF THE WINNIPEG RIVER.

Rainy Lake.—Description of.—Rainy River.—Affluents of Rainy River.—
Fort Frances.—Lac la Pluie Indians.—Valley of Rainy River.—Cha-
racter of the Valley.—The Winter Road to the Lake of the Woods.—
Arrangement for crossing the Swamps to Red River direct from the
Lake of the Woods.—Fertility of Rainy River.—The Manitou Rapids.
—Obstructions to Navigation.—The Long Rapids.—Indian Encamp-
ments.—Tumuli.—Graves.—Banks of Rainy River.—Caterpillars.—The
Lake of the Woods. — Beauty of the Lake of the Woods.— Confervæ.
— Garden Island.— Refraction.— Indians.— A Council. — Its Results.
—Grasshoppers.—Shoal Lake.—North West Corner of the Lake.—Monu-
ment Bay. — Route to Rat Portage. — Indians.—Sturgeon. — Polished
Rocks.

In 1826 a map of Rainy Lake, as part of the survey
under the seventh article of the treaty of Ghent between
Great Britain and the United States, was constructed by
David Thompson, Astronomer and Surveyor.   No labour
was spared in producing a correct delineation of the
geographical features of this part of the country, and the
portion of the map accompanying this narrative, which
includes Rainy Lake, Rainy River, and the Lake of the
Woods, besides the Pigeon River referred to in the previous
chapter, is reduced from an authorised copy of those parts
of the survey.   Dr. Bigsby, who accompanied the Commissioners as Geologist, communicates the chief facts in
the following enumeration of the geographical position,
&c., of Rainy Lake, in the Quarterly Journal of the
Geological Society for May, 1851.*

* On the Geology of Rainy Lake, South Hudson's Bay.   By Dr. J. J.
Bigsby, F.G.S., &c.

Rainy Lake, or Lac la Pluie, as it is more frequently called by the voyageurs, is 225 miles west of Lake Superior, and from the head of Rainy River eighty-five southeast of the Lake of the Woods. It is fifty miles long by thirty-eight and a half broad, and is 294 miles round by canoe route. Its form is that of three equal troughs, with deep lateral indents, the main one running in an east and west direction, the other two northerly from it. It is through the main trough that the canoe route lies from the mouth of Nameaukan River, in latitude 48° 30′ N., longitude 92° 40′ W. to the source of Rainy River, thirty-eight miles distant, in a direction a few degrees to the north of west.

The shores of Rainy Lake are generally low, and often consist of naked shapeless masses of rock, with marshy intervals, or they rise in ridges which become hills 300 to 500 feet high, half a mile to four miles from the lake. The timber seems to be very small and thin in the marshes, and on the islands, which exceed 500 in number, the largest growth was observed. Taken as a whole, the general aspect of the shores is forbidding, and furnishes on the ridges and hill flanks a picture of hopeless sterility and desolate waste. Dr. Bigsby says that there is but little loose débris about Rainy Lake, the earth or gravel banks being scarce, and seldom exceeding a few feet in thickness. Whenever the land rises, for the most part bleached and naked rocks occur for many square miles together.

Colonel Lefroy made Rainy Lake 1,160 feet above the sea, by barometrical measurement. Its height deduced from the levels taken at the portages, and the estimated rise and fall in the current of the rivers along the line of route, was 1,035 feet. In this calculation the level of Lake Superior is taken at 600 feet above the ocean.

FALLS ON RAINY RIVER OPPOSITE FORT FRANCES.

Printed by Spottiswoode and Co.]

[New street Square, London.

Major Long estimated it to be 1,200 feet above the same level.

Rainy Lake freezes about the 1st December, and is open about the 1st of May ; as is usually the case where large rivers issue from spacious lakes, the discharging stream is not frozen for a number of miles from its source. The warm waters coming from beneath a shelter of ice in their capacious feeding lake, retain their heat so as to enable them to resist the cold of these regions for many miles below the Chaudière Falls.

At the entrance of Rainy River on the evening of August 19th, the delightful odour of the balsam poplar (*Populus balsamifera*) loaded the air, and seemed to welcome our arrival in a region differing altogether from those through which we had lately passed. Where Rainy River issues from Rainy Lake, it is a broad and rapid stream, with low alluvial banks, clothed with a rich second growth. The fine forests with which they were once covered had long since been stripped of their ornaments by the occupants of the old North-West and the present Hudson's Bay Company's Fort.

The general course of Rainy River is a few degrees to the north of west for a distance of eighty miles by the windings of the river, and in an air line sixty miles. The rapids at its source offer no impediment to skilful navigation, nor do the whirlpools which usually accompany the passage of such a large body of water, in consequence of their being distributed over a wide area. Two miles below the head of the river, Fort Frances is situated on a high bank, just below the Chaudière Falls. These magnificent cascades let the river down 22·88 feet, and at their foot is a famous fishing ground, from which the Lac La Pluie Indians obtain an abundant supply of their staple food. Three miles from Fort Frances, the river takes a sudden southerly bend, which it maintains for a distance of four miles ; it then again assumes a course due west for about sixteen miles, and receives the Pekan, or Little Fork, the Missatchabe, or Big Fork, and the Kakmaskatawagan Rivers, on the south or United States side ; the course then turns abruptly due north, and continues for a distance of six and a half miles, when it again resumes a westerly direction for eighteen miles, its otherwise gentle and uniform current is here broken by the Manitou Rapids and Long Rapids, which let the river down about two and a half feet, and three feet respectively. Six miles from the Long Rapids a short northerly bend again occurs, after which the river, with slight meanderings, pursues a north-west by west direction, until it debouches into the Lake of the Woods. In this part of its course it receives on the British side, small sluggish streams known by the names of the Kiskarko or Pine River, the Kahlawakalk, and Kawawakissinweek, and from the territory of the United States, the Muttontine, or River of the Rapids, the Wishahkepekas, and Kapowenekenow, or Winter Road River. Its affluents, on the British side, are insignificant outlets to the swamps which occupy the region north of Rainy River valley ; but some of those on the United States side are of important dimensions.

Fort Frances, two miles from the source of Rainy River, is situated on the right bank, in lat. 48° 36′, and longitude 93° 33′ W. Mr. Pether, the gentleman then in charge, stated that the river never freezes between the falls and the Little Fork, a distance of twelve miles, nor between the falls and its source in Rainy Lake. Wheat is sown at this establishment of the Honourable Hudson's Bay Company, from the 20th to the 23rd of May ; it ripens about 1st September. Potatoes, turnips, carrots, and indeed all common culinary vegetables succeed well.

Potatoes are dug in the first week of October, and barley is ripe by the middle of August. Snow falls here to the depth of four feet.

The great enemies to extended cultivation are the Lac la Pluie Indians. They are not only numerous, but very independent; and although diminishing in numbers, they frequently hold near Fort Frances their grand medicine ceremonies, at which 500 and 600 individuals sometimes assemble. The number of Indians visiting this

Ojibways at Fort Frances, Rainy River.

fort for the purpose of trade, reaches 1,500. They do not scruple to jump over the fences, and run through the growing crops, if the ball in their games is driven in that direction.

In the immediate neighbourhood of Fort Frances, the swamp or morass bounding the valley of Rainy River on the right bank, is about half a mile in its rear. This swamp, which extends from Rainy Lake to the Lake of the Woods, is described by Mr. Pether, and the Indians who were questioned about it, as consisting of a springy, movable surface, overlying a vast deposit of peat, through which a pole might frequently be pushed to the depth of thirty feet without reaching the bottom. The surface sustains low bushes, with here and there islands of small pine. Its borders approach and recede from Rainy River with the windings of that stream ; the breadth of the dry wooded and fertile valley varying from half a mile in the rear of Fort Frances, to six or eight miles in the direction of the Lake of the Woods. The average breadth of superior land for a distance of seventy miles might perhaps, with propriety, be assumed to be not less than four miles, giving an area of available soil of high fertility, exceeding 170,000 acres ; and there can be little doubt, that with the progress of clearing, much that is now included in the area occupied by swamps would be re-claimed without difficulty or great expense.

In 1858 Mr. S. Dawson was instructed to examine the country on the right bank of Rainy River, with a view to ascertain the extent of surface available for the purposes of an agricultural settlement. He reports as follows :—

" The land immediately bordering on Rainy River, on the British side, is of an alluvial description, and almost as uniformly level as the prairies at Red River. For a mile or so inland from the main stream the ground is dry, and a dense growth of large timber, consisting of poplar, elm, oak, basswood, and occasional white pines, indicates a productive soil. For a mile or two beyond this, however, swampy ground predominates, while beyond that again the land gradually rises to a range of hills of no great eminence, which, as far as we could observe, seemed to run parallel to the river, at a distance of from four to eight miles back. The distance from Rainy Lake to the Lake of the Woods, following the windings of the stream, is about eighty miles, and throughout the whole of this extent the land fronting on the river is fit for settlement."

At Fort Frances the party was separated into three divisions, with a view to explore different routes. Mr. Napier was furnished with a guide to conduct him by the Rivière du Bois, from one of the north-westerly bays of Rainy Lake to the Lake of the Woods. The result of this exploration established the fact, that however advantageous this route may be for Indians, in their small canoes, it is far inferior to that by Rainy River, and the Lake of the Woods, as a boat communication. In describing it Mr. Napier says :—

" This is the winter road, and is preferred to the route by the Rainy River, as being more sheltered and free from the long open traverses necessary in crossing to the Rat Portage from the mouth of Rainy River. From Rainy Lake this road follows a chain of small lakes and connecting creeks, with occasional portages, until the north-east corner of the Lake of the Woods is reached, where the route continues through the numerous islands to the Rat Portage. The land throughout is rugged, rocky, and timbered with spruce and birch."

Mr. Gladman, accompanied by the remainder of the party, with the exception of Mr. S. Dawson, myself, and two French Canadian voyageurs, took the old route by Rainy River, the Lake of the Woods, and the Winnipeg, to Red River.

On Saturday, August 22nd, I started from Fort Frances at noon, in company with Mr. S. Dawson, for Muskeg River, a small stream flowing easterly from the swamps which occupy the summit of the water-shed between the

valley of Red River and the Lake of the Woods; our
object being to ascend the Muskeg River, cross the
swamps, and descend the Roseau River to its junction
with Red River. We were provided with two small
canoes fit for transportation through the swamps, and
supplied ourselves with provisions to last for ten days,
one change of clothing, a small tent, and a pair of
blankets each.

Fort Frances, Rainy River.

In Mr. Dawson's canoe were a French Canadian
(François) and an Iroquois (Pierre). In my canoe an
Indian guide from Garden Island, Lake of the Woods,
and Lambert, who acted as interpreter.

In describing the general aspect of the banks and
valley of Rainy River, it will be advantageous to sketch
with considerable minuteness the features of the soil and
vegetation at the different stopping places, where very
excellent opportunities were offered for acquiring information on these particulars.

The ground at our camp, twelve miles below Fort Frances, was covered with the richest profusion of rose bushes, woodbine, convolvulus in bloom, helianthii just beginning to flower, and vetches of the largest dimensions. Fringing this open interval of perhaps 280 acres in extent, were elms, balsam-poplars, ashes, and oaks. One elm tree measured three feet in diameter, or nine feet eight inches in circumference; and there is no exaggeration in saying that our temporary camping place was like a rich, overgrown, and long neglected garden.

Similar intervals to the one just described were noticed occasionally as we descended the river; the banks preserving an average altitude of about forty feet, and sustaining a fine growth of the trees before enumerated. No part of the country through which we have passed west of Lake Superior can bear comparison with the rich banks of Rainy River which everywhere preserves a very uniform breadth, varying only from 200 to 300 yards. The soil is a sandy loam at the surface, much mixed with vegetable matter, but where the bank has recently fallen away, clay may be seen stratified in layers of about two inches in thickness, following in all respects the contour of what appears to be unstratified drift clay below. Basswood is not uncommon, and sturdy oaks, whose trunks are from eighteen inches to two feet in diameter, were found in open groves with luxuriant grasses and climbing plants growing beneath them. The lodge poles of an Indian camp of former seasons were covered with convolvulus in bloom, and the honeysuckle twined its long and tenacious stems around the nearest support, living or dead.

The banks of the river maintain for twenty miles an altitude varying from fifteen to sixty feet. Occasionally, they show the abrupt boundaries of two terraces, the lower boundary having the form of a sloping bank or an abrupt cliff from fifteen to thirty feet in altitude; in its rear the upper terrace rises gradually or abruptly from fifteen to twenty feet higher, according to its position with reference to the river. There is every appearance in places of fire having destroyed a former larger growth of trees than those which now occupy these areas.

The extraordinary height of the water in August 1857, was seen by the lodge poles of former Indian encampments at the foot of the bank. They were under water to the depth of one and even two feet. The river does not appear to rise high in the spring, as the trees fringing the banks to the water's edge show no action of ice. The difference between the highest and the lowest water levels may be seven feet, and no record of recent higher levels meets the eye.

The Manitou Rapids let us down about two and a half feet, and appear to be caused by a belt of rock crossing the river at nearly right angles to its course. On the American side the hill range has an altitude of about eighty feet, on the British side it is much lower, and appears to subside rapidly in gentle undulations. The Manitou Rapids are capable of being ascended by a small steamer of high power without difficulty, and cannot be considered as presenting an obstacle to the navigation of this important stream as long as the water maintains its present altitude, which is about three feet higher than is usual at this season of the year, but often exceeded in the spring and fall. Two locks of ten feet lift, with one guard lock, would overcome the falls at the mouth of the river, and thus form a splendid water communication between the head of Rainy Lake and Rat Portage, Lake of the Woods, by the north-west coast, a distance of 190 miles, or between the head of Rainy Lake and the north-west point of the Lake of the Woods, a distance of 170 miles.

High clay banks are exposed above and below the Rapids, and many hundred acres are very scantily timbered with second growth.

Ascending the bank two miles below the Rapids, I was much surprised at the number of birds of different kinds chirruping and singing in the light and warmth of a bright morning sun. I heard more birds in ten minutes here than during the whole journey from the Kakabeka Falls on the Kaministiquia to the mouth of Rainy River.

At the second or Long Rapids an extensive area denuded of trees presents a very beautiful prairie appearance. Here we landed to examine two immense mounds which appeared to be tumuli. We forced our way to them through a dense growth of grasses, nettles, and helianthii, twisted together by the wild convolvulus. On our way to the mounds we passed through a neglected Indian garden, and near it observed the lodge poles of an extensive encampment. The garden was partially fenced, and contained a patch of helianthii, six and seven feet high in the stalk, and just beginning to show their flowers. The wild oat attained an astonishing size, and all the vegetation exhibited the utmost luxuriance. The mound ascended was about forty feet high and one hundred broad at the base. It was composed of a rich black sandy loam, containing a large quantity of vegetable matter. On digging a foot deep no change in the character of the soil was observable. The Indian guide called them underground houses; he informed Lambert that a tradition existed regarding the origin of these mounds which he had often heard his father repeat in the spring of the year when his tribe assembled at the foot of the Rapids to catch sturgeon. About two hundred

and fifty years ago, so runs the tradition, a large party of
Sioux had penetrated from the south-west into the hunting
grounds of the Ojibways, to make war upon them in the
heart of their country and at their best fishing station.
The Sioux were driven back, but with terrible loss to the
Ojibways. A grand council was held after the defeat of
the enemy, and it was resolved to erect a number of
mounds or underground houses, wherein a store of provisions might be laid up and the women and children
retreat in case of a sudden invasion. These mounds were
the result of that determination, and are regarded by the
Lac la Pluie Indians, according to my guide, as the for-
tifications which their ancestors erected to protect their
families from the invading Sioux, to enable the warriors,
freed from the embarrassment occasioned by the presence
of their wives and children, to harass the enemy in all
directions, and possibly cut off his retreat.

The Rainy River mounds are far larger than the burying-
places or ossuaries which are scattered over Canada, and
especially on the south-eastern shores of Georgian Bay,
Lake Huron. It might be supposed by many familiar
with those curious mounds containing the bones and
relics of the barbarous people who occupied Canada some
centuries since, that, like the smaller ossuaries distributed
in the forests of the Lake Region, the gigantic mounds of
Rainy River were places of sepulture. The custom of
burying the remains of many individuals in one spot and
heaping over them a mound of earth was common in
remote times among the wandering tribes who hunted
over the rocky and barren plateau north of Lakes Huron
and Superior. The dead were laid upon the bare rock
and covered with stones to protect the body from wild
animals. After a certain number of years the tribe made
a gathering of their dead, and bore the bones to a suitable resting-place where earth existed in sufficient

abundance to admit of a mound being made without difficulty. Over very extensive areas on the north shore of Lake Huron this could not be easily accomplished, in consequence of the rocky character of the country and the general absence of loose earth, except in the valleys of rivers liable to annual overflow.

The modern graves of Indians are numerous on Rainy River; they have each a little birch bark roof placed over them, and facing the south a small opening is left, through which the relatives introduce tobacco, rice, or other offerings at their periodical visits.

About three hundred yards below the second rapids, which have a fall of three feet at the present high stage of water, twenty-three skeletons of Indian lodges were seen, all clothed with the wild convolvulus, and now serving as records of the love of change which seems to form a leading characteristic in the habits of the barbarous race who possess, without appreciating or enjoying them, the riches of this beautiful and most fertile valley. Limestone fragments and boulders, more or less water worn, with pebbles of the same rock, are found everywhere on the beach, at the foot of the clay or loamy banks.

As we approached the Lake of the Woods the river increased in breadth, and at each bend a third low terrace was in process of formation, often from 200 to 300 acres in area, and elevated above the present high water level from one to three feet. Coarse grasses grow in great abundance upon many of these rich outlying alluvial deposits, and it appeared very probable that in ordinary seasons they would furnish some thousand acres of rich pasture land, as the grasses they sustain are like those which on the Kaministiquia the settlers cut for their winter supply of fodder for cattle. Near the mouth of the river the tall tops of a few red and white pine may be seen, which rise far above the aspens occupying the lower plateau, while a vast reedy expanse, probably in ordinary seasons available for grazing purposes, marks the junction of Rainy River with the Lake of the Woods.

Rainy River flows upon an alluvial bed partly of its own formation, the materials being probably derived in a great measure from the cutting away of the clay and sand which constitute the higher of the two terraces by which its boundary is so well defined. The first or lowest terrace being generally from twelve to fifteen feet above the present water level, frequently terminates on the river in abrupt, low, clay bluffs, capped with loam and sand, or rich alluvial deposits; sometimes both terraces come upon the river together in one bold bluff, often forty feet in altitude, and again the lower terrace is found to occupy the bank without the higher one in the rear being visible from a canoe.

The separation of these terraces is an important item in the description of the topography and general characteristics of Rainy River. Where the lower terrace is alone visible, the vegetation it sustains is often characteristic of a poor and sandy soil. Red pines, some of them of fair dimensions, red cedar and small poplars occupy it, and if any passer by were to draw an inference from the prevailing timber which, in such situations, meets the eye, he would at once form the opinion that the land was comparatively worthless. But let him cross the lower terrace until he reaches, at a distance of 200 yards or perhaps a quarter or half a mile, the higher one, and the magnificent growth of poplar, elm, and basswood would quickly reverse such judgments. As far as I penetrated in different places back from the river, the soil of the higher terrace was of admirable quality, and supported a heavy growth of timber. The clay upon which it rested was often exposed by the steep banks of numerous sluggish streams which cut the terrace to nearly the level of

Rainy River, and evidently form channels by which the swamps in the rear are drained.

On the 6th June, 1846, Mr. Kane* saw the trees on each side of Rainy River, and part of the Lake of the Woods, for full 150 miles of the route, literally stripped of foliage by myriads of green caterpillars, which had left nothing but the bare branches. In such extraordinary numbers were these destructive insects that Mr. Kane's party found it impossible to breakfast on land unless they submitted to the unpleasant addition of numbers of these caterpillars to their meal, dropping as they did from the trees without intermission and covering the ground.

Sir John Richardson relates that in 1847 multitudes of caterpillars spread like locusts over the neighbourhood of Rainy River. They travelled in a straight line, crawling over houses, across rivers, and into large fires kindled to arrest them. Throughout the whole length of Rainy River, on the Lake of the Woods, and on the River Winnipeg, they stripped the leaves from the trees and ate up the herbage. They destroyed the *Folle avoine* (wild rice) on Rainy Lake, but left untouched some wheat that was just coming into ear. This was the first time that Fort Frances had experienced such a visitation. The following year Sir John Richardson found the still leafless trees covered with the cocoons of the previous year, in each of which there remained the hairy skin of a caterpillar. In 1858 we noticed the trees in the Bad Woods, on the north bank of the Assinniboine, covered with an incredible number of small green caterpillars, resembling the palmer worm, so destructive in the United States in some seasons.

The Lake of the Woods is about seventy-five miles in

* Wanderings of an Artist among the Indians of North America, by Paul Kane.

length, and the same in breadth. It is 400 miles round by canoe route*, and is broken up into three distinct lakes by a long promontory, which in periods of high water becomes an island. The southern part is termed the Lake of the Sand Hills, the eastern portion White Fish Lake, and the northern division the Lake of the Woods. White Fish Lake and Lake of the Woods are separated from Sand Hill Lake by the broad promontory before referred to, respecting which little is known. The name of the latter division is derived from vast numbers of low sand hills, which occupy its south-western coast. The distance of the Lake of the Woods from Lake Superior, is north-west 325 miles by the Pigeon River route, and 381 by the route from Fort William, followed by the expedition. The north-west corner of the lake is only about ninety miles from Red River, in an air line. Its elevation above Lake Superior is 377 feet, or 977 feet above the sea. Major Long made it 1,040 feet above the ocean level.

The scenery among the islands towards the north-west corner of the lake is of the most lovely description, and presents in constantly recurring succession every variety of bare, precipitous rock, abrupt timbered hills, gentle wooded slopes, and open grassy areas. Some of the islands are large and well timbered, others show much devastation by fire, and often a vigorous second growth of a different kind of tree under the blackened trunks of branchless pines.

The ordinary course of the canoe route to Red River lies in a north-easterly direction, following the trend of the coast towards Turtle Portage, which leads from the Lake of the Sand Hills to White Fish Lake. In pursuance of our intention to endeavour to pass from the west

* See vol. viii of the Quarterly Journal of the Geological Society, for an account of the Lake of the Woods, by Dr. Bigsby.

side of the Lake of the Sand Hills across the country, in
as direct a line as possible to Red River, we made a
traverse in a north-westerly direction towards the south
point of Keating Island, a distance of sixteen miles. The
surface of the lake was perfectly smooth, reflecting the
sun's rays with extraordinary power and brilliancy. As
we receded from the shores the low sand dunes to the
south-west were refracted into the similitude of distant
mountain ranges, and the rocky coast of the eastern side,
as seen through a glass, into high, precipitous, half wooded
cliffs.

The origin of the sand dunes is interesting; they probably point to the existence in this neighbourhood of the
Chazy formation (Lower Silurian) so characteristic of the
west coast of Lake Winnipeg.

About four miles from land the water became tinged
with green, deriving its colour from a minute vegetable
growth (confervæ), which increased as we progressed,
until it gave an appearance to the lake like that of a vast
expanse of dirty green mud. On lifting up a quantity of
water in a tin cup, or on looking closely over the side of
the canoe, the water was seen to be clear, yet sustaining
an infinite quantity of minute tubular needle-shaped
organisms, sometimes detached, and sometimes clustered
together in the form of small spherical stars, varying from
a quarter to half an inch in diameter. Five miles from
the shore the lead showed thirty-five feet of water, and
four miles further on thirty-six feet; the green confervæ
increased in quantity, and the little aggregations assumed
larger dimensions, some of them exceeding one inch in
diameter.

The temperature of the lake near the mouth of Rainy
River was 67° at half-past eleven, A.M.; yet five miles
from land it was found to be 76°, six inches below the
surface; an hour afterwards, repeated and careful obser-

vations showed the temperature to be $77\frac{1}{2}°$. At one, P.M. the temperature two feet below the surface was 71°, and at the surface 78°. The depth of water was here thirty-six feet, and the green confervæ uniformly abundant, so that it was impossible to obtain a tablespoonful of liquid free from their minute forms. The presence of this " weed," as the voyageurs termed it, was the probable cause of the unusual temperature of the lake. Occasionally grasshoppers were seen resting on its calm glistening surface, and as we approached Keating Island they increased in number, all of them preserving, with singular uniformity, a direction towards the southeast. The Indians think the " weed " proves destructive to fish; they had seen it on Lake Winnipeg, where also we recognised it in September of the following year.

After passing the south point of Keating Island we steered for Garden Island, distant from us about nine miles. On the west side of Keating Island the Indian guide pointed out one of their fishing grounds, where he stated the water was thirty fathoms deep, and illustrated the manner in which he arrived at that estimate of the depth by explaining, through the interpreter, the mode of fishing during the winter months, showing the length of a fathom and the number of these in the lines his people employed to reach with their nets the feeding grounds of the white fish at that period of the year. He also described the thickness of the ice through which they had to break before they arrived at the water as sometimes exceeding five feet.

Wild rice (*Zizania aquatica*) grows very abundantly in the marshes bordering the Lake of the Woods. It is an important article of food to the Indians. They gather it about the end of August and beginning of September, and lay up a store for winter consumption. A soup made of wild rice and blue berries is a very palatable dish, and eagerly sought after by those who have been living on salt food for several weeks.

On approaching and receding from Keating's Island, the effects of refraction were most astonishing, elevating low detached island rocks into huge precipitous promontories, and giving to a shore, a few feet above the level of the water, the appearance of a high rock-bound coast. On nearing a small island about four miles east of Garden or Cornfield Island, the grasshoppers on the surface of the lake became more numerous, the green confervæ was visibly less in quantity, and before we landed to dine it had disappeared altogether, but the grasshoppers were found in great numbers on the shore. The island on which we rested for an hour was about three acres in extent, and sustained some fine old oaks and elms, with a profusion of long grass, not much destroyed by the grasshoppers, which had evidently only just arrived there, as was afterwards inferred, while those which had been observed scattered over the surface of the lake were probably stragglers from a vast flight of these insects, whose main body we saw subsequently on Garden Island.

During the morning the sky had been cloudless, the air still, and the sun oppressively hot; but in the afternoon a long gentle swell began to rise upon the lake, and when we put off for our destination, a wind arose which gradually increased to a gale before we landed during the evening of the 24th August, on a low gravelly beach, at the north-west corner of Garden Island.

We camped near a well-cultivated field of Indian corn, and a rapid exploration of the island revealed to us a large potato patch, and a small area devoted to squashes and pumpkins of different kinds. We ascertained that the island had been cultivated by the Lake of the Woods Ojibway Indians for generations.

My companion and the Iroquois, Pierre, both complained that evening for the first time of being unwell.

Our camp fire evidently soon attracted the attention of a number of Indians, who were then living on a neighbouring island about four miles from us, for at midnight we were aroused by the sudden appearance at the door of the tent of two of these people, and in half an hour twenty or more had arrived. In the morning we answered their inquiries, and were requested to visit their chief, who remained with his tribe on the island already referred to. Declining their invitation, as we were anxious to hasten to the mouth of the Muskeg River, they told us they would send for their chief, who would arrive as soon as the wind fell. We made the necessary preparations for a long council, and about noon the chief's son, who was one of the first arrivals on the evening previous, announced that the canoes were coming.

We counted thirteen canoes, and found that they contained in all fifty-three men and boys, there being seven of the latter; the others were the chief and warriors of the tribe. A portion of them had just returned from an expedition against the Sioux, and were decorated or disfigured, according to taste, with whatever advantages paint, feathers, and ornaments could confer. As the object of their visit was to ascertain the reasons why we wished to pass through this part of their country, a long council or " talk " was the result of the visit.

The council terminated by a distinct refusal, on the part of the chief, to allow any of the tribe to guide us through the swamps which separate the Lake of the Woods from the prairie country to the west. The replies and objections of the chief were often couched in very poetical language, with a few satirical touches, which were warmly applauded by the audience. The following is a specimen of the colloquy : —

" What reason can we offer to those who have sent us, for your having refused to allow us to travel through your country ? "

Chief.—" The reason why we stop you is because we think you do not tell us why you want to go that way, and what you want to do with those paths. You say that all the white men we have seen belong to one party, and yet they go by three different roads, why is that ? Do they want to see the Indian's land ? Remember, if the white man comes to the Indian's house, he must walk through the door, and not steal in by the window. That way, the old road, is the door, and by that way you must go. You gathered corn in our gardens and put it away ; did you never see corn before ? Why did you not note it down in your book ? Did your people want to see our corn ? Would they not be satisfied with your noting it down ? You cannot pass through those paths."

" We ask you now to send us one of your young men to show us the road ; we shall pay him well, and send back presents to you : what do you wish for ? "

Chief.—" It is hard to deny your request ; but we see how the Indians are treated far away. The white man comes, looks at their flowers, their trees, and their rivers ; others soon follow ; the lands of the Indians pass from their hands, and they have nowhere a home. You must go by the way the white man has hitherto gone. I have told you all."

At the close of the council, the chief said to the interpreter, " Let not these men think bad of us for taking away their guides. Let them send us no presents ; we do not want them. They have no right to pass that way.

We have hearts, and love our lives and our country. If twenty men came we would not let them pass to-day. We do not want the white man ; when the white man comes, he brings disease and sickness, and our people perish ; we do not wish to die. Many white men would bring death to us, and our people would pass away ; we wish to love and to hold the land our fathers won, and the Great Spirit has given to us. Tell these men this, and the talk is finished."

A hasty consultation with my companion as to what we should do in this dilemma, was abruptly closed by being informed that the Iroquois Pierre was very ill at the back of the tent. Without his paddle, without guide, and Mr. Dawson feeling much worse than on the previous evening, we determined at once not to attempt to cross the swamps between the Lake of the Woods and Red River alone, but decided to go to Red River by the Rat Portage and Winnipeg.

We told this determination to the chief, and asked for assistance to take the canoes to the settlements.

He pointed out two young men, who received orders to take us down the Winnipeg. One was to return from Islington Mission, the other to go on to Red River. We then told the chief that we would send him presents from Red River, at which he expressed satisfaction, and suggested tea and tobacco.

Garden Island is about a mile and a half long and a mile broad at its widest part. Its western half is thickly wooded, but the greater portion of the eastern half is cleared, cultivated, and planted with Indian corn. Near the centre of the field were several graves before referred to, with neatly constructed birch bark coverings. Only one lodge was seen on the island, and that was placed about 100 yards from the graves. Near the space devoted to Indian corn, were several small patches of potatoes, pumpkins, and squashes. An air of great neatness prevailed over the whole of the cultivated portion of the island, and in the part still remaining in its natural state, thickets of raspberry, black currant, and gooseberry bushes grew in the intervals between groves of elm, basswood, and oak; and on the sandy beach an abundance of the sand cherry (*Cerasus pumila*), the favourite Nekaumina of the Indians. Large flocks of passenger pigeons (*Columba migratoria*) flew backwards and forwards over the island, occasionally alighting in dense masses in the small groves. The shores were covered to the depth of two or three inches with countless millions of grasshoppers, which had been washed there during the gale of the preceding night. The greater number of the grasshoppers were alive, and as the rising sun warmed and invigorated them they spread with much regularity over the field of Indian corn and the potato patches; their progress across the potato patches was like that of an invading army of insects, eating and destroying every living green thing in their way. Before we left the island they had advanced, here and there, some thirty or forty yards from the beach, in a well defined and undulating line, leaving behind them nothing but the bare and blackened stalks of the plants over which they had spread themselves and destroyed. By inclining the head, and seeking shelter from the wind under the lee of a bush, the noise of their jaws could be distinctly perceived; and had it been calm, I have no doubt it would have been heard with the greatest ease for a distance of several hundred yards. The Indians had seen the grasshoppers before, but never in such alarming numbers; they appeared, however, quite indifferent to their progress, and quietly amused themselves as they squatted or lay on the ground, by jerking the intruders

off their arms and legs with a thin piece of wood, bent by
the fingers so as to act as a spring. In another chapter
it will be shown that this grasshopper is a true locust
(*Acrydium femur-rubrum*). The term grasshopper is
retained in the narrative because it is universally applied
by the half-breeds to this destructive insect.

From Garden Island to the north-west corner of the
lake, in longitude 95° 15′, is about twenty miles, but the
westerly limits of navigation are not yet found there. It
is possible to proceed without difficulty some miles in a
due west direction, through a narrow, shallow channel,
into a small lake called Shoal Lake. Although no facts
derived from personal observations can be here adduced
respecting the general features of Shoal Lake, yet the importance which it derives from its position requires special
mention to be made of it. From our Indian guide, permitted to take us to Rat Portage by the chief on Garden
Island, I learned that Shoal Lake is a reedy expanse of
water, eight or ten miles long, connected with the Lake
of the Woods. The north and west sides of Shoal Lake
were represented to be blended with a vast marsh or
muskeg, which stretches from near Rat Portage to far
south of the Lake of the Woods, and is the source of
numerous rivers which flow from it both eastward and
westward. It is this great muskeg or marsh which
forms the barrier between the Lake of the Woods and
Red River valley.

On part of the south shore of Shoal Lake, and all along
the west coast of the Lake of the Woods, there is a considerable area of dry land timbered with spruce and small
pine. Shoal Lake is about eighty-seven miles in a direct
line from Fort Garry, but by the very dangerous and circuitous Winnipeg route, it is at least 320 miles. The South
shore of Shoal Lake is in latitude 49° 23′, and the same
parallel cuts Red River at a spot twenty-five miles north of

the boundary line. The importance of the north-west corner of the Lake of the Woods, and possibly also of Shoal Lake at the terminus of a communication by land with Red River cannot fail to be appreciated.

The north-west corner of the lake is styled Monument Bay; it marks the point where the boundary line between British America and the United States, after passing down the middle of Rainy River and striking across the Lake of the Sand Hills reaches the north-west corner of Monument Bay; from this point the boundary line pursues a due south direction, intersecting the 49th parallel in longitude 95° 15′, according to Dr. Bigsby's map published in the Quarterly Journal of the Geological Society, but on the map of the Boundary Survey in the Crown Lands Office at Quebec, the longitude of the Monument at the extremity of the north-west corner of the Lake of the Woods is represented to be 95° 24′.

Shoal Lake, or Lac Plat, as it is also termed, was examined during the winter of 1857-8 by some of the assistants attached to the Surveying Department of the Expedition. The exploration having been made during the winter months, when the deep bogs and wide spreading marshes were frozen and covered with snow, loses much of the interest it would doubtless have possessed had it been executed during the summer season. An effort to penetrate from Fort Garry to the Lake of the Woods in the direction indicated by the exploring party of the previous year, was wholly unsuccessful in the autumn of 1858. A number of half-breeds from the settlements at Red River also made the attempt to reach the Lake of the Woods in September 1858, but were defeated in the attempt by unpassable swamps. In 1859 some half-breeds belonging to Mr. Dawson's party succeeded in passing through with horses.

The route we pursued, from a point near the north-west corner of the lake, lay through a labyrinth of islands in a north-east by north direction for a distance of twenty-eight miles. Six miles more, nearly due north, through scenery of the same description but of a bolder character, brought us to Rat Portage, on one of the numerous mouths of the rocky Winnipeg. Much good pine timber was seen on the larger islands, near the northern part of the Lake of the Woods, and if conclusions may be drawn from the accounts which the Indians gave us of their gardens, it is very probable that extensive areas of excellent land exist on the great promontory, and on some of the large islands. They spoke of growing Indian corn to a far greater extent than seen by us on Garden Island.

During our voyage from Garden Island to Rat Portage we met with six small canoes one after the other; four of them contained only women and children, who had been gathering berries on the islands. We exchanged tea and tobacco for a birch bark basket of blue-berries (*Vaccinium Pennsylvanicum*), of which they had collected a large supply.

Sturgeon are very numerous in the Lake of the Woods: they were repeatedly seen leaping out of the water in their gambols at the approach of evening. The Indian in my canoe told me that in calm weather they could be seen lying on the bottom in deep water, among the beautiful islands with which the narrows between the Lake of the Sand Hills and the true Lake of the Woods are studded. Whenever we landed to dine or camp before reaching Rat Portage we found hosts of grasshoppers; they covered the bare rocks, and rose in small clouds from the grass at my approach. We succeeded in securing a large pike which lay basking in the sun at the surface of the water,

by cautiously approaching and striking him on the head
with a paddle ; he proved to be a splendid fish, probably
not less than ten pounds in weight. A thunder storm of
great violence detained us for a few hours on a small
island, where the rock was so smoothly worn and polished
by the action of ice, that we found some difficulty in keep-

Falls at Rat Portage.

ing our feet. The recent action of ice is often beautifully
shown on the islands in this lake, and instances of ancient
glacial action are by no means uncommon here as elsewhere on the canoe route. We arrived at Rat Portage,
where the Great Winnipeg issues from the Lake of the
Woods, on the morning of the 27th of August.

CHAP. V.

THE WINNIPEG RIVER.

Character of the Winnipeg.— Rat Portage.— Thunder Storm:— Thunder Storms in the North-West.—A View on the Winnipeg.—Islington Mission.—Cultivable Areas on the River.—Rev. Robert Macdonald.—Church Service.—State of Islington Mission.—Indian Superstitions.—Farm at the Mission.—The School-House the Hope of the Mission.—En route for Red River.—James' Falls.—Animal Life.—Rice Grounds.—Mr. Clouston.— Otter Falls.—The Pennawa River.—Scarcity of Food on the Winnipeg.— Bonnet Lake.—Indian Cache.— The Silver Falls.— Fort Alexander.— Lake Winnipeg.— Character of the Coast.—Camp in the Marshes.— Mouth of Red River.—Indian Village.—Christian Indians.

ISSUING from the Lake of the Woods through several gaps in the northern rim of the lake, the River Winnipeg flows through numerous tortuous channels for many miles of its course in a north-easterly direction. Some of the channels unite with the main stream ten to fifteen miles below Rat Portage, and one pursues nearly a straight course for a distance of sixty-five miles, and joins the Winnipeg below the Barrière Falls. The windings of this immense river are very abrupt and opposite, suddenly changing from north-west to south-west, and from south-west to north-west, for distances exceeding twenty miles.

In its course of 163 miles, it descends 349 feet by a succession of magnificent cataracts. Some of the falls and rapids present the wildest and most picturesque scenery, displaying every variety of tumultuous.

cascade, with foaming rapids, treacherous eddies, and huge swelling waves, rising massive and green over hidden rocks. The pencil of a skilful artist may succeed in conveying an impression of the beauty and grandeur which belong to the cascades and rapids of the Winnipeg, but neither sketch nor language can portray the astonishing variety they present under different aspects; in the grey dawn of morning, or rose-coloured by the setting sun, or flashing in the brightness of noon day, or silvered by the soft light of the moon.

Hudson's Bay Company's Post at Rat Portage.

The river frequently expands into large deep lakes full of islands, bounded by precipitous cliffs or rounded hills of granite. The Fort in the occupation of the Honourable Hudson's Bay Company at Rat Portage is beautifully situated on an island at one outlet of the Lake of the Woods. It is surrounded with hills about 200 feet high, and near it some tall white and red pine, the remains of an ancient forest, are standing amidst a vigorous second growth. The rock about Rat Portage is chloritic slate, which soon gives place to granite, without any covering of drift, so that no area capable of cultivation was seen until we arrived at Islington Mission. We did not pursue the usual canoe route, but in the hope of overtaking the other members of the expedition, followed an Indian route for some miles, which was said by our guide to be half a day's journey shorter than that by the main river.

In descending this branch of the Winnipeg, a terrific thunder storm, accompanied by a hurricane of wind and an extraordinary fall of hailstones, approached us from the south as we reached the high portage which connects this route with the great river. Turning round in my canoe I saw, about half a mile in our rear, a white line of foam advancing rapidly towards us. Directing the attention of Lambert who occupied the stern, to the approaching squall, he instantly changed the course of the canoe to the opposite bank. The river here was about 300 yards broad, and we were swiftly paddling close to the west side, which was bounded by high and precipitous granite rocks. I had barely time to stretch a gutta-percha cloth over the canoe, before the hurricane came down upon us. Lying, as directed, at full length on the floor of our small craft, I left her to the dexterity of Lambert and the Indian. They met the shock skilfully, and paddled before the storm with great rapidity, as did also our fellow-travellers in the other canoe. We continued on our way for some minutes, gradually drawing near to the right bank, where we intended to land. Suddenly, however, large hailstones began to descend with such force as to bruise my hands severely in endeavouring to retain the covering in its place. The Indian in the bow laughed heartily at first, but having no covering on his head beyond his thick and matted hair, he soon crouched, and drew a part of the gutta-percha cloth over him. Lambert being provided with a thick fur cap held bravely on, although he loudly exclaimed that the hail-stones were bruising his hands, and he would not be able to paddle much longer. Fortunately we were now close to the bank, and Lambert called out to the Indian to keep the canoe from striking against the rocks. A few strokes of the paddles brought us within a yard of the shore, when the Indian, lightly springing out of the canoe, caught her bow as she was about to strike the rock. I succeeded in disentangling myself from the covering, which was pressed down by an accumulation of hail-stones, enough to have filled at least three buckets, and looking over the side of the canoe I saw the Indian in Mr. Dawson's tiny craft leap out on the bank, and catch the fragile vessel as it was about to strike with wonderful agility. Lambert's hands were much bruised, and soon became swollen and very painful; he wrapped them in wet cloths, and on the following morning, with his usual lightheartedness, declared they were as sound as ever. We made a large fire and a comfortable tent of spruce bows, soon forgetting, as the storm cleared away and allowed us to enjoy a most gorgeous sunset, the imminent danger from which we had escaped. Violent thunder storms are very common during the summer months on the Winnipeg; hailstones have been known to descend with such force as to pass through the thin birch-bark of the small canoes constructed by the Indians who hunt on this river. On the grand Coteau de Missouri, the tough buffalo-skin tents of the Crees and Sioux are sometimes penetrated by the small angular masses of ice which fall on that elevated plateau. In the summer of 1858 the canoes we carried across the prairies to the elbow of the

south branch of the Saskatchewan were indented, and in
some instances penetrated by hailstones during a severe
storm; in several places, many miles apart, we observed
the long grass laid smooth and uniform over wide areas
as if pressed by a heavy roller.

Just before sunset I ascended a hill about 250 feet
high, and obtained from its summit a very extensive
view of the surrounding country. The broad river,
with its numerous deep and spreading bays, was seen
stretching far to the north, and all around dome-shaped
hills, similar to the one on which I stood, showed
their bare and scantily-wooded summits; generally they
seemed to be thinly covered with stunted pine, but in
the hollows or valleys between them, pine and spruce
of large dimensions, interspersed with fair sized aspen
and birch, flourished abundantly. The pine on the granite
hills grew in little hollows, or in crevices of the rock.
The general surface was either bare, and so smooth and
polished as to make walking dangerous, or else thickly
covered with cariboo moss and tripe de roche. The
aspect of the country was similar in its outline to the
region about Milles Lacs, but the vegetation was far
inferior. Until we arrived at Islington Mission, the
general features of the country maintained an appearance
of hopeless sterility, and inhospitable seclusion.

Islington Mission, or the White Dog, or Chien Blanc,
for by these names it is known to the voyageurs, occupies
an area of what seems to be drift clay extending over
250 acres, surrounded by granite hills. The soil of
this small oasis is very fertile, and all kinds of farm
and garden crops succeed well. Wheat sown on the
20th of May was reaped on the 26th of August; in
general it requires but ninety-three days to mature.
Potatoes have not been attacked by spring or fall frosts

during a period of five years ; Indian corn ripens well, and may become a valuable crop on the Lower Winnipeg. Spring opens and vegetation commences at Islington about the 10th of May, and winter sets in generally about the 1st of November. These facts are noticed in connection with the small cultivable tract at the Mission, on account of the occurrence of other available areas, varying from fifty to three hundred acres in extent, between the Mission and Silver Falls, about eighteen miles from the mouth of the river. From Silver Falls to where the river flows into Lake Winnipeg, poor and rocky land is the exception, alluvial and fertile tracts, bearing groves of heavy aspens and other trees, prevailing.

The cultivable areas on the river banks may yet acquire importance, for they may be regarded in the light of productive islands in a sterile waste of rock and marsh.

The Rev. Robert Macdonald, the missionary in charge at Islington, informed me that the Mission was formerly held by the Roman Catholics for a period of several years, but was abandoned on account of the opposition of the Indians, who drove them away in consequence of the death of a young girl in the nunnery at Red River settlement. The heathen Indians persuaded the converts that all who embraced Christianity would soon die. The Mission was left vacant for a period of six years, after which Mr. Philip Kennedy was appointed catechist in 1850, a post which he held until the 20th October 1853, when the Rev. Robert Macdonald was enabled to revive the Mission by the generous and Christian liberality of an English lady.

This Mission is at present sustained by a munificent gift from Mrs. Landon of Bath of 1,000l. sterling for its establishment, and 100l. a year for its maintenance. Its present prospects are favourable, and it will eventually become an important station in the wilderness by which it is surrounded. I attended divine service in the schoolhouse, where it was celebrated in the Ojibway language; at my request the heathen Indians, who had assisted in conducting us from Garden Island, were present. They maintained a respectful silence during the service, and a favourable impression may have been produced. Mr. Macdonald has given away ten hundred weight of flour and forty bushels of wheat since May last, to the wandering Indians who occasionally touch at the White Dog. Their attention and sympathy must first be enlisted by continually renewed presents; this shows the advantage of agricultural operations being associated with spiritual labour at remote stations. In seasons of scarcity the Indians assemble at the Mission and expect to be assisted by material contributions; these failing, they lose confidence and heart, and the influence which the missionary may have succeeded in acquiring is soon lost when they disperse in search of the means of sustaining life. The present congregation numbers about forty-five when the entire body of Christians attached to the Mission are assembled. They belong to the Swampy Crees, and hunt on the lower Winnipeg.

Mr. Macdonald divides the Indians who hunt north-east and north of the Lake of Woods from those who inhabit the shores, islands, and the country east and south-east of that beautiful lake. The former belong to the Muskeg nation (Muscaigoes) or Swampy Crees, the latter to the great Ojibway nation. The Swampys have a tradition that at a remote period they drove the Ojibways from the lower Winnipeg to the country bordering on the Lake of the Woods, and since that time they have maintained their footing in the conquered territory.

ISLINGTON MISSION—WINNIPEG RIVER.

Printed by Spottiswoode and Co.

[New-street Square, London.

The heathen Muskegs or Swampys address their invocations to the Evil Spirit, but they acknowledge the existence of a Supreme and Good Being. The Bishop of Rupert's Land had an opportunity of witnessing the idolatrous worship of the conjurors of the Swampys on the Lower Winnipeg in 1852. He describes the scene in the following words : — " There were two or three tents. I entered the largest, and there found the son of Wassacheese sitting in solitary state. I was about to sit down where I saw some articles expanded, and where at first I thought he had prepared a seat for me ; but I found, on a second look, that these were the idols of the chamber of imagery, the instruments of his· art as conjurer, and the feast spread out for spirits. I asked him to explain his magic art, which he said he would if I would give him some flour. I gave him instead a little tobacco, and then heard his tale. He showed me, as a special favour, that which gave him his power — a bag with some reddish powder in it. He allowed me to handle and smell this mysterious stuff, and pointed out two little dolls or images which, he said, gave him authority over the souls of others ; it was for their support that flour and water were placed in small birch-rind saucers in front. . . . The altar was raised a little on some Indian matting, and on it, ranged in order, the bags and images and all the instruments of his craft. O what is man without the grace of God ! " *

At Islington there are now (1857) five houses, besides the mission-house and a school-house, which is used as a chapel. With the exception of the hill of drift clay, on which the Mission is situated, the country is wholly granitic in the immediate neighbourhood, and resembles the

---

* The Net in the Bay, p. 19.

bold but dreary region through which the Winnipeg flows from Rat Portage to this, the first small area on or near the banks of the river where farming operations on a small scale can be conducted with any prospect of success. The river is about 250 yards broad, and flows past this little oasis with a rapid current. Mr. Macdonald has about six acres of wheat and four acres of potatoes in a very promising condition. In a garden surrounding the dwelling of one of the Christian Indians, I observed an acre and a half of potatoes, some Indian corn, and a patch of wheat. Missionary labour here, as elsewhere, is not encouraging; the adults are generally very indolent, unsettled, and improvident; they make but little improvement, and their conduct tends to confirm the impression which I find to prevail among missionaries, that many embrace and adhere to Christianity for the sake of the temporal benefits it secures, often relapsing when material advantages diminish or cease, into their original condition of barbarous idolatry. Mr. Macdonald looks to the schoolhouse as the main hope for the future. When the Mission was established the Indians who visited this favourite camping ground made many promises; but even in 1852, when the Bishop of Rupert's Land visited Islington, he had occasion to remark, "that the Indian promised more at that time than he has since performed will not appear surprising to those who know the Indian character, and therefore all would stand prepared for some little disappointment in the carrying out of the plan."*

I started from Islington Mission on the 31st August, with Lambert, a young Indian lad, and the Ojibway from the Lake of the Woods whom I had persuaded to accompany me to the settlements at Red River, leaving Mr.

* The Net in the Bay, by the Bishop of Rupert's Land.

Dawson in the charge of the Rev. Robert Macdonald and his sister, in a fair way of recovery, although far too much indisposed and feeble to attempt the rough and dangerous voyage down the Winnipeg in a small canoe, weakly manned. I arranged with Mr. Macdonald to send supplies, nourishing food, and a north canoe, as soon as I arrived at the settlements, which I hoped to reach in five days.

The poles of wigwams are numerous in the flat country bordering the shores of Grand Turn Lake; the timber is aspen and poplar, with a few Banksian pine. Eagles and fish-hawks were the only birds visible in this part of the Winnipeg, the other species of the feathered tribe being in the rice-fields. Near James' Falls there is a small area of clay sustaining aspen and poplar; it may embrace 200 or 300 acres. Here I observed some gigantic ant-hills close to the river. The largest were from five to seven feet in diameter at the base, and from four to five feet high. I counted thirteen of these curious habitations of the "wise" ant. James' Falls are about thirteen feet high, and are always approached with great caution in descending the stream; they are considered to be among the most dangerous obstacles on this broken and tumultuous river. The foaming torrent passes over three steps, each about four feet in height and ten apart, caused by a ledge of rock stretching across the river in a direction W. 10° S.

Below James' Falls the poles of wigwams are numerous, and many Indians were seen at the foot of the different rapids engaged in fishing. The scarcity of animal life of all kinds was very remarkable, eagles and fish-hawks, ducks and rabbits, being the only representatives seen. This scarcity is, however, confined to the autumnal months, as to the time, and to the Great Winnipeg River in respect of area. Some distance from the river the extensive rice grounds cover many thousand acres, and continue for miles on either bank. Here the game congregates, and reveling in the midst of such an abundant supply of nutritious food, vast flocks of ducks, geese, and all kinds of aquatic birds common in these regions, are to be found. It is here too that the Indians assemble at stated periods amidst the rice-grounds, procuring without any difficulty, in favourable seasons, a large supply of food for winter consumption.

The falls of the Portage du Bois, where we arrived on the following day, are exceedingly beautiful; they let the river down ten and a half feet, and are divided into four parts by three prettily wooded islands ; the whole breadth of the river here is about one-third of a mile. The portage path passes over a fertile oasis sustaining oak, ash, and cherry, with rosebushes, honeysuckle, and a great variety of sweet scented flowering plants, which fill the air early in the morning with delicious fragrance. Succeeding this romantic break in the river, the Point aux Chiens falls and rapids let the river down nineteen feet. It is succeeded by the Roche Brûlé Portage and then by the Slave Falls, the scene of one of those terrible incidents in Indian life formerly of frequent occurrence in these inhospitable wilds. Tradition tells that a slave of a ferocious master, maddened by long continued cruelty, calmly stepped into a canoe above these falls in the presence of the tribe, and suddenly pushing off from shore, wrapped her deer-skin robe round her face and glided over the crest of the cataract, to find rest in the surging waters below.

Above the Barrière Falls, which are next approached, a clay cliff comes on the river, and exposes a section about fourteen feet high ; its breadth is not great, as in the rear bare rocks can be seen from the river ; it occupies, how-

ever, a considerable area, being also observed below the
falls, where it attains an altitude of eighteen feet. At this
spot we met Mr. Clouston, of the Hudson Bay Company's service, in charge of supplies for Rat Portage
and Fort Frances. He kindly consented to open his cases
and allow Mr. Macdonald to select some medicine and
whatever necessaries he might require for Mr. Dawson's
use. The motley crew of Indians and half-breeds were
engaged at the time of our meeting in hauling a bateau

Slave Falls, Winnipeg River.

over the falls, having deposited the cargo on the rocks.
Twenty to twenty-five men were pulling at a rope attached
to the bateau, and with the utmost exertion slowly lifting
the unwieldy craft up the wall of water, five feet high,
which constitutes the Barrière Falls. Late in the evening
we arrived at the mouth of the Pennawa, a small branch
of the Winnipeg, which joins it again at Bonnet Lake. At
and below the Otter Falls were a number of Indians engaged in fishing; as we hurried past them they shouted

to us to stop and camp with them, but not being provided with presents, and having only a very small stock of provisions, I thought it advisable to decline the invitation; leaving a few pieces of tobacco on the portage path we hastened to the mouth of the Pennawa, and camped about half a mile down the river.   Being anxious to reach Fort Alexander, I awoke the men an hour before daylight, and whilst breakfast was preparing, strolled over the rugged and fissured rocks through which the little river finds its way.   The almost oppressive silence was broken only by the occasional splash of a pike, or the distant howling of a wolf, or the subdued roar of the Winnipeg wafted by a very gentle breath of wind which now and then stole from the west.   The musk rats were busily engaged swimming across the little ponds which separate the rapids into which the Pennawa is broken, near where it branches off from the Winnipeg.

Instead of following the course of the Great Winnipeg, after arriving at the Otter Falls, I passed down the Pennawa River into Bonnet Lake, in order to avoid the dangerous " Seven Portages," and save several miles of route.   Near the entrance of the Pennawa into Bonnet Lake, the little river winds through an immense marshy area covered with wild rice, and I succeeded in collecting a considerable quantity as the voyageurs paddled through its light and yielding stalks with undiminished speed.   There, too, were seen vast numbers of different species of duck, and many other kinds of birds, such as herons, pigeons, woodpeckers, cedar birds, jays, &c.

The Indians we met lamented the failure of the rice this year ; they described the appearance in favourable seasons of the ground through which we were hurrying, as a vast expanse of waving grain, from which they could soon fill their small canoes, by beating the heads with a stick. The waters of the river and marshes were unusually high, so as to check the growth of the rice to an extent which, when coupled with other deficiencies, threatened them with famine during the coming winter.

The same cause which originated the partial failure of the wild rice led also to a great scarcity in fish. In general, the Winnipeg teems with fish, among which are sturgeon, pike, two kinds of white fish, perch, suckers, &c., affording a bountiful supply to the Indians who hunt and live on or near the lower portion of this majestic river. The extraordinary height of its waters during the summer of 1857 had so extended the feeding grounds of the fish, that they were with difficulty caught in sufficient numbers to provide the Indians with their staple food.

The unlooked-for short supply of wild rice and fish were more severely felt in consequence of the unaccountable disappearance and death of the rabbits, which are generally found in vast multitudes in the region of the Lake of the Woods and Winnipeg River. During the spring and summer, large numbers of rabbits were found dead in the woods, owing probably to the exhaustion which followed a severe winter, prolonged this year to an unprecedented length in these regions. With a partial failure in the rice, great scarcity of fish, and the prospect of a very limited supply of rabbits, the anticipations of the coming winter on the part of those who cared to think of the sufferings of the wretched Indians of the Winnipeg, were gloomy indeed.

The Pennawa enters Bonnet Lake between high and rugged rocks, the termination of a range of dome-shaped hills, which decline from an altitude of 250 feet to small island bosses in the lake. The strike of the range is nearly north and south, curving slightly to the south-east; the summit

of the range is bare, rounded, and apparently polished.
Camping on an island in the lake to escape an approaching squall, I found an Indian cache elevated on a stage
in the centre of the island. The stage was about seven
feet above the ground, and nine feet long by four broad.
It was covered with birch bark, and the treasures it held
consisted of rabbit-skin robes, rolls of birch bark, a ragged
blanket, leather leggings, and other articles of winter
apparel, probably the greater part of the worldly wealth
of an Indian family. At the great Bonnet Portage there
are many acres of good farming land, and, indeed, from
this point to Lake Winnipeg, a strip on the river, widening
as we descend, possesses all the requirements as far as soil
and timber is concerned for a large settlement. The
Silver Falls, the last but one, are perhaps the most imposing and beautiful of all the cascades on the Winnipeg.
The volume of water precipitated here is immense ; all
the inosculating branches of the Winnipeg uniting some
distance above the magnificent Silver Falls. The vast torrent descends a slope about 200 yards long with an
inclination of nearly sixteen feet, in the form of five or
six gigantic swells. The observer may stand close to the
huge heaving waves and watch them rush past him with
astonishing velocity and ever-changing form. Sometimes
they send a thin sheet of water over the smooth rock on
which he is standing at the edge of the torrent ; in another
minute there may be a gulf ten or fifteen feet deep, with
a terrible whirlpool raging below, between him and the
crested swell fifty feet from the shore ; suddenly the gulf
is filled, and the turbulent waters, dashing against the
rocks, send a shower of spray far and wide over the
polished gneiss which confines them. We reached Fort
Alexander at four in the afternoon, stayed half an hour to

procure some flour and potatoes, and then hurried on to
Lake Winnipeg, camping at the mouth of the river.

Fort Alexander is situated within one mile and a half
from Lake Winnipeg. In fields near the fort I saw wheat
in process of being harvested on the 3rd of September,
and obtained some new potatoes of great size and excellent quality. I was informed by the officer in charge of
the fort that Indian corn succeeded well in many parts of
the south-eastern rim of the lake, and that it was very
rarely touched by late spring frosts ; it is cultivated by the
Indians. On the following morning I aroused the men at
three, and made the first lake traverse by moonlight. The

Fort Alexander, Mouth of the Winnipeg River.

west shore of Traverse Bay is high, and shows an excellent soil, thickly covered with balsam-poplar, aspen and
birch. The lodges of Indians are very numerous on the
west point, which forms one of their most important fishing stations. When we landed to breakfast or dine, opportunities were afforded of examining the precipitous but
unstable cliffs which were occasionally exposed. At a

point on the east coast of the Grand Traverse, a section showed one feature of interest, which is common to all the great lakes of the St. Lawrence basin. The summit of the cliff, clothed with an inch or two of sandy loam, exhibits an ancient lake beach, composed of water-worn boulders, pebbles, and stratified sand, two feet thick. This is underlaid by sixteen feet of stratified sand, containing limestone fragments, and boulders of the unfossiliferous rocks; it is flanked by a talus of shingle, slabs and boulders, among which, bright yellow, cream-coloured, and beautifully variegated limestones are numerous. This talus is the present shore of the lake, and the shingle, slabs, and boulders have probably been washed out of the unstable cliff. Its breadth may reach sixty feet, with an inclination of three to five feet from the level of the lake, giving to the ancient beach at the summit an elevation of twenty-one feet above the present level of Lake Winnipeg.

About five miles further south, I ascended a cliff fifty feet high, consisting of stratified sand and marl, in which were embedded primitive boulders of most gigantic dimensions: some of them measured twelve to fifteen feet in diameter; they were all water-worn, and distributed throughout the cliff. On the surface, walking was exceedingly difficult on account of their numbers and size. Many of them were covered with the Virginian creeper (*Ampelopsis quinquefolia*). The base of the cliff was well protected by an immense accumulation of these erratics, which had fallen from the loose sand of the cliff. The temperature of the lake, six miles beyond this point, was 64° 5'. A heavy squall from the north-west compelled us to approach the shore when within three miles of the mouth of the Red River; the waves rose with great rapidity, as is usual in large, open, and shallow sheets of water, compelling a hasty retreat among the willows and rushes, where, notwithstanding exposure to the discomfort of the waves washing over our camp during the night, we were compelled to remain in a damp maze of reeds until the wind and waves subsided. Here I had an opportunity of observing the vast number of duck, geese, and plover, which congregated amongst the rushes at the approach of night. Early in the morning, flights swept backwards and forwards close to our camp in constant succession.

Red River enters Lake Winnipeg by six distinct channels, which will be described in the second volume. Its junction with the lake by the branch through which we entered is marked by a low spit of sand, which was the only piece of land visible amidst the tall reeds extending far to the south.

Land which is dry during the summer months at the stage of water in the river on the 5th September, 1857, begins five miles from the mouth of the main channel. Half a mile above this point, Netley Creek comes in from the west, and by means of this small affluent, much of the water from the upper country during floods reaches Lake Winnipeg. Large numbers of haystacks were seen here in September 1857. An immense area is flooded during the spring, and produces a very rank profusion of those grasses which delight in a rich and marshy soil.

A little below the Indian village, fourteen miles from the mouths of the river, the whole country rises, the banks are about thirty feet high, the timber is imposing, and all the aspects of a level, fertile region, gradually invest the scene; but the sameness in the general appearance of the banks at this season of the year soon becomes monotonous, after the wild and varying beauties of the Winnipeg. The sight of clearings, however, with the neat white houses of settlers at the Indian missionary village, speedily creates other impressions, aroused by such fair comparisons between the humanizing influence of civilization, and the degraded, brutal condition of a barbarous heathen race.

These familiar and suggestive signs of improvement in moral and social position, rapidly create a healthy tone of feeling in passing from the cascades and rapids of the Winnipeg, where half clad savages fish and hunt for daily food, to the even flow of Red River, where Christian men and women, once heathen and wild, now live in hopeful security on its banks.

I arrived at the Stone or Lower Fort about six P.M., and after a little delay succeeded in procuring an Indian horse, which carried me to Fort Garry, a distance of twenty miles, in a little less than three hours, where I found the other members of the expedition camped under the walls of Fort Garry.*

* The Blue Book, published by order of the Legislative Assembly, dated 10th May 1858, contains an account of the steps which were taken to send supplies to Islington Mission.

CHAP. VI.

RED RIVER SETTLEMENTS.

The Red River of the North.— Its Tributaries.—The Red Fork.—The Red River within British Territory.— Its physical Features.— Objects seen on ascending the River.— Section of the River and Prairie.— Objects on the Banks. — The Settlement.— The King's Road.— Character of the Country north of Fort Garry.—Aspect of the Prairies.—Beauty of the Prairies.—The Assinniboine River.—Effect of Evaporation on the Volume of Water in the Assinniboine.—Description of the Assinniboine.—Prairie Portage.—Mud and Sand Flats in the River.—Timber.—John Spence. —Lignite reported to exist on the Assinniboine.—Sioux.—Indian Corn. —The Big Ridge.—An Overturn.—The Prairies of the Assinniboine.— Mr. Lane.—Mr. George Flett.—Mr. Gowler.—Mr. Gowler's Farm.—His Opinions respecting the Prairies on the Assinniboine. — Melons. — Old Associations.—Independence. — Mr. Gowler's Success. — The Nor'wester. — A Newspaper published at Red River Settlements.

THE Red River of the North rises in Ottertail Lake, State of Minnesota.* The north-east end of Ottertail Lake is in latitude 46° 24′ 1″, and the general course of the river is westerly, through an attractive undulating country, until it makes its great bend to the north, in latitude 46° 9″. It then meanders through a boundless prairie, which gradually declines in elevation until it forms a vast level plain, elevated above the water of the river only about one and a half to two feet, at its ordinary stage in June. The distance of this great bend is 110 miles from Ottertail Lake, and the vast low prairie through which the river subsequently flows, in an exceedingly tortuous

* The description of that part of Red River which lies within the territory of the United States is abbreviated from Dr. D. D. Owen's account in his geological survey of Wisconsin, Iowa, and Minnesota.

channel, is level as a floor.   In latitude 46° 23' 30" a
belt of timber sets in, and continues with some interruption along its banks, on one side or the other, to Pembina,
near the boundary line.   In latitude 46° 41' 12" the level of
the prairie above the river is thirty feet; this depression is
probably due to the gradual erosion of the river channel
in soft clay.   Red River receives some important tributaries south of the 49th parallel: on the left bank the
Shayenne sweeps round the north-east flank of the Grand
Coteau de Missouri, and joins Red River in latitude 47° 3'.
Numerous streams, draining a lower terrace of tableland
than the Grand Coteau, cut the fertile prairies through
which Red River meanders at right angles to its course,
but no affluent of importance is received between the
Shayenne and the Pembina, near the international
boundary line, or 525 miles from Ottertail Lake, by the
windings of the main stream.

On the right bank, the Red Fork, issuing from Red
Lake, and joining the main river in latitude 47° 50', is the
most important; in consequence of its being on the line
of water communication between Rainy River and Red
River, and also separated by a low-water parting from
the Mississippi and Lake Superior, it may hereafter
acquire some degree of prominence.

Dr. Owen remarks of the country through which Red
River flows in the United States territory, that it possesses
features both geologically and physically of great sameness and flatness, without the least indication of minerals
of any value, except salt, which may be crystallised out
of saline springs.

The length of Red River within British territory is
about 140 miles by the windings of the stream.   It
debouches into Lake Winnipeg, in latitude 50° 28', longitude 96° 50'.   Its most important affluents on the east

side are Roseau River and German Creek. On the west side it receives, in latitude 49° 53′ 24″, and longitude 96° 52′, the Assinniboine* River; at the confluence of these streams Fort Garry, the capital of Assinniboia, and the head quarters of the fur trade in British America, is situated.

The following description in detail of Red River, within British territory, supposes the observer to ascend the stream from Lake Winnipeg in a bateau or canoe, and is confined to those objects which come under observation during the voyage.

Fourteen miles from the mouths of the river, the Indian missionary village occupies a terrace thirty feet above the summer level of the stream. Above the village the banks are fringed with oak, elm, and maple, which soon give way to aspen, and then to open prairie land, the trees of larger growth appearing at intervals on the points and on the insides of the bends.

About four miles above the Indian missionary village, a remarkable bend in the course of the stream gives rise to a sharp projection of the level plateau of the prairie, called Sugar Point, from the groves of maple which cover it. It has been preserved from the abrading action of the stream by numerous fragments of limestone which lie at the bottom of the river bank and continually increase in number and size in its ascending course, as far as the exposed strata of rock in position, at and above the Stone Fort, where their place is supplied in part by the parent rock.

The maple, which at one time grew in considerable quantities near Sugar Point, is not the true sugar maple (*Acer saccharinum*) so common in Western Canada, but

* Assinniboine, from " assinni " a stone — Cree. Howse, in his grammar of the Cree language spells the name of the tribe, " Assinne-boigne."

another species, generally known as the ash-leaved maple (*Negundo fraxinifolium*), also furnishing an abundance of juice from which sugar is made as far north as the Saskatchewan. Near to Sugar Point is an Indian school, in connection with the Indian Mission below, situated north of the line which divides the parish of St. Peter from that of St. Andrew, and marking the northern limits of the Red River Settlement. The banks on both sides are very heavily timbered close to the river; but between this place and the Stone Fort there are very few farmhouses. The general direction of the river from Sugar Point to Fort Garry, is a few degrees to the west of south. In an air line the distance is twenty miles; by the road on the left or west bank twenty-one, and by the river itself twenty-three and a half miles. The scenery and objects which meet the eye in ascending the river between the Lower Fort and 49th parallel are uniform, but singular and interesting.

In order to arrive at a true conception of its physical features, it is merely necessary to imagine a river from 200 to 350 feet broad, with a moderately rapid current, having in the course of ages excavated a winding trench or cut to the depth of from thirty to forty feet in tenacious clay; through a nearly level country for a distance exceeding 100 miles, and the general physical aspect of Red River, within British territory, is reproduced. Here and there local diversities occur which give some appearance of variety. Such are noticed at the Grand Rapids, where the even flow is broken and disturbed by a ledge of limestone, which may occasion a fall of six feet within a mile. A lower terrace has here and there been excavated perhaps ten feet below the general level of the prairie banks. Occasionally sand, mud, and gravel bars are formed at the numerous sharp turns in the general

Printed by Spottiswoode and Co.]

RED RIVER FROM ST. ANDREW'S CHURCH.

[New-street Square, London.

course of the stream, before the volume of its waters is augmented by the Assinniboine which comes in at Fort Garry. These projecting bars or points are often covered with fragments of limestone, primitive boulders, and vast numbers of large fresh-water shells. The current round them is rapid, and during ·low stages they present an obstacle to the navigation of the river by means of steamers exceeding 120 feet in length. Often, too, on one side or the other, and sometimes on both sides, a narrow belt of heavy forest timber closes upon the river, and seems suddenly to narrow and darken its abrupt windings. The most uniform character, however, and one which is more frequently found on the west side, is a clean and steep line of bank about thirty feet in altitude, perfectly level to the eye, and forming the boundary of a vast ocean of prairie, whose horizon or intermediate surface is rarely broken by small islands of poplar or willow, and whose long, rank, and luxuriant grasses show everywhere a uniform distribution, and indicate the character of the soil they cover so profusely. A subsequent closer inspection of the soil never failed to establish its fertility and abundance, as well as its distribution over areas as far as the eye can reach, both eastward and westward, from the banks of this remarkable river.

The objects which arrest attention in ascending the river between Sugar Point and the Stone Fort, are limited to precipitous clay banks, fringed with elm, poplar, maple, oak, and ash, all of large growth, but not fair representatives of the forest which once skirted the stream, so long subjected to a destructive culling process, in order to supply the necessities of the settlement above. Among the underbrush, the Virginian creeper, and occasionally a wild grape, with a profusion of convolvulus twining round hazel, alder, and rose bushes are most conspicuous. At the Stone Fort, massive layers of limestone crop out, which have been extensively quarried, and their application is seen in the walls and bastions of the fort built upon the bank, here about forty feet in altitude, and forming the abrupt termination of the prairie stretching westward, which for some distance sustains a small but dense growth of aspens. At each turn of the river above this point the houses of the inhabitants of Red River Settlement come in sight, and occupy the banks, at short intervals, all the way up to Fort Garry, a distance of twenty-three and a half miles by the windings of the river. Two miles above the Stone Fort, is the so-called Whirlpool Point, and immediately above it the Big Eddy; these are obstacles to further progress, formidable only in name, and like most other local descriptive titles on this river, must be accepted with the mildest interpretation, and only understood to designate marked differences from the general even flow of the river. A small brook on which a water-mill is situated enters the river at the Big Eddy. A short distance above the same locality limestone crops out in heavy layers on the west bank, and detached fragments in great abundance protect the base of the cliff, which in no instance, from the mouth to the 49th parallel, as far as my observation permitted me to judge, rises above forty feet from the water level. Some very substantial illustrations of the adaptation of the limestone for building purposes occur here, and particularly at the Grand Rapids, two and a quarter miles farther up the stream. The east side of the river is wooded to a depth varying from a few yards to a mile, and generally this feature prevails along the eastern bank as far as Fort Garry; the timber is similar to that already described. At the Grand Rapids, which even during the low stage of water in

September, offer no formidable obstacle to the Company's and freighters' boats carrying four and five tons, an assemblage of well-built stone buildings are grouped, which create a very favourable impression of Red River resources and comfort, not unfrequently repeated in ascending the stream. There is erected a very substantial stone church, capable of seating 500 people, and surrounded with a stone wall enclosing an extensive burying-ground. About 300 yards south of the church, the parsonage house is seen from the river, and a visit to its interior, to be more fully noticed subsequently, proved that every desirable comfort was enjoyed by the kind and hospitable incumbent, Archdeacon Hunter. Adjoining the parsonage is the residence of the curate, Mr. Kirkby (now on Mackenzie's River), and next to it a capacious and well-built school-house of wood. Four miles above the Grand Rapids, Water-mill Creek enters the river, having cut its way through the yielding clay substratum of the prairie to a depth of twenty-five feet, half a mile from its mouth.

Above Mill Creek the river banks break off abruptly from the prairie level, and are well wooded on the east side, the houses of the inhabitants occurring at regular intervals upon the west bank. At a short distance above the very commodious and comfortable residence of Mrs. Bird, a lower terrace, caused by denudation, commences, and its prairie boundary passes in the rear of the house occupied by the Expedition, and comes upon the river again before reaching the Presbyterian church. The following section shows the relation of the lower terrace to the general level of the Great Prairie, the relation of the Big Swamp, noticed hereafter, to the river, and also of the ancient beach or ridge of Lake Winnipeg to the general level of the country.

Distance from water mark, West.	Height above level of river, 18th September.	
Water mark 0 feet.	0 feet.	
66 „	18·48 „	
109 „	11·36 „	
152 „	20·74 „	Bench-mark.
233 „	20·06 „	
830 „	16·52 „	
1230 „	19·07 „	
1330 „	25·76 „	
1853 „	27·52 „	King's Road.
2482 „	23·80 „	Small shallow bed of creek.
2667 „	27·38 „ }	Grand Prairie Level.
2988 „	27·30 „ }	
4212 „	26·31 „	Commencement of marsh.

East.

4 miles nearly N. E. }	. . . 8 feet.	{ Ancient beach of Lake Winnipeg.

Above St. Paul's church, in the Middle Settlement, eight miles north of Fort Garry, the river winds between high prairie banks, which generally maintain an altitude of about thirty feet; houses and windmills occur at regular intervals, until the steeple of St. John's church, the peaked roof of St. John's College, the school-house, the Bishop's residence, &c., offer the appearance of a large village, which is again re-produced after a sharp turn at Point Douglas, by the imposing Roman Catholic church, dedicated to St. Boniface, the spacious nunnery and the parish school, with other buildings on the left, and a group of several commodious private dwelling houses just below Fort Garry, on the right. About half way between these small centres of population, German Creek, a small meandering stream, comes in from the south-east. A quarter of a mile above the Roman Catholic church, the Assinniboine enters Red River, and a short distance up this stream the bastions of Fort Garry come into view. Above the mouth of the Assinniboine the course of the river is exceedingly tortuous. An idea of its meanderings may be obtained from the comparison between

distances by the river from Fort Garry to the mouth of
La Rivière Sale, and the relative position of the same
places by the road, the former being sixteen, and the latter
nine miles. The houses of settlers appear at intervals on
the banks for several miles above La Rivière Sale, the last
house being situated thirteen miles from Fort Garry, or
fifty-seven from the 49th parallel. Above this the river
windings are fringed with forest, varying in depth from a
few yards to half a mile. Here and there naked bends
are exposed to the prairie, the peninsula portion on the
opposite side being generally clothed with trees of large
dimensions; this character is preserved far south of the
49th parallel.

Returning to the Indian village, and following the road
to Fort Garry, thence to the 49th parallel, the following
description refers to the west or most thickly inhabited
parts of Selkirk Settlement.

From that part of the Indian village which lies on the
west bank of the river to the Stone Fort, little can be
seen of the surrounding country, as the road traverses a
forest of small aspens, and farms are few in number and
small in extent.

The Stone Fort covers an area of about four acres, and
encloses within its walls numerous buildings of which
several excellent photographs were taken in 1858. The
main or King's Road does not follow the windings of
the river, but stretches from point to point, sometimes
approaching it at these places within a quarter or half a
mile. Where the river windings throw it back to a
distance exceeding a mile, inner roads, as they are
termed, branch off to the river for the convenience of
settlers, and there is a bridle path all the way from the
Lower to the Upper Fort, on the immediate bank of the
river. Aspen woods continue to shut out the view until

we arrive within a mile or two of Water-mill Creek, when a scene opens upon the right which discloses on the one hand the white houses and cottages of the inhabitants, with their barns, haystacks, and cattle yards, grouped at short distances from one another, and stretching away in a thin vanishing line to the south; while on the other hand, a boundless, treeless ocean of grass, seemingly a perfect level, meets the horizon on the west. The same kind of scenery, varied only on the left hand as the road approaches or recedes from the farmhouses, on the river banks, or passes near neat and substantial churches, which at almost regular distances intervene, prevails without interruption until within four or five miles of Fort Garry. Here, stretching away, until lost in the western horizon, the belts of wood on the banks of the Assinniboine rise above the general level, while from the Assinniboine toward the north again is an uninterrupted expanse of long waving prairie grass, sprinkled with herds of cattle, and in the fall of the year with clusters of stacks of hay. This is the ordinary aspect of the country comprising that portion of Red River Settlement which lies between Water-mill Creek and Fort Garry. Remove the farmhouses and churches, replacing them on the river banks by forest trees of the largest growth, and the country between Fort Garry and the 49th parallel, as seen along the road to Pembina, a distance of seventy miles, is continually reproduced in its ordinary aspect of sameness and immensity.

The vast ocean of level prairie which lies to the west of Red River must be seen in its extraordinary aspects, before it can be rightly valued and understood in reference to its future occupation by an energetic and civilised race, able to improve its vast capabilities and appreciate its marvellous beauties. It must be seen at sunrise, when

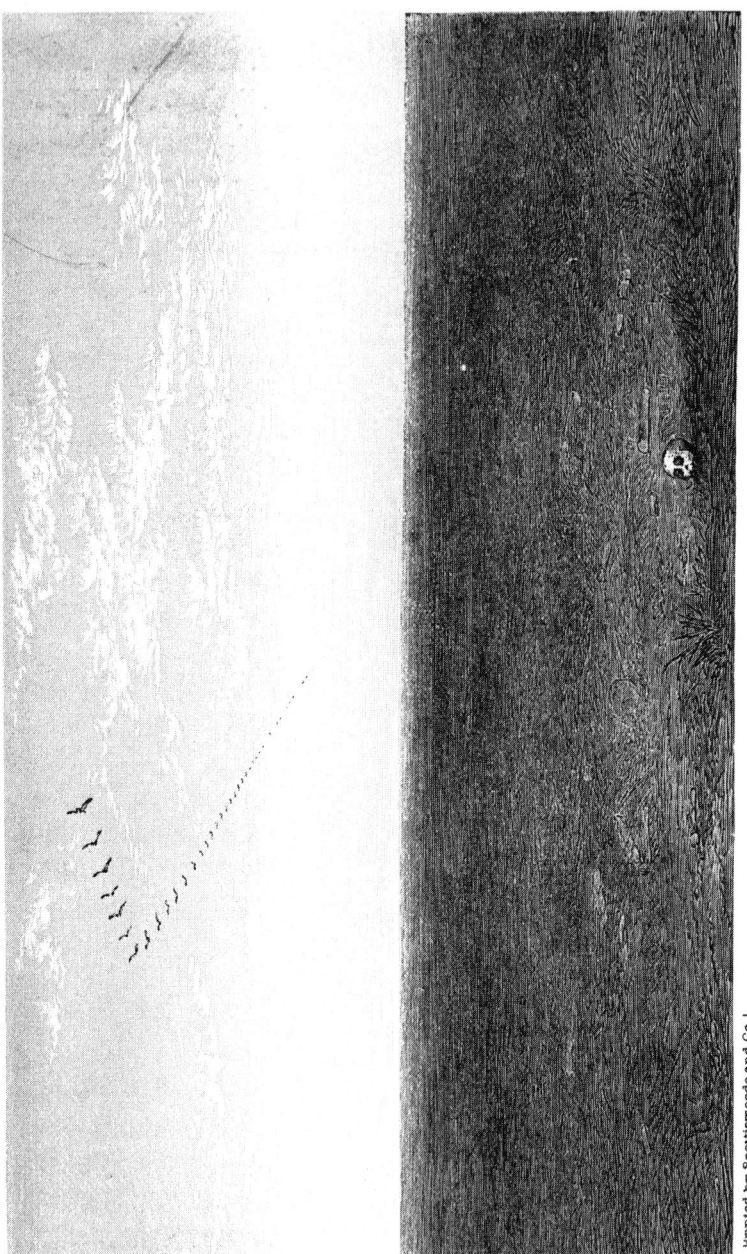

Printed by Spottiswoode and Co.]

[New-street Square, London.

THE PRAIRIE LOOKING WEST.

the boundless plain suddenly flashes with rose-coloured light, as the first rays of the sun sparkle in the dew on the long rich grass, gently stirred by the unfailing morning breeze. It must be seen at noon-day, when refraction swells into the forms of distant hill ranges the ancient beaches and ridges of Lake Winnipeg, which mark its former extension; when each willow bush is magnified into a grove, each distant clump of aspens, not seen before, into wide forests, and the outline of wooded river banks, far beyond unassisted vision, rise into view. It must be seen at sunset, when, just as the huge ball of fire is dipping below the horizon, he throws a flood of red light, indescribably magnificent, upon the illimitable waving green, the colours blending and separating with the gentle roll of the long grass in the evening breeze, and seemingly magnified towards the horizon into the distant heaving swell of a parti-coloured sea. It must be seen, too, by moonlight, when the summits of the low green grass waves are tipped with silver, and the stars in the west disappear suddenly as they touch the earth. Finally, it must be seen at night, when the distant prairies are in a blaze, thirty, fifty, or seventy miles away; when the fire reaches clumps of aspen, and the forked tips of the flames, magnified by refraction, flash and quiver in the horizon, and the reflected lights from rolling clouds of smoke above tell of the havoc which is raging below.

These are some of the scenes which must be witnessed and felt before the mind forms a true conception of the Red River prairies in that unrelieved immensity which belongs to them in common with the ocean, but which, unlike the ever-changing and unstable sea, seem to promise a bountiful recompence to millions of our fellow-men.

On the 10th September I started for Prairie Portage,

on the Assinniboine, about sixty-five miles west of Fort Garry, with Mr. Napier and a Cree half-breed named John Hallet.   Our conveyance, which resembled a very shaky old fashioned light cart, was furnished by Mr. M'Dermott, the most enterprising and wealthy merchant and freighter in the settlements.   Mr. M'Dermott became a kind of purveyor-general to the Expedition, and supplied us with whatever was to be procured for money in this remote region.   Hallet provided two horses, one of which he declared to be an excellent buffalo runner, but not to be trusted in shafts, as we found to our cost.   In the following general description of the Assinniboine, a few facts are incorporated which were acquired during the exploration of the following year.

The Assinniboine rises in latitude 51° 40′, and pursues a south-easterly course for a distance of about 260 miles parallel to the basins of the Great Lakes on the east of the Riding and Duck Mountains. Within eighteen miles south of the 50th parallel it takes a sudden bend to the east, which general direction is preserved until it falls into Red River, a distance of about 240 miles from the great bend.   At Lane's Post, twenty-two miles from Fort Garry, the Assinniboine is 120 feet broad (June 28th, 1858), with a mean sectional depth of six feet. Its greatest depth here is seven and a half feet, and the rate of its current is one and a half mile an hour.   Near Prairie Portage, sixty-seven miles from Fort Garry, the speed of the current is two miles an hour, and its fall, as ascertained by levelling, is 1·18 feet in a mile.   At its junction with the Little Souris, an affluent which it receives 140 miles from its mouth, the breadth of the river is 230 feet, its greatest depth twelve feet, and its mean sectional depth 8·6, the speed of its current being one and a quarter mile an hour; this river is apparently considerably larger 140 miles from its outlet than twenty-two miles from the same place. Even at Fort Ellice, 280 miles from its junction with Red River, the Assinniboine is 135 feet wide, 11·9 feet deep in the channel, with a mean sectional depth of eight feet, and a current flowing at the rate of one and three quarters of a mile an hour; in other words, this river, 280 miles from its mouth, carries a larger body of water than at a point twenty-two miles from it.

The following table shows the quantity of water which the Assinniboine carries at three different points, distant respectively in round numbers, twenty-two miles, 140 miles, and 280 miles from its outlet by the windings of the river valley, but not by the windings of the river itself, which will be at least double the length of the river valley.

	Cubic feet per hour.	Distance from outlet at Fort Garry.
Lane's Post	5,702,400	22 miles.
Mouth of Little Souris . .	12,899,040	140 „
Opposite Fort Ellice . .	9,979,200	280 „

It thus appears that the volume of water in the Assinniboine is nearly twice as large at Fort Ellice as 258 miles lower down the river, if the foregoing table affords sufficient data on which to rest an opinion. It is very probable that the character of the season would modify these results in different years. The measurements were not made simultaneously, and the rainfall in the neighbourhood of the Touchwood Hills and in the region about Fort Pelly in 1858 was represented to be more in the extreme, than is usual during the summer months. But judging from the appearance of the river bank, and the statements of Indians and half-breeds, familiar with the summer level at the localities where the sections were made, there is no reason to suppose that its waters were in excess of their ordinary summer level. It is, therefore, very probable that evaporation during a long and tortuous course through an open valley, is adequate to diminish the volume of water in the Assinniboine very much in excess of the supply which it receives from tributaries or springs during its course to Red River.

East of Prairie Portage the Assinniboine flows through a flat, open, prairie country, not more than sixteen or twenty feet above the general level of the stream, until within a few miles of Fort Garry. The whole country rising in steps west of Prairie Portage, the Assinniboine has excavated a deep and broad valley through it, in which it meanders with a rapid current.

At the mouth of the Little Souris, or Mouse River, this valley is 880 yards across and eighty-three feet below the general level of the prairie. At Fort Ellice the valley is one mile and thirty chains broad, and 240 feet below the prairie.

The Assinniboine receives numerous and important affluents. On its eastern water-shed are the Two Creeks, Pine Creek, Shell River, Birdstail River, and Rapid River or the Little Saskatchewan. The distances of the rivers from Fort Pelly, which may be considered as lying at the head of bateau navigation, will be noticed hereafter when the country they drain is described. From its western water-shed it receives the White Sand River from the Touchwood Hills; the Qu'Appelle or Calling River, inosculating with the south branch of the Saskatchewan; Beaver Creek, a small rivulet on which Fort Ellice is situated; and the Little Souris or Mouse River, sweeping round the flanks of the Grand Coteau de Missouri. The Crees of the Sandy Hills on the south branch, state that Elbow Bone Creek, an affluent of the Qu'Appelle River, inosculates by a deep valley with the Mouse River, or an arm of it, and is connected continuously with the Assinniboine.

For a distance of several miles above Fort Garry the Assinniboine flows in a trench excavated through a level prairie to the same depth as the river it feeds, or from twenty-five to forty feet. Differences due to local variations in the height of the bank are often referable to very slight undulations in the level of the prairie, and to the occurrence of ancient lake beaches or ridges, the first of which is cut by the river, near St. James' church. This ridge continues in a direction nearly due north, until it dies away in the general rise of the prairie. It is near this spot that the rapids occur which, in the summer months, when the water is low, offer the only impediment to the continuous boat navigation of the Assinniboine for many miles.

Some short distance above the rapids the river widens; at its mouth it may be 130 feet in breadth, and four miles from its mouth, 150 feet, a breadth which it preserves with considerable uniformity for a distance of 130 miles. About six and a half miles from Fort Garry the Assinniboine receives a small affluent called Sturgeon Creek, coming from the north-west. The general direction of the river up to this point is nearly due west, and its course comparatively straight. The south bank is heavily timbered to a small depth, but the north bank is chiefly taken up by farms, and devoid of timber.

From Sturgeon Creek the course against the stream continues still westerly, but with more decided meanderings, and the wooded points on both sides of the river rarely penetrate a quarter of a mile into the vast prairie on either side. The distance from Fort Garry to where it makes its north-westerly bend is twenty-three miles by the river's windings, but by the road through the prairies and settlements only sixteen miles. The river banks are here about eighteen feet high, and their height imperceptibly diminishes until, at Prairie Portage, they were found by measurement not to exceed sixteen feet, during the stage of water, on the 7th of September, 1857.

After making its north-west turn, the Assinniboine is so remarkably crooked that a straight line drawn through the tract of country in which it meanders for a distance of twelve miles, would be cut eighteen times by the river, and these windings are confined within such a limited breadth that in a strip of the same length, and 1,000 yards broad, the curves of the river would just overlap this boundary four times.

The physical features of the Assinniboine as far as Prairie Portage, resemble in every important particular those of Red River. The tortuous sinuosities of the larger stream are reproduced with curious fidelity in the magnificent prairies through which its western rival flows.

There is little or no variety in the character of the banks either of Red River or the Assinniboine; they consist of Post Tertiary stratified clays and marls, overlaid by vegetable mould. At Lane's Post, twenty-two miles west of Fort Garry, a fresh exposure of the bank, which, by the way, is continually breaking down in small patches and changing, during the lapse of many years, the channel of the river, exhibited stratified whitish marly clay, and dark drab coloured clay from the water's edge to within five feet of the prairie level, which here, as is frequently the case, comes abruptly upon the river. Dark alluvial clay succeeds, having an average thickness of about four feet; this is followed by from six to eighteen inches of black prairie mould.

Beyond Lane's Post the river course is westerly for a few miles, it then makes a bend towards the north-west, until Long Lake, an old bed of the river, is reached, after which it turns towards the south-west for about sixteen miles, thence westerly, ten miles further to Prairie Portage. Nine miles beyond Lane's Post the settlements cease; they recommence thirty miles further up the stream by the road, and although the distance from Lane's to Prairie Portage is not more than forty-three miles, the course of the winding Assinniboine would probably exceed ninety miles. The river banks are heavily timbered, and sustain trees of very large dimensions. The distance between the top of the banks on either side of the river is variable, but it appears to be generally between 600 and 800 feet, at sharp turns it was often not more than 400, and whenever it exceeded that distance one side was steep and washed by the water, the other occupied by a sand spit or mud-flat at the foot of the opposite bank.

During my stay at Prairie Portage, in September, 1857, I had an excellent opportunity of examining the relation of the sand and mud flats to the river banks, as well as the forest which fringed it to the depth of half a mile. The river is here about 180 feet broad, and with a rapid current sweeps under the south bank, which forms the outer arc of a very beautiful curve extending over 120 degrees. The cord of this arc is well defined by the old north bank of the river, under which, probably, it once swept, but now only touches when the channel is full during spring freshets ; the length of this cord is perhaps 700 yards, and at each extremity the river is seen sweeping between steep banks, sixteen feet high, until, a little lower down or a little higher up, similar curves, with their accompanying sand and mud-flats recur. These sand and mud-flats are arranged in the order of the specific gravities of the materials which compose them, but with such singular regularity, and with such curious and interesting admixtures, that I have considered it worth while to describe them with some degree of particularity.

At the western extremity of the curve, a few rounded boulders were seen, not exceeding eight inches in diameter;

these were followed by gravel spits as the area opened; beyond the gravel spits, which extended perhaps over a quarter of the segment, flats of coarse sand showed themselves; these were strangely filled and strewed with the decayed and broken horns of the elk, the bones and horns of the elk, buffalo, deer, and just beyond these a human skull, with two or three scattered and water-worn skulls of buffalo; the sandy areas ceased in curved lines, with a small steep descent of about two feet, and were succeeded by mud partly covered here and there with fine sand, probably drifted by wind. The sandy mud was followed by fine compact mud, with numerous deep cracks partially filled with fine sand. Another fall of about three feet occurred in the form of a bank, and recent mud, smooth and treacherous, occupied the remaining portion of the segment a few inches above the present water level. This arrangement of mud, sand, and gravelly spits was noticed elsewhere, and probably frequently occurs in the Low Prairie region through which the Assinniboine is continually changing its course by excavating new channels in the soft and yielding clays.

The timber on the banks of this river is perhaps not so heavy as on Red River, nevertheless some very fine oak and elm, with white wood and poplar of extraordinary dimensions, were seen near the Prairie Portage. A fair quantity of sugar is made by the Assinniboine half-breeds, but not in comparison with what might be easily obtained if systematic habits and a proper appreciation of the fruits of industry existed here. A species of grape grows in profusion on the banks of this river. I suppose it to be the frost grape (*Vites Andifolia*). The fruit when first gathered is not very palatable, but after hanging in the open air for forty-eight hours it acquires a sweet taste and a very delicious flavour.

The name Prairie Portage is derived from the existence of a carrying place nine miles long, between this part of the Assinniboine and Lake Manitobah. It is stated by half-breeds at the settlement, that at seasons of extraordinary high water, canoes can approach each other from the Assinniboine and Lake Manitobah, so as to leave but a very short distance for the portage ; and instances have occurred of water, during periods of high floods, flowing from the Assinniboine into Lake Manitobah by the valley of Rat River.

I had an opportunity of meeting, at this isolated settlement, with one John Spence, a Cree half-breed of great experience in Rupert's Land. He drew a small chart for me, showing the position of what he called " coal " on the Assinniboine. I saw and conversed with a half-breed who had brought " a few bushels " of this coal to the settlement, for the purpose of ascertaining its fitness for the forge ; he stated that he was a blacksmith, had used the coal and found it answer, but it required a strong draft. I procured from another half-breed several specimens, and ascertained that it was lignite, and not the true coal of the coal measures. On the Little Souris, a tributary of the Assinniboine, the lignite was described as cropping out in bands exceeding a foot in thickness, and occupying a large area on its banks, a statement which the exploration of the succeeding year did not verify, at least in the locality pointed out to me. Dr. Hector, however, found lignite in the valley of the Souris, three hundred miles west of Prairie Portage. I endeavoured to induce John Spence to accompany me and point out the locality where the lignite cropped out on the Assinniboine; he expressed perfect willingness to do so, if I could procure for the trip ten men in all, so that watches might be established by night, in consequence of the pre-

sence of several bands of Sioux Indians on the trail of the
buffalo hunters, who were then coming in from the Great
Prairies, after their summer hunt. The Sioux had succeeded in driving off ten horses from the tail of the caravan, about half a day's journey from Prairie Portage the
night preceding my arrival there ; and this incident led
John Spence and others to decline going with me, unless
the number of the party amounted to ten in all. So large
an addition I found it impossible to procure at Prairie
Portage, and after my return to the settlements, the time
at my disposal was too short to admit of the exploration.
In carefully searching the recent mud-flats of the Assinniboine at, and a little above Prairie Portage, I found
numerous small fragments of lignite, from which it might
be inferred that an exposure of the parent rock was situated some distance up the river ; but beyond this and the
reiterated statements of many who had descended the
stream in a batteau, I found no proof of the existence of
lignite in available quantities.

In the settlements on Red River and the Assinniboine,
small specimens of lignite were frequently shown to me
by different people, who stated that they procured them
from the crossing-place on the Little Souris ; and an Indian
had a bag containing about half a bushel of the same material, together with specimens of silver mica, carefully treasured up in many folds of dressed buffalo skin. Many intelligent people in the settlements appeared to be much im-
pressed with the importance of ascertaining the true nature
and extent of the lignite beds on the Little Souris. The
great scarcity of wood in the prairie country, and all
through the valleys of Red River and the Assinniboine,
making the question of a permanently increasing settlement in a measure dependent upon the supply of fuel
which may be obtained from other sources than those

offered by the aspen-covered ridges, or the thin strips of timber on the immediate banks of the rivers.

In order to reach John Spence's house, I passed through a field of Indian corn, and from the proprietor I obtained the following statement respecting the cultivation of this valuable grain. The kind of Indian corn most common in the settlements is called the Horse-teeth corn, and it does not always ripen. The variety sown by Spence he termed the Mandril corn, the seed was procured from the Indians near the head waters of the Missouri; probably the " Mandan corn " would be the correct name. He had cultivated it for two years; it ripened

Prairie Portage, Assinniboine River.

well both years. One of his neighbours, a Cree Indian, had cultivated it for four years and had not met with any failure. Spence sowed his corn on the 1st June, and gathered it 10th September, or after a period of 102 days. In dry seasons it ripens earlier, and is planted about the 20th of May; the wet spring of the present year retarded all agricultural operations. A small house adjoining the one in which Spence resided I found filled with a portion of his corn crop.

The road from the village of Prairie Portage follows

a general north-easterly direction for a distance of twenty-nine miles, before it turns south-westerly towards Fort Garry. This deviation is necessary in order to avoid Long Lake, an ancient bed of the river, now converted into a narrow, winding lake of great length.

On each side of the road is a very magnificent prairie, bounded on the right by the wooded banks of the Assinniboine, and on the left by the horizon; a few scattered clumps of poplar are seen here and there, but no trees, until the " Big Ridge" comes in sight. The ridge is probably an ancient beach of Lake Winnipeg; its elevation does not appear to be more than sixty feet above the prairie level. Where the road touches Long Lake, a spur of the Big Ridge is distant about three miles. I made a diversion from the main track for the purpose of examining the character of the ridge. It rises almost imperceptibly from the prairie, and at its base are numerous small fragments of limestone. Ascending the ridge, the limestone débris was found to increase in quantity, and near its summit the slabs were of large dimensions.

Our cart-horse beginning to show signs of breaking down soon after we left the Big Ridge, Hallet proposed to put his buffalo runner into the shafts, thinking that the journey had so far subdued his spirit, that he would submit to the indignity of drawing a light cart. He travelled for a few hundred yards quietly enough, but when we came into the open prairie, he started off at a gallop, and swerving suddenly, overthrew the cart, projecting its contents into the grass. Mr. Napier received a serious sprain in the wrist, but I fortunately escaped without injury, having fallen on my feet. The buffalo runner soon broke loose from the shafts, and after a short gallop stopped to survey the ruin he had made. Hallet caught him without much trouble, but did not venture to harness him again. A few strips of buffalo hide placed our cart in travelling condition, and permitted us slowly to resume our journey.

Regaining the main road, well marked by the deep ruts formed by the buffalo hunters' carts, we soon arrived at the White Horse Plain, a vast, slightly undulating prairie, bounded by the horizon in every direction but the south, where the distant wooded banks of the Assinniboine afford some relief to the eye. The grass is long and rank, and the soil a black mould of great depth, often exceeding eighteen inches. In many places it is thrown up into conical heaps by moles, and uniformly displays the same rich appearance, truly represented by the bountiful profusion of verdure it sustains. In 1857 the edges of the White Horse Plain unfortunately teemed with another kind of life. The grasshoppers (locusts) appeared in countless millions just before my arrival; every bare patch of ground in the road was filled with their eggs, the living insects were leaping through the tall grass in infinite multitudes, yet, notwithstanding, failing to change the appearance of the country in the midst of so great a profusion of food. What the next year's brood may do remains to be seen, their progenitors had come in swarming clouds from the south side of the Assinniboine, but no one could tell of their origin, or of the devastations they must have created before they took their flight, and alighted on the White Horse Plain.

The last house of the settlement, westward of White Horse Plain, is about thirty-three miles from Fort Garry, and between it and the Company's Post, in charge of Mr. Lane, there are nine houses and farms. The Prairie Portage road, however, does not pass near them, it touches the river only at those bends which do not necessarily compel much deviation from a straight course. The farmhouses are similar to those on Red River, but the soil appears to be, if possible, of a better description.

We were very hospitably entertained by Mr. Lane, the gentleman in charge of the Hudson's Bay Company's Post on the Assinniboine, twenty-two miles west of Fort Garry.

Mr. Lane informed me that Indian corn did not always ripen on that part of the river. Spring frosts rarely affect it, but autumn frosts sometimes cut it off. He thought that careless cultivation was the reason why it did not progress fast enough to escape the early autumnal frosts. Indian corn sown on dry points of the river arrived at maturity much sooner than that which was sown on the rich and moist prairie mould.

Leaving Lane's Post, the river is touched again at the Roman Catholic Mission of St. François Xavier. The road now follows the general course of the river, in the rear of the farms which, from this point to Fort Garry, are not far apart.

On the night of the 15th September, we stayed at the house of Mr. Geo. Flett, fifteen miles west of Fort Garry. Mr. Flett's turnips had been altogether consumed by the grasshoppers, but his wheat was safe and good ; he says that Indian corn succeeds well, and almost always ripens ; it is his opinion that it may always be relied upon when care is taken ; it does not progress quick enough on the open prairie to escape every season the early·autumnal frosts ; on the points of the river where the soil is lighter and dryer than in the open prairie, and where some shelter may be obtained from the neighbouring timber, he has never known it to fail. Mr. Flett finds the cut worm the great enemy to his turnips ; his potatoes for the summer crop are planted 1st June, and ready for eating from the 10th to the 15th August; the winter supply he does not lift until October. Over the whole of the White Horse Plain district, thirty bushels to the acre is an average crop of wheat, but on new land forty bushels is not only common but generally expected.

On the morning of the 16th we paid a visit to Mr. Gowler, whose farm is situated on the immediate banks of the Assinniboine, about nine miles from Fort Garry. Nearly all farming operations were over, but an inspection of his farmyard and garden enabled me to form an opinion of his success and prospects as an agriculturist on the Assinniboine.

A small stack-yard was filled with stacks of wheat and hay; his barn, which was very roomy, was crammed with wheat, barley, potatoes, pumpkins, turnips and carrots. The root crops were shortly to be transferred to root houses, which he had constructed by excavating chambers near the high bank of the Assinniboine, and draining them into the river. The drain was supplied with a close and tightly fitting trap, which was closed when the water rose during the spring above its mouth, at that time eight feet above the level of the river. The chambers were about nine feet high, and their ceilings three feet below the prairie level. Access was obtained through a hole in the ceiling, which was covered with a neat little movable roof. There were three of these cellars or root-houses before the dwelling-house, and between it and the river. Frost never entered them, and he found no difficulty in preserving a large stock of potatoes and turnips through the severe winters of this region.

Mr. Gowler farmed fifty acres in white and green crops, hay and pasture being furnished by the prairie. He owned much more land, but found it useless to crop it, as no market for surplus produce existed. In 1856 he had sold many bushels of potatoes at sixpence per bushel, and had carted them nine miles. I had been previously in-

formed of the extraordinary success of Mr. Gowler in
growing wheat, but I found upon inquiry that the practice he employed was simply not to grow wheat after
wheat; he had grown fifty-six measured bushels to the
acre. The price of wheat at the time of my departure was
4s. 5d. sterling a bushel, but last year at the same time
it had been 3s. 6d. sterling. His turnips (Swedes) were
magnificent; four of them weighed 70 lbs., two weighing
39 lbs., and two others 31 lbs. Whatever manure his yard
and stables supplied he gave to green crops and the
garden. A portion of the potato crop was still in the
ground; they far surpassed in quantity, quality and size,
any I had ever seen before. Mr. Gowler very kindly
turned them up out of the soil wherever I pointed
out. I counted thirteen, fourteen, and sixteen potatoes,
averaging three and a-half inches in diameter, at each
root respectively. They were a round white-skinned
variety, like those known in Canada as the "English
White." The potatoes were planted on the 1st June, and
were ready for eating on the 16th or 18th August. The
winter supply was rarely taken out of the ground before
the beginning of October. The greatest enemy to the
turnip crop is the cut-worm (the grub of an elater).

Indian corn succeeds well on Mr. Gowler's farm, and
onions of rare dimensions were growing in his garden.
He had had this year a splendid crop of melons, the seed
being sown in the open air at the end of May, and the
fruit gathered about the 1st September. At the time of
my visit the melons had all been consumed, but I had
several opportunities of tasting and enjoying this fruit, at
Fort Garry and elsewhere, on the Assinniboine and Red
River. In every instance they were grown in the open
air, without any artificial aid beyond weeding, from the
time the seed was planted to the maturation of the fruit.

Mr. Gowler insisted on my tasting his wife's cheese, and smoking his tobacco, before I departed. The cheese was tolerable; the tobacco, which was grown in the neighbourhood and highly prized by Mr. Gowler, was dreadfully strong, and would involve long training in order to acquire a taste for its qualities. Nevertheless, Mr. Gowler preferred it to some excellent fig-leaf which I offered him; he remarked that he had grown and prepared it himself, and knew what it was.

I may here relate, with a view to show how long old associations linger in the recollections of the European portion of the population in this remote region, that when I sat down to table Mr. Gowler turned inquiringly to his wife, saying, "And where is my plate?" "Oh, John! you would not think of sitting at table with gentlemen?" Mr. John seemed puzzled for a moment; his son-in-law and children were looking in silence from different corners of the room. He cast a hasty glance around, and the true feelings of independence and manly right showed themselves, as he exclaimed, "Give me a chair and a plate; am I not a gentleman, too? Is not this my house, my farm, and these my victuals? Give me a plate."

As Mr. Gowler accompanied me to the gate, where my horse was tied, he expressed, with much warmth of feeling and manner, the following opinion of husbandry and its prospects in Assinniboia: —

"Look at that prairie; 10,000 head of cattle might feed and fatten there for nothing. If I found it worth my while, I could enclose 50, 100, or 500 acres, and from every acre get 30 to 40 bushels of wheat, year after year. I could grow Indian corn, barley, oats, flax, hemp, hops, turnips, tobacco, anything you wish, and to any amount, but what would be the use? There are no markets, it's a chance if my wheat is taken, and my potatoes I may have to give to the pigs. If we had only a market, you'd have to travel long before you would see the like of these prairies about the Assinniboine."

The substantial character of the barn, stables, and piggeries, constructed of wood, their neatness and cleanliness, the admirable arrangements of the hammels for cattle, and the sheds for sheep, all showed how far a little energy and determination, instructed by the experience of earlier years, would go in reproducing amidst the boundless prairies of Assinniboia, the comforts and enjoyments which are by no means the rule among the small farmers of Great Britain. I regretted to find that a few days before my visit the grasshoppers had arrived from the south-west, and consumed in a single day every green leaf in the garden which remained exposed to their attacks.

The "Nor'-Wester," a newspaper published for the first time at the Red River settlement on the 28th December, 1859, mentions Mr. Gowler's success as an agriculturist in the following terms : —

" At seed-time of the present year (1859), all traces of the pestilence (the grasshoppers) had disappeared, and Mr. Gowler having before his eyes the pretty sure prospect of a good market, brought under cultivation a greater breadth of land than any year previously. He sowed 63 bushels of wheat, 36 of barley, 24 of oats, and 101 of potatoes, and from these he realised 700 bushels of wheat, 350 of barley, 480 of oats, and 2,100 of potatoes. The cost of the seed was 50l. ; in preparing and tilling the soil, about 25l. more were expended ; and the cost of gathering in and thrashing the crops is set down at 100l. — making a total expenditure of 175l. Place against that the sums representing the sale of the wheat at 6s., the barley at 3s. 9d., the oats at 2s. 6d., and the potatoes at 1s. 3d. per bushel (average prices, which the produce will easily command), and an argument more strong and convincing than could be wrought out by any other process of reasoning, stands stubbornly forth in favour of the claims of the settlement as being one of

Confluence of the Assinniboine and Red River.

the best agricultural countries on the face of the globe. It should be added that Mr. Gowler's profits have already enabled him to enlarge the bounds of his estate to 600 acres ; to stock it with a noble herd of cattle and horses, and to make the necessary preparations for erecting thereon, next summer, a snug and comfortable mansion."

CHAP. VII.

As soon as I returned from Prairie Portage preparations
were made for an exploration of the Roseau or Reed-grass
River, with a view to ascertain whether a communication
might be effected between the Lake of the Woods and
Red River, as well as to ascertain the limit of the palæozoic
rocks, and their junction with the metamorphic series on
which they rest. On the 21st September, accompanied
by my assistant, Mr. Fleming, I started from our quarters
in the Middle Settlement. Our equipment consisted of
three men, five horses, and a Red River cart. We crossed
the Assinniboine by the ferry at Fort Garry, and took the
road on the west side of the river to Pembina.

The country lying to the west of Red River was
examined by Mr. Dickinson in 1858, and, for the sake of
uniformity, his observations are incorporated in the description which follows, of La Rivière Sale, and an extensive range of table-land called Pembina Mountain.

Nine miles above Fort Garry, La Rivière Sale joins with
the main stream. The buffalo hunters' trail to the great

south-western prairies on the Grand Coteau de Missouri passes up the south side of this river for a distance of thirty miles, cutting across the large and winding bends of the valley.

The country lying between it and the Assinniboine is very marshy, and is covered with willows and clumps of small aspen. In the valley of La Rivière Sale, and along both sides, grow oak and elm and some fine ash, many of the trees being two feet in diameter; this narrow forest extends the whole way up the river on the north bank. On the south side there is a prairie apparently as level and boundless as the ocean; the grass on it is most beautiful and luxuriant, indicating the richness of the soil.

The valley is about a quarter of a mile wide and forty feet deep, abounding in salt springs, which make the water in the river quite brackish, from which it derives its name. The river higher up opens out into small lakes, and rises from an extensive marsh. The track here joins the hunters' track from the White Horse Plain; it then turns to the south, in which direction it goes for about twenty-five miles through open prairies, until it crosses "La Rivière des Iles des Bois," a river fifteen feet wide and two deep, flowing into the Scratching River. This portion of the country is all a level prairie, the greater part of it being wet and marshy, except near the last-named river, where it is quite dry for five miles; the land is a rich sandy loam, yielding most luxuriant grass. On both sides of the river there is a skirting of trees, chiefly of oak, averaging one foot six inches in diameter.

The buffalo hunters, when they have crossed this little river, begin to keep a sharp look-out for the Sioux, and to take their usual precautions, such as setting watches by night and placing their carts in a ring.

The track, continuing in the same direction, crosses a

prairie twenty miles wide, of light sandy soil, with clumps of aspen and willows growing here and there; it is intersected by many small valleys, in all of which, with one exception, the brooks that formed them are now dried up. The valley of " La Riviere Tabac " is seven chains wide and twenty feet deep, with very little water in the fall of the year, where in spring time there is a rapid flow.

The prairie on the south and west is bounded by what is generally called the " Pembina Mountain," which is rather a series of steps rising up from the prairie below to one above. There are three steps, from twenty to twenty-five feet high, together with a gradual ascent for two miles; the whole of it is thickly strewn with granitic boulders. This " mountain," which consists of clay, gravel and sand, runs in a south-easterly direction, from a little above Prairie Portage to Pembina. Where we crossed it there is no timber, but on both sides it is well covered, particularly on the south, where the trees seemed large and good. Here the forest is said to begin which reaches to the Assinniboine, but with the exception of some oak on the mountain there is no good timber, nothing but young aspen from twenty to thirty feet high, growing very close together, and forming a dense thicket.

Scratching River joins the main stream thirty-seven miles from Fort Garry. The postman who carries the mail between Pembina and the settlements lives here, and has established an apology for a tavern and a ferry. Scratching River winds for many miles through a boundless prairie, without a tree or shrub on its banks.

We arrived at Pembina Fort on the 24th, just in time to partake of an excellent dinner with Mr. Murray, the gentleman in charge. In the afternoon we crossed Red River, passing through the miserable cluster of houses which bears a prominent position on maps of the north-western states, figuring under the name of the town of Pembina. Most of the inhabitants of Pembina have moved to St. Joseph, so that the population of this frontier village does not now number more than 100 souls. On the evening of the 25th we camped on the banks of the Roseau, after a hard march of thirty-one miles through a very fine and promising country.

The general course of this stream from its confluence with Red River to Roseau Lake is a few degrees to the south of east. It enters Red River about ten miles north of the 49th parallel, and it is probable that Roseau Lake is on the boundary line between Rupert's Land and the State of Minnesota. The course of the Roseau is very tortuous, and for the first twenty miles it meanders through a beautiful prairie, with a belt of heavy forest trees on its banks. Near the mouth of the river, on the south side, there is a large area of low land, but above that point the banks vary from fifteen to twenty feet in height until, at the crossing place, the ancient Lake Ridge is reached. Here the banks are from fifty to fifty-five feet above the level of the river. Near the crossing place, the ridge has probably an elevation exceeding sixty feet above Red River; it, with its offsets, form a very singular and most interesting feature in the topography of the valley of this river.

The ridge once past, the whole face of the country changes. The soil becomes poor and sandy, although still preserving a prairie or plain character. The timber on the banks of the river fast dwindles to small sized oak, elm, birch, and poplar, until it gives place, about forty-six miles from the mouth, and perhaps seventy or eighty by the winding of the stream, to extensive marshes, in which there are islands of small pine. At the commencement of these marshes the Roseau River moves sluggishly, and its stream soon becomes dead water. with a vast ex-

panse of flooded land on either side, extending, according
to our guide, fifty miles to the right hand and to the left.

Having found it impossible to proceed further on horseback than the beginning of the great marshy tract of the
Roseau, and not being provided with a canoe, the following
description of the country rests upon the authority of the
guide who accompanied us, and who had resided at Roseau Lake for a year and a half when in the service of the
Hudson's Bay Company. The river channel can be traced
through a marsh ten miles long, nearly on a level with the
water in the river. The depth of the marsh does not exceed three feet, and it is quite possible to wade on horseback through it. The Hudson's Bay Company's route to
their post on Roseau Lake (in 1851) retired from the
banks of the river when the waters began to flow sluggishly, and pursued a direction some miles to the south of
the channel, probably within the United States territory.
In 1847, a very dry season, it was possible to proceed
with carts in a direct line near the banks of the river from
the beginning of the marsh to the post, one mile and a
half from Roseau Lake.

An idea of the character of the country about this post
may be inferred from the guide's description of his attempts to destroy the monotony of his life, when stationed
at Roseau Lake. He informed me that when he wished
" to see anything " beyond the four walls of his log shanty,
and the rushes by which it was surrounded, he was in the
habit of mounting to the roof, and from the top of the mud
chimney enjoying the view; which consisted of reeds to
the north, reeds to the south, and reeds to the west, as far
as the eye could reach, and to the east, Roseau Lake,
fifteen miles long by ten broad, with a deep fringe of reeds.
On the bosom of this retired sheet of water, in the spring
and the fall, he was enabled to watch countless millions

of ducks and geese; and the noise of their shrill cries, with
the flapping of wings as they would rise to take their morning flight to the north or south, according to the season of
the year, were almost the only sounds he heard, save the
sighing of the wind through the reeds, during his dreary
abode in the waste of Roseau Lake. The altitude of Roseau
Lake above Lake Winnipeg probably does not exceed 170
feet; and as the elevation of the Lake of the Woods is at
least 370 feet above the same level, there must still be a rise
of 200 feet to be overcome before reaching the height of
land. Our guide described the Roseau River, before it
enters Roseau Lake, as stretching to the south in the territories of the United States. He also said, that issuing from
the Great Muskeg, or swamp, occupying so much of the
height of land between Red River and the Lake of the
Woods, was a narrow rapid stream fifty miles long, emptying
into Roseau Lake, thus forming a route by means of which
none but the smallest sized canoes can pass from Roseau
Lake, through the Great Muskeg, to the Lake of the Woods.

The ancient Lake Ridge is a continuation of the one referred to on page 132; it extends in an unbroken line,
except where the river from the higher level in the rear
has cut channels through it, from near Lake Winnipeg,
far beyond the international boundary. At the crossing-
place on the Roseau, about forty-six miles from Red
River, its height was estimated to be the same as at the
Middle Settlement; it forms a beautiful dry gravel road
wherever traversed, and suffers only from the drawback
of being the favourite haunt of numerous badgers, whose
holes on the flank, and sometimes also on the summit, are
dangerous to horses; it is, apparently, perfectly level for
a hundred miles, and everywhere, as far as my observation enabled me to judge, shows the same even rounded
summit; it may yet form an admirable means of com-

munication through the country, and it marks the limit of the good land on the east of Red River. This ridge is a favourite resort of the prairie hen (*Tetrao cupido*), when they perform their curious circular dances in the early spring months. We frequently met with a ring of sticks, placed in a circle about ten feet in diameter, to each of which a noose of sinew was attached. Our half-breeds informed us that they were snares which the Indians set to catch the prairie hens In the spring the males congregate on dry gravelly ridges, frequenting the same spot year after year, and march round and round, with feathers erect and wings rubbing the ground as a preliminary to a general combat. The Indians observe the spot where the birds congregate, and after night-fall set their snares on the edge of the ring, which the male birds have selected to try their strength, and to attest their claims to the favour of the females who are perched on the neighbouring bushes. In the battle which ensues, or during their solemn march, some of them are caught and strangled. The following spirited description from Audubon's delightful "Birds of America" will, doubtless, be read with interest:—"Their love season commences, and a spot is pitched upon to which they daily resort until incubation is established. Inspired by love, the male birds, before the first glimpse of day lightens the horizon, fly swiftly and singly from their grassy beds, to meet, to challenge, and to fight the various rivals led by the same impulse to the arena. The male is at this season attired in his full dress, and enacts his part in a manner not surpassed in pomposity by any other bird. Imagine them assembled to the number of twenty by daybreak; see them all strutting in the presence of each other; mark their consequential gestures, their looks of disdain, and their angry pride as they pass each other. Their tails are spread out and inclined forwards, to meet the expanded feathers of their neck, which now, like stiffened frills, lie supported by the globular orange-coloured receptacles of air, from which their singular booming sounds proceed. Their wings, like those of the turkey cock, are stiffened and declined so as to rub and rustle on the ground, as the bird passes rapidly along. Their bodies are depressed towards the ground; the fire of their eyes evinces the pugnacious workings of the mind; their notes fill the air around, and at the very first answer from some coy female, the heated blood of the feathered warriors swells every vein, and presently the battle rages. Like game cocks, they strike and rise in the air to meet their assailants to better advantage. Now many close in the encounter; feathers are seen whirling in the agitated air, or falling around them tinged with blood. The weaker begin to give way, and one after another seek refuge in the neighbouring bushes. The remaining few, greatly exhausted, maintain their ground, and withdraw slowly and proudly, as if each claimed the honours of victory. The vanquished and the victors then search for the females, who, believing each to have returned from the field in triumph, receive them with joy."

At noon on the 26th September, when discussing with the guide the possibility of proceeding further up the banks of the Roseau River on horseback, we heard the sound of a gun, proceeding apparently from the river. Having fired one in return, we were not surprised some time afterwards to see an Indian approach. He had just arrived with his family from the Lake of the Woods, by the route proposed to be taken by Mr. S. Dawson and myself some weeks before. He described the route in the same way as the guide, and in no material respect differed from the accounts we had before received from the Lac la Pluie Indians, who had been engaged to convey us through it, before the intervention of the tribe at Garden Island, narrated in Chapter IV. He had been ten days on the road, but might have accomplished the journey thus far in shorter time, had he not found it necessary to hunt for his family, who accompanied him. At my request he drew a chart of the route, which was, in almost all particulars, similar to that furnished by the Indian at Fort Frances. He ascended a small river, marked on the map Reed River, from the Lake of the Woods, for a distance of thirty miles to the Great Muskeg at the height of land. He was two days dragging his canoe through the Muskeg, which is here nine miles broad. He then descended the rapid stream, forty to fifty miles long, before noticed, which is called by the Indians Muskeg River, and found himself among the rushes or reeds of Roseau Lake. In his canoe we found his wife and two children. The half-naked little savages were busily engaged in plucking a goose for their noon-day meal. I offered him some tea in exchange for the bird, and when the transfer was made, asked him what they intended to eat for their own dinner; he replied by pointing to the bow of his canoe, addressing at the same time a word or two to his wife, who raised a piece of birch bark and disclosed two more geese, which he had shot a few minutes before we saw him. Having bartered for them also, with a small plug of tobacco, I asked the guide what he would take for a new stone pipe which one of the children was playing with; to my astonishment the Indian replied, three beaver skins (about five shillings), but at the same time casting his eyes upon our cups and saucers which lay on the grass, he said he would prefer a cup, worth about four-pence. He really knew nothing of the value of money or of cups, although he was quite aware of the worth of a beaver-skin in ordinary articles of trade, such as powder, shot, tobacco, or tin-ware, but a painted earthenware cup was something new to him; and his wife expressed great delight as she examined with much minuteness the addition to her household goods.

Returning nearly in our steps to the crossing-place, we went over to the right bank of the Roseau, and after threading through a forest of fine oaks, about one quarter of a mile deep, found ourselves emerging upon an open, dry prairie, bounded on the east by the low wooded ridge before noticed as occurring on the south side of the river. The distant belt of woods, fringing Red River, might just be seen in the far western horizon, the whole intervening space being a rich and level prairie, without shrubs or willows. On the bank at the crossing place the skeletons of Indian wigwams and sweating-houses were grouped in a prominent position, just above a fishing weir where the Ojibways of this region take large quantities of fish in the spring. The framework of a large medicine wigwam measured twenty-five feet in length by fifteen in breadth; the sweating-houses were large enough to hold one man in a sitting position, and differed in no respect from those frequently seen on the canoe route between Lakes Superior and Winnipeg, and which have been often described by travellers.

Six miles from the Roseau, Still Water Creek occurs; its waters are deep, and, as its name implies, sluggish, or almost stagnating. Between Still Water Creek and Rat River some marshy spots occur, while on the right the ridge, wooded with aspen, continues in the direction of the Rapids of Red River, near which spot it is found within four miles of the banks of the main stream. Rat River is an insignificant brook coming from the Great Muskeg, which occupies the height of land to the east of the Valley of Red River. At the crossing place it is fifteen feet broad.

Four miles from Big Rat River, Little Rat River was crossed, and the trail then led to the point of junction of the two streams, until it came upon a ridge, which it followed for a distance of ten miles, after which the great Nine Mile Swamp occurs, where water lodges in marshy intervals, for the distance which has given its name to this wet prairie. A strong Scotch plough, drawn by a stout team of oxen, would soon effect the drainage of the Nine Mile Swamp. It partly originates from the excessive luxuriousness of the grasses growing upon the level expanse, which, in a humid season, hold up sufficient water to give permanency to the wetness of this portion of the prairie. Hay, in considerable abundance, as exemplified by the stacks which were seen in all directions, is made on the dry intervals of the Nine Mile Swamp. A French settlement commences immediately on the northern extremity of this characteristic illustration of Red River indifference and unconcern in regard to the improvement of the country. A very little well-directed labour would convert these extensive marshy areas into the richest pasture and hay privileges, and drive to more congenial haunts the myriads of snipe and plover we disturbed in our passage through them.

We arrived at Mr. Pierre Gladieux's house an hour after sunset on the evening of September the 29th. We were soon provided with an excellent supper, and our horses, seven in number, well supplied with hay in the yard. Before starting next morning an almost sumptuous breakfast was given to us, and while the horses were being saddled, I begged permission to see the farmyard, &c. Under a small shed there was a neat, light, four-wheeled carriage, which as we passed Mr. Gladieux very politely and kindly placed at my disposal during the remaining period of my stay at Red River. He remarked that on the morrow he was going to the plains to hunt buffalo, and should not require the carriage for several weeks after my proposed departure. I requested the guide to ask what I had to pay for the entertainment of the party. The polite answer returned was as follows : " Nothing ; it is not the custom of the people of this country to charge strangers who may honour them with a visit."

Mr. Gladieux is a French "native," he resides on the right bank of the Red River, five miles south of Fort Garry. He showed me his farmyard, barns, garden and cattle.

The Red River at Pierre Gladieux's.

Four pea stacks, several wheat stacks, and five or six hay stacks, all of fair dimensions, were neatly arranged in the stack yard, while the cattle yard was tenanted by a number of cows, pigs, horses, and poultry. His peas were sown on the 7th May, and reaped on the 25th September. Before Mr. Gladieux's house, the trunk of an immense liard lay ready for splitting into firewood ; the size appeared to be so unusual that I measured it carefully, and found it to be four feet ten inches in diameter six feet from the base, and four feet eight inches in diameter ten feet from the base ; at the base it measured 16·5 feet in circumference, and showed 150 well defined rings.

German Creek, or La Rivière Seine, flows into Red River about two miles below Fort Garry. In 1858 this river was explored by Mr. Dickinson, whom I had requested to try and penetrate to the Lake of the Woods. Mr. Dickinson set out on the 16th September : his description of the country is published in my Report on the Assinniboine and Saskatchewan Expedition, and the introduction of an abstract here, while completing the view of the Valley of Red River within British territory, will not materially affect the narrative of our explorations in the succeeding year. I give the following description nearly in Mr. Dickinson's own words :—

" As the country east of Red River, extending to the Lake of the Woods, is quite unknown, except for a few miles back from the river, to any but Indians, I was anxious to procure one of them as a guide. Having succeeded in doing so after some little delay, I was obliged to examine this part of the country first, as the Indian guide was about to leave the settlement in a few days for his winter quarters.

" Considering that one of the objects of this exploration should be that of ascertaining where a *summer road* could be most easily made from Red River to the Lake of the Woods, that being now a subject of great interest among the settlers, who were about sending a party out for that special purpose, I thought it advisable first to go along the straight picket line made by Mr. Dawson last winter, in which direction, I understand, he reports that a road can be made for some miles, in order that I might be able to institute a comparison between this and any other portion of the adjacent country through which the Indian might guide me.

" The first day I was able only to go about fourteen miles — two-thirds of this distance at least being through marsh and wet prairie. The general course was along the picket-line, from which I was obliged to diverge frequently — sometimes a mile or more, but always keeping it in view — in order to avoid, when possible, the wide marshes through which it passes. The next day I continued in the same direction, and having reached a point opposite the 22nd mile-post, on the picket-line, I could go no further, being stopped by a swamp or quagmire, impassable for horses, or even men, extending in front for many miles, and on both sides as far as the eye could reach. Though taking advantage of all the dry places within reach, ten miles of the course I took lay through marsh and wet land, and five miles at least through swamp. There are a few small clumps of young aspens along the line, and low willows in some of the marshes; but far away towards the north may be seen some clumps of larger trees.

" The land is, for the most part, a rich loam, with a subsoil of sandy clay; but the difficulty, or rather the impossibility of draining the numerous swamps and marshes, and the want of timber, render this tract of country unfit for settlement; and for the same reasons, the difficulty of constructing a suitable road through it would be very considerable, and the expenses enormous.

" Judging, then, that I had seen enough of this part of the country for my purposes, I retraced my steps to the settlement, from which I set out again, under the guidance of the Indian, who promised to conduct me by the only dry path towards the Lake of the Woods, as far as the boundary of his hunting grounds.

" On the morning of the 23rd, I proceeded along the south side of ' La Riviere Seine,' or German Creek, which flows into the Red River a little below its junction with the Assinniboine. There are farmhouses and a good road along it for a distance of five miles, when an Indian track begins, which keeps close to the valley of the creek for eight miles, between it and the marsh.

" This dry space varies from half a mile to a quarter mile wide, and is crossed by two small sluggish creeks, which, if widened and deepened, would effectually drain the marsh. There is plenty of good timber along the valley, consisting of poplar, elm, and black ash, with small oaks. Leaving German Creek here on our left, we went along a low ridge about one foot above the level of the marsh, and varying in width from fifty to a hundred yards; it runs in a south-easterly direction for about three miles, and then widens out on the left as far as I could see, and on the right to half a mile. At this point we were about three miles from German Creek, which we lose sight of now for some time. Continuing in the same direction for three miles more through beautiful rich grass, with clumps of aspens on the left, and high willows on the right, we came to Oak Creek, which is about two chains wide, but so still and sluggish that it rather resembles a long lake. Our course lay along it nearly due east for two and a half miles, when the creek turns to the south. This would be an admirable place for a settlement, the land being as rich as any in the whole country, and there being a large supply of oak, averaging one foot six inches in diameter, and poplars suitable for fencing.

" On the south side of Oak Creek the open prairie stretches away to the horizon, the greater part of that which was within view being dry, there being only a few patches of wet land. Leaving Oak Creek we went through a country of this character for about nine miles in a south-easterly direction, our track winding, however, a little to avoid the wet places, a few of which we had to cross ; but none of them were more than seven or eight chains wide, and easy of crossing. There are numerous clumps of small aspens and willows in every direction. We then proceeded nearly due east for about seven miles, German Creek being from one and a half to two miles on the north, a beautiful and rich prairie lying between us and it, and on the south, one mile distant, runs a well wooded ridge, parallel with our course; then turning to the south-east we wound round numerous large clumps of aspen from five to thirty feet high, and willows for seven miles, when we came to a rising ground so densely covered with young aspen and fallen timber, that it was impossible for carts to go further; we therefore left them here, and made packs of a few things for the horses to carry. Here the land becomes of a lighter description, being of a light sandy and clay loam. The timber has been all burnt, and the ground was so thickly strewed with the fallen logs, that it was with much difficulty the horses could travel. Two miles further on we came to the banks of German Creek ; its valley here is from fifteen to twenty chains wide, and about forty feet deep ; it is full of excellent timber, elm, oak, poplar, and black ash, all large enough for building purposes. The creek, which is here very rapid, is thirty feet wide, and about one foot six inches deep. We followed its course for twenty-seven miles, never being more than half a mile away from it. The country through which we passed is for the most part covered with trees of various kinds growing in large clumps, such as balsam poplar, aspen, tamarack, balsam spruce, cedar, and oak. The whole country has been burnt some years ago; the remains of the timber everywhere to be found indicate that there was once a vast forest of large trees.

"The Indian guide now said he had come to the boundary of his own country, and could not bring me further; and though I tried to induce him by every available means, he remained firm to his resolution. He was unwilling for some time even to give me a description of the country beyond; but finally I procured from him the following account : —

" At half a day's journey on snow shoes, or a distance of fifteen miles from where we were, there is a mountain or ridge thickly covered with trees stretching towards the Lake of the Woods. A part of this intervening space is a swamp in which grow tamarack, cedar, and spruce; the remainder is dry ground covered with small aspens and willows. Passing along the ' mountain ' you come to a marsh which extends to the 'Lake of the Woods;' but through it there flows a river up which large canoes could come within the hearing of a gun-shot, or about two miles from the mountain. The entire length of the way I had come was seventy miles; fifty miles, at least, of this distance, being fit for settlement, and throughout the whole of it a road could be made without the slightest difficulty, and at little cost. If time and means had permitted, I would have pushed through to the lake, but under the circumstances I considered it better not to attempt it.

" From the description given by the Indians of the country, and which I think may be relied on as correct, I am of the opinion that a road can be easily made through it."

On the 29th September I arrived at the settlement again, and during the week following, while Mr. Fleming was making preparations for our journey to Canada, *viâ* St. Paul, and Mr. Napier and his assistants were engaged in preparing maps and a report on the canoe route, I traversed the settlement from the Indian Mission to the last French house on the road to Pembina, gathering whatever information was accessible respecting the capabilities, condition, and resources of the Valley of Red River, and its singular inhabitants.

CHAP. VIII.

BRIEF HISTORY OF THE COLONY. — STATISTICS OF POPULA-
TION. — ADMINISTRATION OF JUSTICE. — TRADE AND OCCU-
PATIONS.

Lord Selkirk.—First Emigrants.—Difficulties of the Emigrants.—The De
Meurons.—Mr. West.—First Missionary.—The Census.—European and
Native Population.—Statistical Table.— Population by Families.—" Na-
tives."—Character of the Half-breeds.—Occupations.— Improvidence of
the Half-breeds.—Aids to Improvement.—Administration of Justice.—
Governor and Council.— Quarterly Courts.— Council of Assinniboia. —
Trade and Occupations. — Absence of Trades. — Mills. — Merchants.—
Freighters.—Land.—Leases.—Unoccupied Area fit for Settlement.

THE first attempt to found a colony in the country now
occupied by the Red River settlements was made in the
year 1812, under the patronage of Lord Selkirk.   A large
tract of country, extending from the sources of the Winnipeg to Lake Winnipego-sis*, and stretching from Lake
Winnipeg far beyond the United States boundary, was
purchased from the Hudson Bay Company by Lord Selkirk in 1811, for the establishment of his contemplated
colony.   The colonists consisted of several Scotch families, who, after they had reached the spot which was to
be their future home, were met by a large party of half-
breeds and Indians in the service of the North-West
Company, and warned not to attempt to establish a permanent settlement.   They were conducted by a number

---

* "Winnipego-sis," Little Winnipeg.   The affix " sis " signifies in Cree
" little."

Plan of Selkirk Settlement.

Scale 4 Miles to an Inch

Engraved by Edw.ᵈ Weller

of these wild and reckless children of the prairies to Fort
Pembina, a post of the Hudson Bay Company, where
they passed the winter in buffalo-skin tents, and soon
adopted the habits of life belonging to the savage and
half-savage natives by whom the 7 were surrounded. In
May, 1813, the emigrants returned to the neighbourhood
of Fort Douglas, about two miles below the present site of
Fort Garry, and here commenced their agricultural labours.
In the fall of the year they again sought refuge at Fort
Pembina, and after a winter of much suffering, revisited,
in the spring of 1814, the scene of the previous year's
attempt to plant themselves on the banks of Red
River, with a determination to make it a permanent resting-place. During the summer, however, their houses
were destroyed by the wandering half-breeds, who were
opposed to the establishment of a colony ; and when, in
October, 1815, the main body of emigrants arrived from
Scotland, they found poverty, ruin, and despondency prevailing where they had hoped to meet with a warm
reception and comfortable homes. The provident care
of Lord Selkirk prevented the colonists from suffering all
the horrors of starvation during the inclement winters of
this region. His lordship had established a general store
of goods, implements, arms, ammunition, clothing, and
food at Fort Douglas, from which the impoverished emigrants were supplied on credit. This store was erected
in the first year of the colony, and regularly replenished
from time to time by shipments from England.*

In 1816 a serious conflict took place between the colonists and the native employées of the North-West Company. Many were killed on both sides, and the settlement

---

* The Red River Settlement; its Rise, Progress, and Present State, by
Alexander Ross. London, 1846.

was again destroyed, the settlers dispersed, and some of them banished, by the half-breeds, to Norway House.

At the time when these disastrous occurrences were taking place, Lord Selkirk was on his way to the Red River with about 100 disbanded soldiers of the De Meuron regiment, composed chiefly of Germans, French, and Swiss. After Lord Selkirk's arrival order was restored, the Scottish emigrants recalled, the De Meuron soldiers rewarded with grants of land on German Creek, a town was laid out on Point Douglas, and such arrangements completed for the government of the colony as the position of the Hudson's Bay Company and the interests of the fur trade would admit of. The social condition of Rupert's Land at this period may be gathered from the following brief description by Governor Semple, who was killed in the unfortunate conflict just referred to : — "I have trodden the burnt ruins of houses, barns, a mill, a fort, and sharpened stockades, but none of a place of worship, even on the smallest scale. I blush to say that, throughout the whole extent of the Hudson's Bay territories, no such building exists."*

On the 16th July, 1818, several French-Canadian families, under the guidance of two priests, Messrs. Provencher and Dumoulin, arrived in the colony, and in the same year, and almost at the same period, innumerable hosts of grasshoppers came from the south-western prairies, and in a few hours destroyed every green thing, threatening the young colony with famine. In 1820 the foundation of a Roman Catholic church was laid near the site of the present Cathedral of St. Boniface ; and in the fall of that year Mr. West, a minister of the Church of England, visited the colony as chaplain to the Hudson's Bay Com-

* Governor Semple. Quoted by Tucker, in the "Rainbow of the North."

pany, aided and encouraged by the Church Missionary
Society. Mr. West's instructions were to reside at Red
River Settlement, and endeavour to meliorate the condi-
tion of the native Indians.* The following extract from
Mr. West's journal shows the state of the settlement at
this period : —

" On the 14th of October we reached the settlement,
consisting of a number of huts widely scattered along the
margin of the river ; in vain did I look for a cluster of
cottages, where the hum of a small population at least
might be heard, as in a village. I saw but few marks of
human industry in ' the cultivation of the soil. Almost
every inhabitant we passed bore a gun upon his shoulder,
and all appeared in a wild and hunter-like state. The
colonists were a compound of individuals of various countries. They were principally Canadians and Germans of
the Meuron regiment, who were discharged in Canada
after the conclusion of the American war, and were mostly
Catholics. There was a large population of Scotch emigrants also, who with some retired servants of the Hudson's Bay Company were chiefly Protestants, and by far
the most industrious in agricultural pursuits. There was
an unfinished building as a Catholic church, and a small
house adjoining, the residence of the priest ; but no Protestant manse, church, or school-house, which obliged me
to take up my abode at the Colony Fort (Fort Douglas),
where the charge-d'affaires of the settlement resided,
and who kindly afforded the accommodation of a room
for Divine worship on the Sabbath. My ministry was
generally well attended by the settlers, and soon after my
arrival I got a log-house repaired about three miles below
the fort, among the Scotch population, where the school-

* The Substance of a Journal during a Residence at the Red River
Colony, &c. &c., by John West, A.M. London, 1827.

master took up his abode, and began teaching from twenty to twenty-five of the children."

In 1821 the North-West and Hudson's Bay Companies united, and from that time the condition and prospects of the Red River settlements became more encouraging, and their progress slow but sure.

In 1823 the population of the colony was about 600, twenty years afterwards it had increased to 5143, and thus assumed an important, though not a prominent, position among Christian communities, in the midst of barbarous and savage races.

POPULATION OF THE SETTLEMENTS.

The census upon which the statements contained in the following pages are founded was taken in the years 1843, 1849, and 1856, and the copies were kindly furnished me by Mr. W. R. Smith, the clerk to the Council of Assinniboia.

The total population at the settlements on Red River and the Assinniboine amounted to 5143 in 1843 ; 5291 in 1849 ; and 6523 in 1856, showing an increase in the first six years of only 148, and in the last seven years of 1232 souls. This great difference in the apparent rates of increase is one which may be easily explained, by enumerating the offsets from Red River Settlement, which have occurred since the period when the census was taken. These consist of a number of families, embracing 120 persons, forming a settlement at Prairie Portage. St. Joseph's at Turtle Mountain has absorbed a very considerable number, exceeding 500 persons, and many families have left the settlement to seek a home in other localities. At the same time the population of Red River has received very few accessions from distant countries ; indeed, the foreign element, as it may be termed, shows a very decided diminution in one important source of supply.

Between the periods of the census taken in 1843 and 1849, there was an increase in the European and Canadian element to the extent of 74 families, and of the half-breed of 113 families. During the seven years which elapsed between 1849 and 1856, a decrease in the numbers of Europeans or Canadians—that is, of people not born in Rupert's Land, although British subjects, and originally coming from England, Scotland, Ireland, or Canada, has taken place to the extent of 102 families. The increase in native or half-breed families during the same period was 132. The diminution in the number of European settlers has already worked a change for the worse in the habits and customs of the half-breeds or natives. The tendency of the native population is gradually to lose many of the humanities of civilisation, and approach nearer to the savage wildness of Indian life. An influx of European or Canadian blood had a very good effect in arresting this tendency, which circumstances, far more than disposition, have induced and fostered.

According to origin, the population of Red River now stands as follows :—

	Families.	Families.	Families.	Period of comparison, 13 years.	
	1856.	1849.	1843.		
Rupert's Land—					
Half breeds . } Natives . . }	816	684	571	Increase in half-breed families	245
Scotland . . .	116	129	110	„ Scotch „	6
Canada. . . .	92	161	152	Decrease of Canadian „	60
England . . .	40	46	22	Increase of English „	18
Ireland. . . .	13	27	5	„ Irish „	8
Switzerland . .	2	2	2	„ Swiss „	0
Norway . . .	1	3	0	„ Norwegian „	1

In 1843, or thirteen years before the census of 1856,

there were twenty-seven more European or Canadian families than there were at Red River in May, 1856. These numbers show that in place of an introduction of emigrants of a character likely to refine and elevate the rough natures of the natives, endowed as they are with many peculiar and valuable qualities, those who have been from their youth familiar with the advantages and blessings of civilisation have gradually left the settlement and sought a home elsewhere. The increase of poverty, or incapability of supporting families, is seen by the average number of individuals belonging to each family.

In 1849 the average of each family was $5\frac{31}{1052}$.

In 1856   „   „   „ $6\frac{43}{1080}$.

The difference in the whole population of 1856 and 1849 being 1232 souls, while the difference in the number of families was 28 only. This very extraordinary discrepancy was stated by Mr. Smith, under whose direction the census was taken, to arise from the generally depressed circumstances in which many families found themselves. Numbers were unable to live in separate houses, and it now happens that two, and sometimes three families, formerly occupying distinct houses, and cultivating distinct farms, are crowded together in one house for the sake of economy. In 1849 there were 137 more males than females in the settlements; in 1856 there were 73 more females than males. The reason of this remarkable change in the relative numbers of males and females in so small a community, and in such a short period of time, was stated to arise from the circumstance that during the past five or six years, many young men have gone to seek recompence for industry in the United States, which the district of Assinniboia has not yet offered to them.

The term " native," distinguishing the half-breeds from

the European and Canadian element on the one hand, and the Indian on the other, appears to be desired by many of the better class, who naturally look upon the epithet "half-breed" as applied to a race of Christian men, scarcely appropriate. There is a strong and growing feeling among the few who have turned their attention to such matters, that in the event of an organic change occurring in the Government of the country, the "native" or half-breed population should not be neglected, or thrust on one side.

The half-breeds of the north-west are a race endowed with some remarkable qualities, which they derive in great part from their Indian descent, but softened and improved by the admixture of the European element. It is, however, much to be regretted that, from the singular necessities of their position, many of them are fast subsiding into the primitive Indian state; naturally improvident, and perhaps indolent, they prefer the wild life of the prairies to the tamer duties of a settled home; this is the character of many, but it belongs more to those of French descent than of Scotch or English origin.

About the 15th of June they start for their summer hunt of the buffalo. There are now two distinct bands of buffalo hunters, one being those of Red River, the other of the White Horse Plain, on the Assinniboine. Formerly these bands were united, but, owing to a difference which sprung up between them, they now maintain a separate organisation, and proceed to different hunting-grounds. The Red River hunters go to the Coteau de Missouri, and even as far as the Yellow Stone River; the White Horse Plain settlers generally hunt west of the Souris River, and between the branches of the Saskatchewan, but also over the same grounds as their Red River brethren.

The improvidence of many of the half-breeds is re-markable. During the winter before the last, those of the White Horse Plain camped out on the distant prairies, and killed many thousand buffalo in wanton revelry, taking only their skins and tongues, little caring that the reckless destruction of these animals must exercise a very important change for the worse in their own condition. As the buffalo diminish and go farther away towards the Rocky Mountains, the half-breeds are compelled to travel much greater distances in search of them, and consume more time in the hunt; it necessarily follows that they have less time to devote to farming, and many of them can be regarded in no other light than men slowly sub-jecting themselves to a process of degradation, by which they approach nearer and nearer to Indian habits and character, refusing to adopt or relinquishing the tame pursuit of agriculture, for the wild excitement and pre-carious independence of a hunter's life. The fascination of a camp in the high prairies, compared with the hitherto almost hopeless monotony of the farms of Red River, can easily be understood by those who have tasted the care-less freedom of prairie life. I was often told that the half-breeds generally sigh for the hunting season when in the settlements, and form but a feeble attachment to a permanent home, which cannot offer to the majority a comfortable maintenance under present circumstances, or secure the consciousness of possessing a free and manly spirit, with rational aspirations and hopes.

But few simple aids are required at Red River to ameliorate and vastly improve the condition of the more improvident and careless half-breeds. They frequently bring in a large quantity of buffalo meat or robes to the trading posts, and receive a considerable sum of money in exchange, or if they insist upon it, a certain quantity of rum. The money is spent at once in simple necessaries, dress, and ornaments. The establishment of a savings bank would have an excellent effect, and doubtless become the source of much permanent good.

There are several hundred half-breeds who, like their ancestors, pass their lives on the prairies, visiting the settlements occasionally, according as they may be in want of ammunition or clothing. It is impossible to arrive at an accurate estimate of their numbers, but there is no doubt that collectively they form a numerous and influential body.

The half-breed hunters, with their splendid organisation when on the prairies, their matchless power of providing themselves with all necessary wants for many months together, and now, since a trade with the Americans has sprung up, if they should choose, for years; their perfect knowledge of the country, and their full appreciation and enjoyment of a home in the prairie wilds during winter or summer, would render them a very formidable enemy in case of disturbance or open rebellion against constituted authorities. The half-breed hunters of Red River could pass into the open prairies at a day's notice, and find themselves perfectly at home and secure, where men not accustomed to such a life would soon become powerless against them, and exposed to continual peril.

The causes which have led to the present condition and prospects of this people are truly a painful subject. It is one which cannot escape the attention and care of philanthropists. Men will inquire how it is, that a race giving evidence of admirable discipline, self-government, and courage, when in the open prairies, should subside into indifferent and indolent husbandmen when in the settlements. Considered as the native population of Red River, how is it, it will be asked, that so few among the many have succeeded in the course of years in acquiring comfortable homesteads, and well stocked granaries and farm yards? and why has the European and Canadian element disappeared? The chances of nearly all have been equal,—land of admirable fertility everywhere surrounds them; with unsurpassed advantages for rearing horses, cattle, and sheep, yet little or no progress has been made; and in respect of sheep, which might soon in a measure supply the place of the buffalo, a serious diminution in numbers has taken place. It is true that within the last few years many hundred head of cattle have been driven across the prairies of Minnesota to St. Paul and sold well there. But this new export trade should have given encouragement to raising stock; yet stock, with unlimited pasture, is diminishing. The distant hunt consumes the time which might be given to far more profitable home industry; and those who really enjoy a settled life, and know the advantages which industry confers, from experience gained in Canada or Europe, leave the country and seek their fortunes elsewhere.

Every stranger is struck with surprise that the houses of half-breed hunters generally show no signs of recent improvement, show no signs of care and attention devoted to gardens or the cultivation of fruit. Plums grow wild in the forest, but none are seen in the settlements. Apple trees are only now beginning to be tried at the Stone Fort. No effort of manufacturing industry is visible beyond the windmills for grinding wheat.

It must not be supposed that this stationary, or rather retrograde, condition, is unnoticed by the mass of the people. They see the comfort by which the retired factors, the clergy, and the traders of the settlement are surrounded, and the comparative luxury which exists at the forts; but they do not rightly understand how their own condition might be remedied, for the majority cannot discover in what way the reward of industry may be won, or where a market for labour is to be found, except that kind of wild labour in the distant prairie, or in the woods, which they love instinctively, and which they have always been taught to consider most profitable, and alone capable of securing their comfort and happiness. Under such circumstances it cannot cause surprise that discontent prevails in the settlements. Much disappointment and dissatisfaction is everywhere seen, and wrongs, real or imaginary, for which they have no redress, form the constant subject of complaint in daily conversation. In these repinings, all who were not in the service of the Hudson's Bay Company, or in some way connected with them, as far as my experience enabled me to judge, uniformly agreed.

Let the condition of the half-breed hunters, generally, be contrasted with the present prosperity of the Gowlers, Gladieux, Fletts, the McKays, and several others that might be named, who farm with industry and economy, and the capabilities of Red River and the Assinniboine will not be overlooked in surveying the paralysed efforts of those who are taught to rely chiefly upon the hunter's precarious gains.

ADMINISTRATION OF JUSTICE.

The mode in which justice has been, and is administered in the settlements, is of rather an undetermined character. In 1839 the first Recorder was appointed, and in some instances the office of Governor of Assinniboia, a district comprised within a circle of fifty miles radius round Fort Garry, has been associated with that of Recorder. The Governor has a council of twelve of the principal inhabitants of the settlement to assist and advise him in performing the duties of his office. All the chief factors of the Company are magistrates, *ex officio*, and although instances have occurred of a court of magistrates trying cases without a jury, yet in general, and probably at the present day, a jury is always empannelled.

The reports in the recently established newspaper at Red River exhibit the mode in which justice is now administered; and as this is the first instance in which a reporter has enjoyed the opportunity of supplying the Red River public with a full account of the proceedings of these courts of justice, it possesses an interest apart from and superior to the subjects to which it refers.

RED RIVER SETTLEMENT, Dec. 28, 1859.

GENERAL QUARTERLY COURT.

A sitting of the above Court took place within the Court House, at the Upper Fort, on Thursday, the 15th instant, before William Mactavish, Esq., Governor; Dr. Bunn, Sheriff; Thos. Sinclair, Esq., Robert McBeath, Esq., and François Bruneau, Esq., Chairmen of the several District Courts. The building was crowded throughout the day, and the liveliest interest appeared to be taken in the proceedings, which, on the whole, as will be seen from the report, were of an animated character. Mr. James Ross acted as French and Mr. James McKay as Indian interpreter.

THE ROBBERY AT THE STONE FORT.

Catherine Daniel and Mary Daniel, aged respectively 13 and 16, were charged, the former with having stolen two several sums of money from the Stone Fort, the property of the Hudson's Bay Company, and the latter with having received part of the cash at the time, well knowing it to have been stolen. They both pleaded not guilty.

Margaret Daniel, a still younger girl than either of the prisoners, stated that on the 10th of October Catherine Daniel stole money from the shop at the Lower Fort, which she entered by the window. Witness saw her get through the window, go to the drawer, and take the money out. When she returned, Catherine told witness she had taken £6. Part of the money was spent by Catherine in the shop, and of the balance she gave witness one pound. Subsequently, Catherine a second time entered the store by the same window, witness being with her. On this occasion Catherine went to the counter drawer and abstracted five £1 notes. The drawer had been left unlocked. They left the shop by the same way that they had entered it. Catherine again spent a portion of the money in the shop, Afterwards, Catherine told witness that she had given some of the money to Mary Daniel; and Mary acknowledged to witness that she had received it. The first time Catherine spoke to Mary about the money she told her she had found it; but when the money had been spent, Catherine confessed to her that she had stolen it. Mary likewise spent her money at the Stone Fort. Witness only once heard Catherine tell Mary that she had stolen the money, and that was after the money had been spent.

There being no further testimony adduced, the prisoners were asked if they had anything to say in defence. They made no reply. Dr. Bunn then summed up, and the jury, after a short deliberation, returned into Court, with a verdict of guilty against Catherine Daniel. Mary was acquitted.

Dr. Bunn, addressing the prisoners, said : Catherine Daniel, after a fair and impartial trial, the jury have found you guilty of felony. The offence you have committed is one of a very serious nature, and in any other country you would in all probability have been condemned to seven or perhaps fourteen years' confinement. The Court, however, taking into consideration the fact that you are young, and hoping that you will never commit such an act again, are inclined to be merciful to you. Already you have been nearly three weeks in prison; we were disposed to have committed you other three months; but the Governor, who has in his power to remit your term of imprisonment to any period he pleases, thinks it will be sufficient punishment to you and warning to others if he treats you with great leniency. Therefore, the sentence of the Court is that you be imprisoned for two weeks from this day. You, Mary Daniel, are discharged. But take care. You have had a narrow escape. There is a strong impression on the minds of every one present that you have acted dishonestly. Avoid being brought up again; for if you come hither a second time, the evidence which has been given to-day will tell heavily against you.

LOWER DISTRICT COURT.

The usual bi-mensal meeting of this Court was held at the Rapids, on Monday, November 28th, before Thomas Sinclair, Esq. (chairman), Donald Gunn, Esq., John Inkster, Esq., and Donald Murray, Esq.

COUNCIL OF ASSINNIBOIA.

The Governor and Council of Assinniboia held a general meeting at Fort Garry, on the 7th instant, at which were present — William Mactavish, Esq., Governor of Assinniboia, President; and the following Councillors of Assinniboia—Right Rev. the Lord Bishop of Rupert's Land, Right Rev. the Lord Bishop of St. Boniface, John Bunn, Esq., John Inkster, Esq., Pascal Berland, Esq., Solomon Emlyn, Esq., H. Fisher, Esq., Maximilian Genton, Esq., Robert McBeath, Esq., Thomas Sinclair, Esq., and John E. Harriott, Esq.

TRADE AND OCCUPATIONS.

Upon making inquiry of Mr. Smith, under whose superintendence the census was taken, why no enumeration of trades and occupations was introduced into the census roll, I was informed that no kind of industry or a distinct trade or occupation existed in the settlements. Almost every man was his own wheelwright, carpenter, or mason; carpenters, blacksmiths, masons, &c., could be found, but they were also engaged in other occupations, either as small farmers or hunters. Mr. Smith did not think that one man could be found in Assinniboia who pursued any particular trade, or limited his industry to one special branch. The present condition of the settlements would not, it was thought, afford a living to any distinct class of artificers. A horse-shoe imported from England could be purchased as cheap as the unmanufactured iron required to make one ; every article, no matter of what description, was imported in its manufactured condition. Even the ponderous and unwieldy grindstone was conveyed across the portages from Hudson's Bay, although material well adapted for grindstones existed on the shores of Lake Winnipeg, not one hundred miles from Red River. Grindstones had, I was informed upon authority I could not doubt, been made from the rock in question, and brought to the settlement, but they could not compete commercially with those imported by the Hudson's Bay Company, which, for a time, were sold little above cost, even after their long and expensive journey.

In 1858 I had occasion to send to a blacksmith near Fort Garry for some hasps which I wished to have made, to replace those which had been broken from the provision boxes in crossing the portages on the Winnipeg.

The hasps were made according to my directions as regards form, but the material was copper instead of iron, the blacksmith alleging that he had no iron from which he could make the hasps. It is not improbable that this was an excuse arising from indolence rather than inability to execute the work, or the want of the material, for I saw a few days afterwards plenty of iron suitable for the purpose in the storehouse at Fort Garry.

The Bishop of Rupert's land says : " After all, our grand want is division of labour. We have no separate trade ; all are engaged in everything, farmers and carpenters at the same time, and so on. At a meeting held two years ago, for the promotion of social improvement, I endeavoured to press this upon them, but they are slow in understanding the 'philosophy of improvement.' We want one skilful in tanning, for the hides of the domestic animals are wasted at present. We want one to instruct them in making soap, to save the importation of this bulky and necessary article from Britain. We want, too, improvement in the fulling of cloth, to bring the wool into use, and provide clothing cheaper than what is imported. We have country cloth now, but the fabric is imperfectly fulled, and therefore not sufficiently warm. Young men coming among us, who could guide and instruct the people in any of these branches, would be a great gain."*

The mechanical force employed in preparing food is represented by sixteen windmills, nine water-mills, and one steam-mill, which is also used as a saw-mill. Articles of pottery, notwithstanding their fragile nature, are imported, although, as if reflecting upon the industry and enterprise of the modern inhabitants of Red River, ancient

* Report on the Red River Expedition of 1857. Letter to the Author. Appendix, No. 9.

articles of pottery, in the form of broken fragments, are sometimes found in making excavations on the banks of the river. Speaking in general terms, it may be said that trades and occupations, as representing special branches of industry, do not exist in Assinniboia.

Under the head of merchant-shops, we find no less than fifty-six enumerated in the last census, a heading which, it will be observed, is not represented in the census of 1849. In fact, the class of merchants, including petty traders, has almost sprung into existence during the last ten years. They obtain their goods from St. Paul on the Mississippi, and purchase them in exchange for gold or peltries. This trade with the United States is fast growing into importance, and from the immense extent of frontier, it is not easily checked by fiscal regulations; its continuance must effect to a most serious extent the position of the Hudson's Bay Company in the valley of Lake Winnipeg.

Some of the merchants at Red River- import largely from England by the Company's vessels, and almost any article of common necessity or ornament can be procured at the stores; which, by the way, are of the rudest description, without the least effort being made by their owners to display the wares, but rather showing an endeavour to conceal from outward view whatever goods they may contain.

Besides being merchant or trader, in the ordinary acceptation of the term, some of the inhabitants are freighters, conveying goods between Hudson's Bay and the valley of Lake Winnipeg. They employ Indians and half-breeds to row their boats of 3 to 5 tons burden, and haul them and their freights over the portages. Fifty-five of these boats are enumerated in the census as belonging to Red River, but whether the Hudson's Bay Company's fleet is included in the number is not stated. The employment of Indians by the freighters has, at times, given rise to some little difficulty between them and the Hudson's Bay Company, as introducing a species of industry not compatible with a hunter's pursuit, and likely to divert attention from the great objects of the fur trade.

Among numerous documents which are in the possession of many of the most respectable people of Red River, treasured up, perhaps, as memorials of bygone but not forgotten difficulties in gaining a livelihood by pursuits not connected with the fur trade or its interests, — the following brief note may or may not possess some little historic interest, and, if rightly understood and interpreted, offer a clue to the present condition of the Red River settlements, and of the Indian missionary stations.

FORT GARRY, June 5th, 1844.

SIR,—I am informed that private freighters from Red River frequently employ and afford passages to Indians along the line of communication to York Factory in their boats, which is highly objectionable in many points of view. I have therefore to desire you will not in future receive as passengers, or employ Indians in your craft, on the line of communication between York and Red River.—I am, &c.

(Signed)   G. SIMPSON.

Mr. Edward Mowat, &c. &c.

Copied, 30th July, 1844.   (Signed)   A.

TENURE OF LAND.

Land in Assinniboia is sometimes sold to purchasers at the rate of 7s. 6d. sterling per acre. The title is conveyed under the form of a lease for 999 years. The conditions in the lease are : 1st. That one-tenth of the land is to be brought into cultivation within five years. 2nd. That trading or dealing with Indians or others, so as to violate the chartered privileges of the Company, be forsworn.

3rd. Obedience to all laws of the Company. 4th. Contributions to expenses of public establishments in due proportion. 5th. All trade or traffic in any kind of skins, furs, peltry, or dressed leather, except under licence of the Company, forbidden. 6th. Land not to be disposed of, or let, or assigned, without the consent of the Company. These are the main features of the lease ; the document is long, otherwise it would have been inserted in full; it is contained in the minutes of evidence taken before the Select Committee on the Hudson's Bay Company.

It is necessary here to remark, that I did not see this lease in the hands of any one of the settlers of whom I made inquiries respecting their tenure. I heard of its existence, and saw a copy, through one of the resident clergy, but in no single instance could I find any half-breed, in possession of a farm, acquainted with its existence. In very many instances the settlers did not know the number of their lots, and had no paper or document of any kind to show that they held possession of their land from the Company, or any other authority. These inquiries were necessary for the purpose of ascertaining the exact position of a line of section across the valley of Red River, which I caused to be made for the purpose of ascertaining the level of the swamps, &c. The required information was obtained through Mr. Smith, the Clerk of the Council, but from the people themselves no information of the kind could be obtained. They knew that they had paid a certain sum for their land, or it had been given them in return for services, or that they had squatted upon it, and that they were now in possession, but of title-deeds or receipts they knew nothing. These remarks refer only to those from whom the information was sought for the purposes mentioned above.

When passing from Fort Garry towards the 49th parallel, with a view to explore the Roseau River, our guide pointed out a number of hay stacks occupying a delightful bend on the west side of Red River, about 25 miles from the settlements; he informed us that the hay stacks were made by himself and some friends, a few weeks ago, and that they intended to "move there" during the winter and form a new settlement. I inquired how much he had paid for his land; the reply was, "Nothing; we are not required to pay anything for land lying beyond the present limits of settlement on the river." I may add, that many million acres of land, which cannot be surpassed for fertility, being composed of rich prairie mould, from 18 inches to 2 feet deep, lie free and unoccupied on the banks of Red River, the Assinniboine, and their tributaries, inviting settlement.

REVENUE AND EXPENDITURE.

The following abstract of the public accounts of the Red River Settlement, ending May 31, 1859, will show the condition of the revenue and expenditure of the colony :—

DISBURSEMENTS.

PUBLIC SERVANTS' SALARIES.

						£	s.	d.
Governor of Gaol and Sheriff	.	.	.	.	.	30	0	0
Executive Officer .	.	.	.	.	.	100	0	0
Presidents of Petty Courts .	.	.	.	.	.	26	0	0
Collector of American Duty .	.	.	.	.	.	15	0	0
Petty Magistrates .	.	.	.	.	.	50	0	0
Constables .	.	.	.	.	.	108	0	0
Total amount of Salaries	.	.	.	.	.	329	0	0

GAOLER AND GAOL.

					£	s.	d.
Gaoler's Salary, Rations, and Advances .	.	.	.	.	32	12	6
Gaoler's Wood	.	.	.	.	4	7	2
Gaol-Expenses	.	.	.	.	3	18	3
Prisoners' Expenses	.	.	.	.	4	1	6
Total, Gaoler and Gaol	.	.	.	.	44	19	5

PUBLIC WORKS.

	£	s.	d
Labour performed	34	11	6
Timber	29	1	3
Roads	282	4	2
Bridges	64	3	6
Total amount expended on Public Works . . .	410	0	5
Sundries	5	0	9
Rent of Court Houses	6	0	0

FERRY AND SCOWS.

Ferry	1	8	0
Scows	10	0	0
Total	11	8	0
Premiums paid for Wolves' Heads	35	10	6
Post Office	154	11	11½
Grand total	996	15	4½

REVENUE.

By Outstanding Creditor Balance, June 1, 1859 . . .	869	9	6½

IMPORT DUTY.

Hudson's Bay Company's European	788	3	10
Hudson's Bay Company's American	129	6	7
Settlers, European	351	11	0½
Settlers, American	*35	15	4
Total amount of Import Duty	1,304	17	4½
Interest on £186 8s. at 4 per cent.	7	9	1
Ferry	6	0	0
Advanced Cash returned	5	0	0
Debtors' Maintenance	0	2	4
Old Materials sold by the Board of Works . . .	2	5	0
Fines	1	0	0
Post Office	149	16	8
Marriage Licence	1	0	0
Grand Total	2,347	0	0

* There is £30 of American duty in the hands of the Collector, which came in too late to be entered in this year's accounts.

		£	s.	d.
May 31, 1859.	To Disbursements	996	15	4½
„	To Balance	1,350	4	7½
		2,347	0	0
May 31, 1859.	By Revenue	2,347	0	0
„	By Balance carried to New Account .	1,350	4	7½

W. R. SMITH, E. O.

JOHN INKSTER, Auditor.

CHAP. IX.

THE MISSIONS AT RED RIVER.

Religious Denominations.—Missionaries, Stations, and Congregations.—The Protestant Congregations.— St. John's Church.— St. Andrew's Church and Parsonage.—The Parish of St. Andrew.—Its History.—St. Paul's Church.— St. James's Church.— Church at the Indian Settlement.— Service.—A Novel Night Bell.—A Contrast.—Peguis.—Prairie Portage.— A Congregation.—Wild Indians.—The Presbyterian Church.—The Presbyterian Congregation.—The Roman Catholic Churches.—The Cathedral of St. Boniface.—St. Norbert.—St. François Xavier.—The Congregations at Red River.—Their Demeanour and Appearance.—Protestant and Roman Catholic Parishes.—Extent of the Charities of the Home Missionary Societies. —Apathy of the wealthy at Red River.— Difficulties of Missionary Enterprise at present.—Privations and Difficulties at remote Stations.

THERE are three religious denominations in Assinniboia —Church of England, Presbyterian, and Roman Catholic. In the census of 1843 and 1849, two divisions only were recognised—Protestant and Roman Catholic—and the numbers of members were stated to be 2798 Roman Catholics and 2345 Protestants. In 1849 the Episcopalian families were stated to number 539, and the Roman Catholic families 513. In 1856, a division in the enumeration of the Protestant element was made, probably on account of the advent of a Presbyterian minister, who responded to the call of a numerous body belonging to that denomination, yet in the absence of a minister, formerly enumerated with the Episcopalians. In 1856, the census, according to religions, stood thus :

FAMILIES AND CHURCHES.

Roman Catholics, 534 families, with 3 churches.
Episcopalian, 488 ,, ,, 4 ,,
Presbyterian, 60 ,, ,, 2 ,,

The settlement at Prairie Portage, and the Indian missionary village, are not included in this enumeration, and, in addition to the churches enumerated, services are performed in two or three school-houses.

Subjoined is a table of the missionaries, stations, congregations, income and sources of income, belonging to the Protestant and Roman Catholic Missions in Assinniboia: —(See Appendix, Vol. II. for the latest statistics.)

PROTESTANT MISSIONS.

	Missionaries	Stations.	Congregations.	Income.	Sources of Income.	Remarks.
				Sterl. £		
1	The Right Rev. the Lord Bishop of Rupert's Land	*Red River.* St. John's	500	700	£300 Hon. Hudson's Bay Company. £400 funded property.	
2	Rev. T. Cochrane	,,	. .	100	Society for Propagation of the Gospel.	
3	Rev. J. Chapman .	St. Paul's	300	200	£150 Hon. Hudson's Bay Company. £50 the Bishop.	The Hon. Co.'s Chaplain.
4	Rev. Arch. Hunter . . .	St. Andrew's	1200	250	Church Missionary Society.	
5	Rev. W. W. Kirkby . . .	,,	. .	200	. . . .	Curate.
6	Rev. A. Cowley .	Indian Settlement.	600	200	. . . .	Indian Missionary.
7	Rev. W. H. Taylor	*Assinniboine River.* St. James	250	200	£100 Society for Propagation of the Gospel. £100 Bishop.	
8	Rev. Arch. Cochrane	Portage la Prairie.	200	200	Church Missionary Society.	

PRESBYTERIAN CHURCH.

Missionaries.	Stations.	Congregations.	Income.	Sources of Income.	Remarks.
Rev. John Black .	*Red River.* Middle Settlement.	400	150	£50 Hon. Hudson's Bay Company. Remainder by the Congregations.	

ROMAN CATHOLIC MISSIONS.

	Missionaries.	Stations.	Congregations.	Income.	Sources of Income.	Remarks.
1	The Right Rev. the Lord Bishop of theNorth-West, and 5 to 7 Clergy.	*Red River.* St. Boniface . St. Norbert. De la Rivière Salé.	1500 Included in the above.	. .	£100 from the Hon. Hudson's Bay Co.	A spacious Nunnery and Schools attached.
2	Rev. M. Thibeault, Grand Vicar	*Assinniboine River.* St. François Xavier	1000	. .	. . . . .	A Nunnery attached.

THE PROTESTANT CONGREGATIONS.

St. John's Church is in a very unstable condition, the walls being supported with wooden props. A large quantity of stone is now lying near it for the construction of a cathedral, which is estimated to cost £5000 sterling.

The Bishop of Rupert's Land returned to Red River in the autumn of 1857 from a visit to England, where he succeeded in obtaining a large sum towards the erection of his cathedral, but not sufficient to defray the entire cost. St. John's Church, and the proposed site of the new cathedral, are within two miles of Fort Garry, near where it is most desirable that a substantial and commodious Church of England cathedral should be erected, in the centre of what will soon become the capital of

Assinniboia, and nucleus of Christianity and civilisation in the great North-West.

St. Andrew's Church, called also the Rapids' Church, is a new and very substantial structure of stone, well buttressed, and very conveniently and neatly furnished : all its interior arrangements are attractive and substantial. It is surrounded by a thick stone wall enclosing a capacious churchyard. The parsonage house, also recently completed, is in every respect fitted for the severities of the winter climate of the country. The size is 50 feet by 30, and two stories high ; the walls, of limestone, are 2 feet 8 inches thick, the rooms lofty and capacious, and in its internal arrangements it leaves nothing to be desired.

The Rev. W. W. Kirkby's (now of Mackenzie's River) house is roomy and comfortable, but its architectural points are far from being attractive. A school-house, constructed of wood, is admirably arranged, and in it I saw sixty children pursuing their studies, under the instruction of Mr. Mayhew, lately from Dublin, with a decorum and attention not often to be surpassed in the primary schools of this or the European continent.

The parish of St. Andrew is the most populous on Red River, and the Church of England element is very largely in excess of the Roman Catholic, there being two hundred and six Protestant and eight Roman Catholic families.

The commencement of the mission of St. Andrew may be dated as far back as the year 1824, when four or five retired servants of the Hudson's Bay Company, with their native wives, first sought a permanent home at the Grand Rapids. To these, year by year, a few families were added until 1829, when the Rev. W. Cochrane (now

Archdeacon), was placed there, and from that time the population of the Grand Rapids' district increased by immigration with great rapidity. After the erection of his own house, the first act of Mr. Cochrane was to build a room about 30 feet by 18, which served the double purpose of school-house and church. As soon as it became known in the far-distant outposts of the Company that a school for the young, and a church, with a resident minister, were established at Red River, many of the retired servants of the Company, half-breeds and Europeans, came from the Saskatchewan, Albany, and even Moose, to avail themselves of the advantages which were now for the first time offered to them, with a guarantee for their continuance.

In two years the congregation had outgrown the school-house, and found themselves obliged to build a larger church, to accommodate their growing numbers.

This they completed in the autumn of 1831. It was a neat wooden building, 60 feet by 24, and cost about £200, which was paid by the people themselves in labour, materials, &c. No sooner was the church built, and things began to wear an aspect of stability and permanency, than others came in from Norway House, and the southern department. The population of the parish went on steadily increasing until every lot of land was taken up for about five miles on either side of the church, and a ten-mile line of white-washed cottages and pleasant homesteads enlivened the banks of the river.

In the year 1844, the church was again too small for the congregation, not more than three-fourths of whom could find admittance. Mr. Cochrane now determined to make an effort to build one of stone, instead of wood, which should be large enough for the increasing wants of the settlement. Fortunately there was abundance of lime-stone about four miles down the river, so that when the people were consulted, all were found to be unanimous on that point. It was commenced in the autumn of 1845, and finished in the autumn of 1849, when it was consecrated by the Bishop of Rupert's Land, who had just arrived in the country. Its dimensions are 88 feet by 44, with a tower 20 feet square, and 100 feet high, that is, to the vane; the stone work is only 50 feet. The tower contains three bells, and it is proposed by the inhabitants to increase their number to five.

The entire cost was about 1600*l*. which, with the ex-ception of 100*l*. kindly given by the Hon. Hudson's Bay Co., 50*l*. by Duncan Finlayson, Esq., of Lachine, 30*l*. from a clergyman in England, were collected on the spot in the shape of money, labour, materials, &c. It is a plain, unpretending, at the same time a solid and substantial building; and one that reflects credit upon the piety and liberality of the inhabitants.

In 1852-3, the substantial parsonage before referred to was built by the Church Missionary Society at a cost of 700*l*., and during the same year the people built the new school-house, at a cost of 120*l*.

The average attendance of divine worship is about 500, and the number of communicants 207.

St. Paul's church, parsonage, and school-house are sub-stantial and serviceable buildings, with no pretensions to architectural display, but well fitted for the object of their construction. They are built a few hundred yards from Red River, and at the edge of a boundless ocean of prairie, which, when illuminated by the setting sun, seems in its bright and gorgeous vastness to be emblematic of eternity according to the hope and faith of a Christian, but when dull and cold and grey in the dawn, it symbolizes the gloomy ignorance of its heathen wanderers.

St. James's Church, on the Assinniboine, is a pleasing object at a distance. I had no opportunity of ascertaining how far its internal arrangements comported with its external aspects. The congregation is the smallest in the Red River settlements.

Birch-bark Tents, west bank of Red River, Middle Settlement.

The church at the Indian settlement is also a new and spacious building of stone, with a wall of the same material enclosing the churchyard, in which is a wooden school-house, where I saw about fifty Ojibway Indian young men, young women, and children, receiving instructions from the Rev. A. Cowley, Mrs. Cowley, and a native schoolmaster. The young Indian women read the Testament in soft, low voices, but with ease and intelligence. During service (Sunday, October 4th, 1857), the church was about three-fourths full. The congregation appeared to be exclusively Indian ; in their behaviour they were most decorous and attentive. The singing was very sweet, and all the forms of the service appeared to be understood, and practised quietly and in order by the dusky worshippers. A seraphino was played by Mrs. Cowley to accompany the singers; the responses were well and exactly made, and the utmost attention was given to the sermon. The prayers were read in English, the lessons in Ojibway, and the sermon was delivered in Cree. After service an Indian child, neatly dressed in white, was baptized. A few of the women and girls wore bonnets, but the greater number drew their shawls over the head.

The minister and part of the congregation suffer under the mutual disadvantage of being separated by the river. The settlement is chiefly on the left, the church, school, and parsonage on the right bank of the river. A good ferry, which will probably soon be procured, would enable the congregation to cross with ease. The Rev. Mr. Cowley enjoys no sinecure; he is not only missionary, but the doctor, magistrate, and arbitrator of the settlement. During my short visit of a day and a half, he was sent for three times to visit sick children, and he says that when the Indians require his services during the night, they come into the parsonage, the door of which is never locked, and tap gently at the stove-pipe, which passes from the sitting-room into his bed-room above, to arouse him. They agreed among themselves that they would adopt this novel kind of night bell, and he has never known them endeavour to call him, after retiring to rest, in any other way; they open the outer door and steal without the slightest noise, in the darkest night, to the well known stove-pipe, give two or three low Indian taps, and quietly await the result.

A wonderful contrast do the subdued Indian worship-

ers in this missionary village present on Sunday, to the
heathen revelers of the prairies, who perform their disgusting ceremonies within a mile and a half from some of
the christian altars of Red River. On two Sundays during
my visit, at the time when Divine service was being
celebrated in all the churches of the settlement, the
heathen Indians held their dog feasts and medicine dances
on the open plain. In one instance five dogs were

Ojibway Tents on the banks of Red River, near the Middle Settlement.

slaughtered, cooked and devoured; in another instance
three; the evil spirit was invoked, the conjuror's arts
used to inspire his savage spectators with awe, and all
the revolting ceremonies belonging to the most degraded
heathen superstition practised within a mile and a half
of the spot where the stones are now gathered for the
Bishop of Rupert's Land cathedral, and nearly the same
distance from two capacious churches, Protestant and
Roman Catholic, where Divine service was at the same
time being solemnized to orderly resident congregations.

The farm attached to the Indian mission is cultivated with more than ordinary care, not only being intended to serve as a model for the Christian Indians settled in the vicinity, but also to provide them with seed and supplies in the event of their own stock failing, a contingency by no means improbable, since habits of forethought or economy are rarely acquired by these people until the second generation. In part of the garden allotted to vegetables, a small area was devoted to wheat in 1857, for the purpose of raising seed from an early variety, which Mr. Cowley had procured from Scotland the year before. The " Scotch wheat" was sown on the 16th and 18th of May. It was ready for the sickle and reaped on the 24th of August, having been 97 days in arriving at maturity. The common wheat of the country was sown May 5th, and harvested August 18th, having required 105 days to grow and ripen. Barley was sown May 28th, and reaped August 18th. Indian corn is planted about the 23rd May, and ripens every year. Potatoes are planted from the 22nd to the 26th of May. The potato crop is here truly magnificent. I was favoured with an inspection of the produce of a small field, afterwards visited, and certainly no finer or more plentiful returns could be desired. All were perfectly clean and sound, and of very unusual size and weight. With the permission of Mr. Cowley I took four potatoes which lay close at hand, on the top of a large heap, containing very many equalling in size those I had taken without special selection ; when carefully weighed, they were found to average ten ounces each (10·1 ounces) ; a practical experiment proved them to be an excellent table variety. I may here mention that in the garden I noticed asparagus growing luxuriantly, beet, cabbages, brocoli, shallots, and indeed most culinary vegetables. In the farmyard were ducks, fowls, turkeys, pigs, sheep, with some excellent milking cows, and, through the politeness of Mrs. Cowley, I was enabled to form a very favourable opinion of several varieties of preserve from the wild strawberry, cranberries and plums, which grew in profusion not far from the village. Among many kinds of wild fruits common here and much sought after by the Indians, are red and black currants, high and low bush cranberries, two kinds of raspberries, two kinds of gooseberries, mossberries, blueberries, summer berry, choke cherry, stone cherry, &c.; these are the common names by which they are known in the settlements.

In the garden around the house some flowering shrubs and annuals were still in bloom on the 3rd October. The air was fragrant with the perfume of mignonette, and the bright orange-yellow eschscholtzia shone pre-eminent among asters and sweet peas, which had escaped the autumn frosts.

I was introduced to Peguis, the great Ojibway chief, who at one time commanded three hundred warriors, and about whom so much has been written by the missionaries. He is now a quiet old man, a good Christian, and happy, as he states, in his belief.

Up to the day of my visit, October 4th, there had been fifty-one baptisms, exclusively Indian, in Mr. Cowley's mission, during 1857 ; and in the same period, twenty-six deaths, six of the number being adults. The population of the mission in 1855 was 473 baptized Indians, and 203 heathens; only four adult baptisms were celebrated in 1855.

Next to the Indian settlement, Prairie Portage is the most interesting illustration of a Christian settlement, in a wilderness still inhabited by roving bands of Indians, who, as of old, occupy themselves in barbarous warfare, hunt for daily food, or submit with abject humility to the conjuror's malignant influence. Prairie Portage owes its existence to the untiring energy and undaunted zeal of Archdeacon Cochrane. The church at this most westerly limit of civilization in the Far West is constructed of wood, and contains twenty-five or thirty very substantial family seats, but is capable of holding three times that number; each seat is manufactured by the owner, according to a pattern supplied by the archdeacon. The congregation (on Sunday 13th) was composed of Plain and Swampy Cree Indians and half-breeds: one Plain Cree woman's home was three hundred miles to the west; she was a fine specimen of the race, and neatly habited in the dress of the half-breeds. Near the door of the church, inside the building, a number of heathen Indians from the prairies stationed themselves to indulge their curiosity; they squatted on the floor, remaining quiet and grave, and conducted themselves with the utmost propriety during the service. They were Plain Crees, followers of the Buffalo hunters, with whom they had lately arrived from the high prairies; some were clothed in dressed skins, others robed in blankets, with head and neck decorations; and one young heathen girl, wild, and almost beautiful, triumphed in a robe of scarlet military cloth. Who can say what benign influence the sight of Christian worshipers may have upon many of these savage children of the prairies, who saunter in during the services of the Church, and with characteristic decorum always maintain a respectful demeanour, and a grave and earnest look.

The Upper Presbyterian Church is a neat building of stone, situated in the middle of the settlement. The cost of its erection exceeded 1,000*l.* sterling, and it has sittings for 500. The manse is delightfully placed on the river bank, which here slopes uniformly to the water's edge from the great prairie, some thirty feet above the stream during its summer level.

In the letter referred to on page 219, the Rev. Mr. Black says, " As to church matters, we have here two congregations, or rather a congregation and a mission station belonging to this congregation. In the one where I live there are about sixty families; in the other (situated at Mr. Gunn's, Stone Fort) there are ten or eleven in all. There are somewhat upwards of 120 members in full communion. The people are mostly Scotch, or of Scotch parentage. There are a few Orkney men, whom our Highlanders scarcely recognise as Scotch, a few half-breeds, one Englishman, and one Swiss. We have sabbath schools at both places: here the attendance may just now average eighty-five; below about thirty. Here we have divine service every Sabbath forenoon, and in the afternoon alternately here and below. We have also week lectures on Thursdays, and prayer meetings on Tuesday evenings. In regard to temporalities, the congregation below have no property but their small meeting-house; the one here has about 300 acres of good land, a stone church which cost about 1,000l., and the cottage in which I live. My stipend is 150l. sterling a year, 100l. of which is raised by voluntary contributions, and 50l. is allowed me by the Hudson's Bay Company. My people are mostly all farmers in comfortable circumstances, but none rich. They are, however, allowed to be the most steady and industrious portion of our population."

THE ROMAN CATHOLIC CHURCHES.

By far the most imposing ecclesiastical building in the settlement is the Roman Catholic Cathedral of St. Boniface, near Fort Garry. The external appearance is neither pleasing nor tasteful, although at a distance the two tinned spires, 100 feet high glittering in the sunlight, give an imposing aspect to the building. They can be seen from a great distance, and with the spire of St. James's Church on the Assinniboine, are well-known landmarks. The internal decorations of St. Boniface, for so remote a region, are very striking, and must necessarily exercise a potent influence upon the large and singular congregation who worship every Sunday within its walls. Two or three very sweet-toned bells ring at matins and vespers, and to a stranger just arrived from a long journey through unpeopled wastes, no sight or sound in Red River creates such surprise and melancholy pleasure as the sweet tones of the bells of St. Boniface, breaking the stillness of the morning or evening air.

The body of the cathedral is 100 feet in length, 45 in breadth, and 40 in height. The three bells weigh upwards of sixteen hundred pounds, and their chimes are deservedly listened to with pride and emotion by the Roman Catholic population of Red River.

The parish of St. Norbert is thinly peopled, but the chapel, which is constructed of wood, is 90 feet long, and 33 broad. The chapel of St. François Xavier is in a very dilapidated condition, but there is every prospect that a new and commodious structure will soon be erected on the White Horse Plain. (See Appendix, Vol. II.)

The appearance of the congregations of the different churches in the settlements, produces a very favourable impression upon a stranger. From whatever direction Red River may be approached, the journey has to be made through a wilderness in which no signs of civilization are to be seen for several hundred miles, much less a church in which divine service is celebrated to attentive and intelligent worshipers.

The congregations at Red River consist of resident and retired officers of the Hudson's Bay Company, some merchants, farmers, and the natives or half-breeds of the respective parishes. The services are conducted in strict accordance with customary forms, and the demeanour of the congregations is very attentive and decorous. A fair proportion of the congregations come to and go from church in neat carriages, or on horseback, and the external appearance of the assemblages, taken as a whole, in relation to dress, is superior to what we are accustomed to see in Canada, or in the country parishes of Great Britain. The young men wear handsome blue cloth frock coats, with brass buttons, and round their waist a long scarlet woollen sash; the young women are neatly dressed like the country girls at home, but in place of a bonnet they wear the far more becoming shawl or coloured handkerchief thrown over or tied round the head; sometimes they allow their long black hair to serve the purpose of a covering and ornament for which, from its profusion, it is admirably fitted. In this particular many of the half-breed girls follow the custom of their Indian ancestry, who, as a general rule, never cover the head.

There is a distinct and well-preserved difference in faith between the populations of the different parishes into which the settlement is divided. Some are almost exclusively Protestant, others equally Roman Catholic. In the parish of St. Norbert there is not one Protestant family, but 101 Roman Catholic families. In the parish of St. Boniface there are 178 Roman Catholic families and five Protestant; so also in the parish of St. François Xavier, on the Assinniboine, there are 175 Roman Catholics to three Protestant families. On the other hand, in the parish of

St. Peter there are 116 Protestants and but two Roman Catholic families, and in the parishes of Upper and Lower St. Andrew's there are 206 Protestant to eight Roman Catholic families.

A very short stay in Red River is sufficient to create both admiration and surprise at what may not be inaptly termed, the condition of religion in Assinniboia. Admiration is aroused by the extent and design of the charities of the different societies in England, who sustain such a large ecclesiastical corps in connection with the Church of England as resident missionaries in the settlement, and who have contributed very munificently to the erection of the excellent churches which are now constructed. In addition to these demands upon their liberality, the home societies give large sums towards the maintenance of missions in different parts of Rupert's Land, so that at the present time there are scattered over this immense country nineteen clergymen of the Church of England, costing between 6000*l.* and 7000*l.* sterling, annually. The Church Missionary Society have expended, up to the date of their last report, very nearly the sum of 60,000*l.* sterling upon missionary operations in Rupert's Land. While, however, so much is done by those in England for charity's sake, it is much to be wondered that so little is contributed by the wealthy residents of Red River, such as the retired factors of the Hudson's Bay Company, the merchants, traders, and better class of farmers, towards the maintenance of the clergy, the support and extension of schools, and to the christianising of the heathen Indians, whose medicine-drum, accompanying the monotonous song of the conjuror, can almost always be heard in summer during the hours of service.

The outward appearance of many among the congregations of the episcopal churches as they come and go in

neat little carriages, or on horseback from comfortable
well-furnished homes, enforces the expectation that in
proportion to their means, they would at least endeavour
to prepare the way for the spread of Christianity among
the thousands of heathens who frequent the settlement;
that such is not the case there is too strong ground for
belief.

In the present condition of the country, with the
interests of the fur trade to be upheld, the whole subject of missionary enterprise in Rupert's Land is full of
difficulty. This much appears certain : the Indians must
be induced to settle in one place for a few months of the
year at least; schools must be founded, and their children
taught the truths of Christianity; missionaries must learn
the Indian language; and then the spread of Christianity
among the heathens may be in some degree commensurate
with the charity which animates the different supporting
societies in Great Britain and Ireland. In the settlements
at Red River, and on the Assinniboine, all the services are
conducted in the English tongue; and among the clergy
of the Church of England at Red River, but one only
speaks one Indian language with the fluency and ease necessary to make himself understood by the natives. Of
course the Indian Mission below the settlements is not included in this enumeration.

The Hudson's Bay Company continue to be very liberal
in their support of missionaries, as far as money is concerned; their contributions will be seen in the foregoing
table; but the impression was irresistibly forced upon me,
and I found it strongly felt by some residents in Red River, that the progress of Christianity among Indians would
be rather aided than otherwise, if missionaries were not
to receive any assistance in the form of an annual stipend

from the Hudson's Bay Company. Perfect freedom of action in inducing Indians to settle; in the education of Indian orphan children, and in teaching them and adults the blessings of a settled, Christian home, as opposed to a heathen hunter's life, are essentially necessary before much satisfactory progress can be made. Can the ministrations of the Church in the English tongue to orderly resident congregations of European, Canadian, or half-breed origin, be missionary labour in the sense in which that highest of all duties is understood by those who contribute to the spread of the truths of Christianity among a wandering, degraded, and barbarous heathen race?

Missionary work at the outposts is altogether different to missionary work in the settlements. In no respect do the material means, necessary to make life comfortable, fall short of rational requirements at Red River, if proper forethought be exercised; but at the distant stations, isolated and often alone, the missionary's work becomes a labour which must spring from the heart in order to secure even an outward show even of promise or success. Many country parishes in England are far less attractive, remunerative, or desirable than the missions at Red River Settlement.

Those who have not experienced the privations resulting to missionaries in remote outposts from the non-arrival of their supplies by the customary route and at the expected season, can form but a feeble conception of the troubles and anxieties which chequer the life of a zealous missionary in the wilds of Rupert's Land. It is not mere personal inconvenience which causes him care and embarrassment; it is the impossibility of taking advantage of many opportunities for inducing wandering Indians to settle around the mission, of clothing and feeding the children entrusted to his charge, and of securing, by aid judiciously applied, the respect and affection of those he is endeavouring to christianise and educate, or seeking to draw from their faith in strange and imaginary gods.

The Indian generally, from his habits and precarious mode of subsistence, requires something tangible in the first instance to arrest his attention, and practical encouragement, often repeated, to secure his good-will, before an impression can be made on his heart. If the missionary be cut off from his supplies in the infancy of a Mission, much of his work has to be done over again. Indian wants are few and simple, but they must be supplied without fail at new stations; hence the importance, if success is to be secured, of effecting and sustaining a tolerably regular communication once or twice a year with the settlements at Red River.

It has sometimes happened that this is not convenient, or perhaps quite impossible; it is natural to suppose that when, from missing a season or from other causes, the supplies for the service of the different posts of the Company are in arrears, and the brigade of boats can take only a certain quantity of goods, those for the purposes of the trade will first receive attention. It has happened two or three times that one year's supplies for the whole population, of tea, sugar, articles of clothing and other important necessaries, have been unavoidably left at York Factory, causing no little inconvenience and trouble to the settlers as well as the missionaries. At Red River their wants can be in part supplied from the stores at Fort Garry, but at the missionary outposts such relief cannot be looked for.

Now that rapid communication may be said to be established between Fort Garry and St. Paul by steamboat and stage coach, there will always be an abundant supply of necessaries at the settlements, which was not the case when the chief means of communication with the outer world lay through York Factory. Opportunities may now be embraced for furnishing distant outposts, which did not exist before Georgetown in the State of Minnesota was connected by steam with Fort Garry, and by stage-coach with St. Paul on the Mississippi.

CHAP. X.

EDUCATION IN THE SETTLEMENT. — AGRICULTURAL INDUSTRY.

Schools.—Protestant Schools in the Settlement.—Subjects taught.— Collegiate School.—Distinguished Scholars.—School Attendance.— Sources of Income.—School Wants.—The Presbyterian School.—The Roman Catholic Schools.—Agricultural Industry.—The Farms.—Want of Improvement manifest.—Cause of the Absence of Progress.—Cultivated Crops. —Indian Corn.—Wheat.—Hay.—Barley and Oats.—Root Crops.—Sugar. —Hemp and Flax.—Live Stock.—Agricultural Implements.—Facilities for raising Stock.—Timber.—Country west of Red River.

THE CONDITION OF EDUCATION AT RED RIVER.

EDUCATION is in a far more advanced state in the colony than its isolation and brief career might claim for it under the peculiar circumstances in which the country has been so long placed.

There are seventeen schools in the settlements, generally under the supervision of the ministers of the denomination to which they belong. The following enumeration is nearly accurate :—

STATISTICS AND ENUMERATION OF SCHOOLS.

1. St. John's College, including a boarding school for boys and girls, under the immediate supervision of the Bishop of Rupert's Land.

2. Archdeacon Hunter's School, " Model Training School," conducted by Mr. Mayhew, recently from the Normal School, Dublin.

3. Mr. Gunn's Commercial Boarding School, more particularly in connection with the Presbyterians.

4. The Rev. Mr. Taylor's Parochial School on the Assinniboine.

5. The Rev. Mr. Chapman's Parochial School, near the middle settlement.

6. The Presbyterian School, under the superintendence of the Rev. Mr. Black.

7. Three minor schools, under the supervision of the Episcopal ministers in different parishes.

8. The Roman Catholic schools are three in number, one of them occupying a very spacious and imposing building, near the church of St. Boniface, and providing ample accommodation for female boarders.

9. At the Indian missionary village, an excellent school is under the control of the Rev. Mr. Cowley. All of the foregoing establishments are independent of the Sunday schools, properly so called, in connection with the different churches.

10. A private ladies' school is now established in the parish of St. Andrew, near the rapids. The house is commodious, and the boarders are under excellent supervision.

PROTESTANT SCHOOLS IN THE SETTLEMENT.

The present condition of Protestant education has been so ably described by the Bishop of Rupert's Land, in a letter with which he favoured me, in reply to inquiries on the subject, that I avail myself of some extracts, which embody as full and exact information as could be desired.*

" We may perhaps take the limits of the settlement as extending from Portage La Prairie to the Indian settle-

---

* This letter was published at length in the Canadian Blue Book, "On the Exploration of the Country between Lake Superior and the Red River Settlements," 1858. Also in the Imperial Blue Book on the same subject, June 1859.

ment. Within these boundaries the schools connected with the Church of England are thirteen. They are necessarily more numerous than would under any other circumstances be required by the population ; from the houses of the settlers lying along the banks of the two rivers, and not being in the form of a town or village, the children cannot go to school above a certain distance, and the schools have been in consequence multiplied to suit the convenience of the inhabitants. The thirteen are exclusive of the two higher academies for young ladies and for boys.

" The subjects taught must vary considerably, from the great difference of capacity in the pupils. The two leading schools would be " St. John's Parochial School," in the upper part of the settlement, and the " Model Training School," connected with St. Andrew's Church. In the former, in addition to the usual branches, the upper pupils have the opportunity of studying Latin, French, and mathematics. In the model school, which is taught by a certified master from Highbury, the senior pupils have also the advantage of instruction in Latin, Euclid, and algebra. They are thus an approach to the grammar schools in Canada. In the other schools, of which St. Paul's is the best example, there is an excellent education afforded in British history, grammar, geography, arithmetic, with the elements of general history. Of course we must be content with much less where the pupils are the children of Indian parents. With them it is difficult to go beyond reading, writing, and arithmetic.

" In the Collegiate School many of the pupils make very great progress both in classics and mathematics. Soon after my arrival in the country I was induced to found some scholarships as an incentive to study, and an approximation to what takes place in other countries. To the scholars elected from year to year, was assigned a free board, and the sum of 10l. a year, or in all about 30l. per annum. Of these so elected, some have done well elsewhere, and reflected credit on their early training. I would only specify among these Mr. Colin C. McKenzie, B.A., of St. Peter's College, Cambridge; Mr. Jas. Ross, B.A., who has distinguished himself very highly at the University of Toronto; the Rev. Peter Jacobs, ordained by the Bishop of Toronto to labour among the Indians on Lake Huron; and the Rev. Robert McDonald, ordained by myself to the missionary station of Islington, on the Winnipeg River. With more advanced pupils the higher classics have been read, such as Æschylus, Herodotus, and Thucydides. The turn of the native mind is, however, more towards mathematics. All attain to excellence in algebra, and acquire it with great ease. All, too, have naturally imitative power, and write and draw well. While I have had great pleasure in carrying on these branches of education, my one feeling of disappointment has been, that there is comparatively little opening for those who distinguish themselves in this country in after life. Yet I have felt that the duty is ours; the event was with God. In the young ladies' school the want of adequate motive to excite to study is felt more than in the collegiate school. They have the opportunity of learning every branch usually taught in such establishments elsewhere, such as French and music, and there is a very great change perceptible in the seven years. Their education is all-important with a view to the training of the next generation; and although the progress may not be visible in their case, the effects will, I trust, be fully acknowledged when they are settled in life.

"In the thirteen schools there may be about six hundred, from that to seven hundred children. In one or two there may be above fifty in attendance in winter, but the average will not exceed forty. The students at the Collegiate School have been as many as twenty-four, but as the standard of education rises in the parochial schools, the Collegiate School, as such, will be comparatively unnecessary, and it will ultimately be limited to those who may be under preparation for holy orders. For such, and for the clergy generally, there is a library, possessing now one thousand books of standard divinity, as well as other useful subjects.

" The sources of income vary much; ten out of the thirteen schools are connected with the Church Missionary Society. The masters of such schools have all a salary from the Society. The model training master is entirely paid by them, and also the masters of the pure Indian schools. In the other schools about one half may be paid by the Society, sometimes less, and the rest made up by the parents of the children. In the three parochial schools unconnected with the Church Missionary Society: in St. John's parochial school, a portion of the salary is paid by my own college, Exeter College, Oxford; in St. James's, by some christian friends in Edinburgh; and at Headingly, by the congregation of the Rev. T. M. McDonald, Trinity Church, Nottingham.

" The sum paid by parents is 15s. a year; where Latin is taught, 1l. In some parishes they prefer to pay the pound, or thirty shillings a family, and to send as many as they choose for that sum.

" We want much, school apparatus, books, and maps. A very large quantity of books have been imported, and the Society for the Propagation of Christian Knowledge has given many valuable sets of maps to several parishes; but scattered over thirteen schools, they are still insufficient.   *   *   *

"On the ground of education, let none fear to make trial of the country. The parochial school connected with my own church is equal to most parochial schools which I have known in England; in range of subjects superior to most, though in method and in the apparatus of the school necessarily a little inferior."

The Rev. John Black, the Presbyterian minister at Red River, also favoured me with an account of his church and school. The following extract from that gentleman's letter conveys the necessary information : *—

"First, then, as to the school : This is entirely supported by the people of the district, or rather by those of them who send their children to it. There is no endowment, no public money, nor any allowance by any missionary or other society. The salaries of the different teachers have varied from 22l. to 40l. sterling a year. The branches taught are English reading, writing and grammar, geography, arithmetic, and the elements of algebra and geometry. In the last two branches I think there are no pupils at present. The average attendance will be from thirty-five to forty. The school is kept open for the whole year, excepting a month in harvest, and the usual holidays. The school is not exclusively composed of the children of Presbyterian families, neither do all the children of such families attend it; some of these, residing at the extremities of the parish, attend the Church of England schools at the upper and middle churches, whilst some of the Church of England people who reside amongst us send their children here. You are aware that we have no public school system in the colony, and this, like the rest, is therefore

* This letter was published at length in the Canadian Blue Book, " On the Exploration of the Country between Lake Superior and the Red River Settlements," 1858. Also in the Imperial Blue Book on the same subject, June 1859.

essentially a denominational school. We would like to raise its character, but owing to other burdens lying upon them, and to their being left without assistance, the people are not able to hold out sufficient inducement in the way of salary to secure the services of an able teacher, at least permanently."

ROMAN CATHOLIC SCHOOLS IN THE SETTLEMENT.

The Bishop of St. Boniface furnished Mr. Dawson, as late as February 7th, 1859, with a statistical account of the schools and missions of the Roman Catholic Church.* The parishes on the banks of Red River and the Assinniboine are four in number, St. Boniface, St. Norbert, St. François Xavier, and St. Charles. Fifty-eight children receive education in the school of the brothers of the Christian doctrine, in the parish of St. Boniface. In the convent belonging to the Sisters of Charity, commonly known in Canada as the Grey Nuns, twenty young ladies are boarded, and receive an excellent education, suitable to their station in life. Besides the boarders, the sisters maintain and educate fifteen poor orphan girls, and keep a day school for the benefit of the poorer portion of the parishioners. In the parish of St. Norbert, thirty-one boys and twenty-nine girls attended the schools kept by a priest and the Sisters of Charity. In the parish of St. François Xavier thirteen boys and twenty-six girls receive instruction from the Sisters of Charity. In the parish of St. Charles there is no school or chapel. With reference to the zeal shown by the Roman Catholic population, in matters relating to education, the Bishop says :—

" Considering the sparse character of the settlements,

---

* Letter from the Bishop of St. Boniface. Red River Settlement; published in Mr. Dawson's Report, 1859.

the schools would need to be increased in number in certain districts, but the absence of any law relative to education, and of zeal in the people themselves, renders it utterly impossible to do more. I venture to assert what all reasonable and impartial persons must, in view of what is done, acknowledge, that the result far transcends the means which we can command. The truth is that, but for the unselfish zeal of some who devote themselves without fee or earthly reward to the arduous and meritorious task, it would be absolutely impossible to keep up the schools. So far, scarcely one child in ten has paid for his schooling, although the charge does not exceed ten shillings per annum, and I am certain that if we insisted on the payment of even that trifling sum, many of the scholars would leave the schools; such is the carelessness and indifference of the parents in that respect, notwithstanding our oft-repeated entreaties and the sacrifices which are made in that behalf. This indifference concerning the education of their children and neglect of the many advantages afforded them is a standing reproach which may be justly cast on our population."

AGRICULTURAL INDUSTRY.

Immediately on the banks of Red River and the Assinniboine, and extending rearwards to the public road, thence into the prairies beyond, the farms of the settlers are laid out in narrow strips, so as to give to each a small frontage on the river.

The houses are generally built close to the edge of the prairie, where it is abruptly cut by the winding channel of the river, and is thought to be high enough to protect them from occasional floods; but where the boundaries of the prairie retire from the present river channel, they are sometimes placed near the public road, and rarely in the depression formed by the ancient course of the stream. Above Mill Creek, on Red River, there does not appear to be any rise of land sufficient to afford security against extraordinary floods, such as those of 1826 and 1852, when the waters rose above the road, or more than thirty feet above the present river level. On the west of the road, as already remarked, is a boundless expanse, in which are enclosures, offering to the eye perfectly level fields of waving grain, or luxuriant pasture. Where no enclosures west of the road have been made, the prairie often passes into what are locally termed swamps or marshes ; but which are so susceptible of drainage, and conversion into the richest pasture lands, that they do not deserve the title which has been assigned to them.

Familiarity with the settlements dispels the favourable impression with which a stranger at first regards them. At a distance, the neat whitewashed houses, with their gardens and farmyards, continuing without interruption for twenty miles between the forts, the herds of cattle, horses, and sheep, feeding on the plains, the vast expanse of what seems to be meadow of the richest description, lead one to suppose that universal prosperity and contentment would here be won without anxiety or trouble. Nevertheless, no one can fail to be struck with the indifference to the future, which seems habitually to characterise the people, especially the French portion of the population, and to show itself in their unfinished dwellings, neglected farms, and extravagant indulgence in dress, or in articles they covet. Many of the apparent efforts of industry which, seen from a distance, excite admiration, shrink upon a nearer approach into sluggish and irregular attempts at improvement, often abandoned before completion. The farms and farm buildings in the occupation of the majority afford no sign of recent improvement; and in general it may be said, that the buildings which in Canada would be considered good, roomy country houses are exclusively possessed and occupied by the retired officers of the Hudson's Bay Company, the traders or merchants of the settlement, and the clergy.

The farmers' homesteads, and the hunters' and trappers' cottages, if these classes here can with propriety be separated, bear rather the appearance of slow decay, and a decline in fortune, than a healthy, hopeful, progressive condition.

With few exceptions, and these are chiefly among the Scotch, farming operations are conducted in a very slovenly manner. Weeds abound in most of the fields appropriated to grain; some fields are seen here and there to be altogether abandoned, and the out-houses wear a neglected aspect, or one of ruinous decay. As might be supposed in this primitive part of the world, manure is allowed to accumulate in the front of the stables and cattle sheds, or is sometimes thrown into the river, or heaped in such a position that it may be swept away by spring freshets. All these drawbacks and indications of negligence and imprudence are not uncommon, within certain limits, in every new country, indeed in any locality remote from markets, and wherever ignorance of the first principles of rural economy prevails; but where such marked neglect and seeming dulness abound in the midst of very general intelligence and acuteness, and are limited to the so-called agricultural class, in possession of a soil of unsurpassed excellence, the enjoyment of an admirable summer climate for agricultural purposes, and no greater share of periodical contingencies than those to which every other country is liable, the causes which induce these evils must be sought for in other directions than those which may be said to spring from a dislike for agricultural operations, or a characteristic inability to take advantage of the boundless appliances promoting happiness and comfort which lie within their reach.

The Bishop of St. Boniface, in his letter before referred to, page 220, points to the results of this apathy in relation to agriculture, in the following remarks on the means and resources of a population of about 7000 ; occupying a country possessing a soil of extraordinary fertility, to supply an unexpected demand for the ordinary necessaries of life.

" In the present condition of the Red River settlement, those who have large families are not the persons who should come ; we are more in want of arms than of mouths. A company of soldiers and the exploring parties who visit the settlement suffice, as it is, to create a famine. The price of many of the necessaries of life has doubled since last year, and although the harvest was pretty good, and hunting and fishing abundantly successful, nevertheless, there is an extreme scarcity of everything."

The description which has been given of the general aspect of the farms and farm-houses in the settlements is not such as to create a favourable impression of the condition of husbandry in this remote region ; but it would be very unfair to form an opinion of the agricultural capabilities of the country from the results obtained by the majority, under its present state of isolation, and the direction of the best efforts of the inhabitants to objects the reverse of those which belong to a pastoral life.

The farm, as an exclusive object of industry and attention, is recognised by very few of the people of Red River.

Hitherto remuneration for agricultural industry has been impossible as a general rule, on account of the want of a market. Where, however, due attention has been given to husbandry, it has secured comfort and solid independence. The fruits yet remain to be reaped; for now that immigration is taking place to a limited extent, prices of all kinds of farm productions have doubled; and those who have looked to the soil as their means of support in Red River, will be in a position to benefit by the industry and care of former years, and probably find a remunerative market for all they can produce during the next two or three years, until the market becomes overstocked, which is extremely probable in respect of grain, on account of the remarkable facilities which the country offers for bringing large areas of land into immediate cultivation.

CULTIVATED CROPS.

1. *Indian Corn.*—Varieties of corn exist, which may always be expected to ripen in Assinniboia. In order to secure this result, the rich and moist prairie soil requires draining, which may be accomplished without difficulty or expense, by running deep furrows with a common plough, at certain distances apart through the field devoted to Indian corn. This grain is a sure crop on the dry points of the Assinniboine and Red River, where the absence of superabundant moisture permits it to ripen within a certain period, so as to be secure against the early autumnal frosts. No doubt varieties of Indian corn are to be found in New England and in Lower Canada, which would ripen several days earlier in Assinniboia than

the horse teeth or even the Mandan corn, which are cultivated there.

2. *Wheat.*—This is the staple crop of Red River; its cultivation is so general, and the good quality of the grain is so well and widely known, that very little need be said respecting it. In favourable years, that is, in years which have not been distinguished by so wet and backward a spring for farming operations as that of 1857, wheat ripens and is ready for the sickle three months from the day of sowing. I think it very probable that new varieties from Canada, or the New England States, would ripen in less than three months, and this is the opinion of several of the best farmers in Red River. No fact, however, is more satisfactorily determined than the admirable adaptation of the climate and soil of Assinniboia to the culture of wheat. Forty bushels to the acre is a common return on new land; and I have elsewhere stated that Mr. Gowler has obtained fifty-six bushels to the acre, without the introduction of any artifice beyond deep land furrows to keep the rich vegetable mould of the prairie dry.

The great drawback to the cultivation of wheat is the want of a market. Asking a native to show me his wheat field, he said that he had grown enough the year before to last for two years, and the chances of his being able to dispose of any surplus were so small, that he determined not to trouble himself this year with growing wheat. As it happened, he would have been well repaid for any surplus, the expected arrival of the troops, and other circumstances, created a temporary market for wheat, which, however, could not have been foreseen by the easy going half-breed.

None of those diseases, with the exception of smut or rust, or insect enemies, to which the wheat crops in Canada and the United States are subject, occur, it is said, at Red River. Of this fact I cannot speak from personal experience ; all I can say is that I heard no complaints of rust, nor did I see a single instance of its presence ; yet it would be very unwise to infer from so short an experience that rust is not an enemy to the wheat crops there ; the character of rust leads to the supposition that it will be found wherever wheat is grown, if the climate be favourable to its production. The absence of rust is probably more a question of summer climate than of peculiarities in the soil which prevent its attacks. Although I made numerous inquiries respecting destructive insects, yet I could hear of none similar to the Hessian fly or wheat fly, as having been observed there. The grasshoppers, from 1817 to 1820, were the most destructive enemies known ; in 1857 and 1858 they destroyed the wheat crop at Prairie Portage, and to a small extent in the settlements.

3. *Hay.*—Quantity unlimited, and quality excellent. The prairies for hundreds of miles, through which Red River, the Assinniboine, Rat, and Roseau Rivers flow, offer everywhere a bountiful supply of grass and hay. Hay ground privileges have been established on the banks of the larger rivers, and the right of making hay within particular limits is recognised by the inhabitants.

Barley and oats are not much cultivated ; hops grow wild and in the greatest luxuriance. In 1858 they were seen on the Little Souris River, at the Qu'appelle Lakes and on the Little Saskatchewan. Beet yields very abundantly. Tobacco is cultivated to a small extent, but from trial of its qualities, I infer that it is susceptible of great improvement in the manufacturing process to which it is subjected. The season is, perhaps, too short for it to acquire maturity, and produce a good article.

All kinds of root crops grow well, and attain large dimensions. All common garden vegetables, which are cultivated in Canada, are equalled, if not surpassed by the productions of the rich prairie soil of Assinniboia. Considerable quantities of sugar are made from the ash-leaved maple on the Assinniboine. As no care is taken of the trees furnishing this useful article, it is probable that the supply from this source will soon cease. In cutting wood for fuel, the "natives" do not seem to have any special regard for the valuable trees.

Some years since, at the instance, it is stated, of Sir Geo. Simpson, flax and hemp were cultivated to a considerable extent by the settlers at Red River. The product was of excellent quality, and gave every promise of furnishing a very valuable commodity for home manufacture, and for exportation. The cultivation of these important crops was stimulated for a few years by premiums given by the Hon. Hudson's Bay Company, but when the premiums were withdrawn the cultivation soon ceased. Many settlers with whom I conversed had grown both of these vegetables, but that universal complaint, the want of a market, or of machinery to work up the raw product, led them to discontinue this very important and profitable branch of husbandry.

LIVE STOCK.—The live stock of the settlements are represented by 2,799 horses, 2,726 oxen, 3,883 cattle, 2,644 calves, 4,674 pigs, and 2,429 sheep. Since the census of 1849 an increase has taken place in all of the foregoing items, with the exception of sheep : this useful animal appears to be fast diminishing at Red River, and little wonder, when only one carding mill, and that not in operation, as I was informed, exists in the settlement. In 1856 there were 667 fewer sheep in Assinniboia than in 1849, and 1130 less than in 1843. This decrease is very much to be lamented; it is said to arise from the want of a market for the wool, or means to manufacture it in the settlement. During the winter of 1855—6, the number of animals lost amounted to 184.

The Rev. Mr. Black expresses a strong opinion in favour of the rearing of sheep.* "You saw what a splendid country it is for sheep pasture, and were there means of making wool into cloths, blankets, &c., greater attention would be given to the rearing of sheep; great quantities of such goods are also required for the fur trade, and it would be an advantage to have them manufactured here. Among the emigrants coming up to take possession of the land, it would be a great advantage were there somebody to establish machinery for carding, fulling and dyeing, perhaps spinning and weaving also."

AGRICULTURAL IMPLEMENTS.

The agricultural implements are English and American ploughs, of which 590 are now to be found in the settlement; together with 672 harrows, eight threshing machines, two reaping machines, and six winnowing machines. Produce is hauled in the celebrated Red River carts, of which there were 2,108 in the settlement in 1856. They are admirably constructed of wood; no iron is employed, but sometimes buffalo-hide is made to serve as a tire; these carts will last for several years; and one which conveyed some heavy boxes of geological specimens from Red River to Crow Wing last autumn, had previously been twice near to the foot of the Rocky Mountains, and was still in good condition.

The vast prairies of Red River and the Assinniboine, clothed with a rich profusion of most nutritious grasses,

---

* Vide letter referred to on page 219.

offer unrivalled advantages for rearing stock. The introduction of mowing machines would enable the settlers to
lay in any required quantity of hay for winter consumption. Few of the better class of farmers keep more than
thirty or forty head of cattle, in consequence of the want
of a market for beef, tallow, hides, &c. The answer I
received on all hands to the question, "Why do you not
raise more cattle?" was always the same in substance:
"Find us a market for beef, tallow, and hides, and we will
soon furnish any quantity of cattle you may require."
There does not appear to be any good reason why sheep
and cattle should not supply the place of the buffalo; the
experience of many years shows that no physical impedi-
ments arising from climate or soil exist to prevent the
prairies of Red River from becoming one of the best grazing countries in the world. Two reasons for the neglect of
this important branch of industry are soon apparent, even
to a stranger at Red River. Buffalo meat, pemmican made
from buffalo meat and fat, together with the robes and
sinews, are always a cash article at the Hon. Company's
stores; whereas beef, mutton, hides, tallow, and wool, are
a mere drug in the market; again, the habits of the half-
breeds, who have long been trained to the hunt, are opposed to the quiet monotony of a pastoral life. Introduce European or Canadian emigrants into the settlement,
with the simple machinery they have been accustomed to
employ in the manufacture of homespun, and in a very
few years the beautiful prairies of Red River and the Assinniboine will be white with flocks and herds, and the
cattle trade, already springing into importance between
the settlements and St. Paul, will rapidly increase, or
without much difficulty be diverted into an easterly channel. Such are the ideas of many with whom I discussed
the subject when in the settlements, and my own obser-

vations lead me to the opinion that no real difficulty exists in the least degree likely to hinder Red River from becoming a grazing country of the first class, when other interests shall be permitted to exist in the presence of that all absorbing, all-controlling service—the Fur Trade.

Timber fit for lumbering purposes is only found in narrow strips on the Red and Assinniboine rivers, and in still less quantities on the Roseau and Rat rivers; the timber consists of elm, oak, maple, and poplar of very large growth, as is stated elsewhere; but if the settlements progress,—and why should they not?—these supplies will soon be consumed. The ridges afford small aspen and pine; and in the rear of the great ridge, on the east side of the Red River, good pine is to be found towards the Lake of the Woods; the Winnipeg would doubtless furnish some good pine, but the difficulty would lie in bringing it up Red River, in its unmanufactured state. Saw-mills are unknown in the settlement, but the rapids of the Winnipeg could afford any required power there. The question of a supply of timber for building purposes is not so important as the requirements of the same material for fuel; hence it is that those who interest themselves in the future of Red River are anxiously turning their inquiries in the direction of the upper Assinniboine and the little Souris, to those supposed beds of lignite or tertiary coal which are so often spoken of by the Buffalo hunters who have occasion to cross these rivers in their progress to the high prairies. The value of the expectations of the settlers in this respect will be discussed in the second volume.

Whatever may be the future of Red River, it is quite evident that it will depend to a large extent upon the character of the country lying to the west of the present area occupied by settlements. With a view to supply some data on which an opinion may be based regarding the probable political and commercial status this country may acquire in process of time, the succeeding chapter is devoted to a very general description of the physical character and resources of the region included between the Saskatchewan, from the elbow of the South Branch to the Grand Rapids, the west coast of Lake Winnipeg, and the 49th parallel or international boundary. It is intended to serve as an introduction to the narrative of the Assinniboine and Saskatchewan Expedition of 1858, which follows the description of the Journey to Canada in the autumn of 1857 by Crow Wing and St. Paul.

CHAP. XI.

SKETCH OF THE COUNTRY WEST OF RED RIVER.

General Surface. — Elevation of the Prairies of Red River. — Pembina
Mountain.— Terraces.— Mountains.— Lakes and Rivers.— East of the
South Branch of the Saskatchewan.—North-East of the Assinniboine.—
Riding and Duck Mountains.—The Great Lakes.—Geological Formations.
—The Touchwood Hills.—Turtle Mountain.—Lake Winnipeg. — Lakes
Manitobah and Winnipego-sis. — The Qu'appelle Lakes.— The South
Branch.— The Main Saskatchewan.— The Grand Rapid.— The Little
Saskatchewan.— The Qu'appelle, or Calling River.—The Little Souris.
—Wooded and Prairie Land.—Areas fit for Settlement.—Valley of the
Assinniboine.—Valley of the Saskatchewan.—East of the Riding and Duck
Mountains.

GENERAL SURFACE.

THE prairies of Red River at Fort Garry are about
eighty feet above the level of Lake Winnipeg. They
form the southern portion of a vast region of lake, swamp,
and marsh, which is bounded in a very well defined
manner by the Pembina Mountain, and its continuation to
the Saskatchewan, which river it crosses a few miles
below the Nepowewin Mission, opposite Fort à la Corne.
Pembina Mountain forms the western limit of an ancient
sea or lake coast; its direction is partly shown on the
map as far as the Assinniboine. On the precipitous
eastern flanks of the Riding and Duck Mountains, it
occurs in the form of a ridge, of which a description is
given in the second volume. From the Pasquia Mountain
the course of the ancient sea or lake coast is westerly
towards Fort à la Corne. The whole of the country
east of Pembina Mountain, and its continuation as

described above, with the exception of the Assinniboine and Red River prairies, is low, swampy, and in great part occupied by Lakes Winnipeg, Winnipego-sis, Manitobah, and other bodies of water of less magnitude, having an area exceeding in the aggregate thirteen thousand square miles.

Rising above Pembina Mountain in the form of steps, are two other terraces, best seen on the east and west flanks of the Riding and Duck Mountains, but obliterated in the valley of the Saskatchewan and Assinniboine by the denuding forces which have swept over the whole of this region.   The south-western limit of these table-lands is marked by the boundary of the Grand Coteau de Missouri, which forms the highest terrace or plateau of the series.

MOUNTAINS, LAKES, AND RIVERS.

Surveying the country in the direction in which the great rivers flow, these vast plains slope gently from a low height of land near the south branch of the Saskatchewan with an easterly trend to the Assinniboine.   This slope is continued throughout the valley of the Assinniboine to Red River, after an abrupt declension, where the Assinni-boine descends the flank of Pembina Mountain.

North-east of the Assinniboine the country rises almost imperceptibly for a distance of fifteen to thirty-five miles, as far as the base of a series of hill-ranges lying parallel to the general direction of the river valley, before it makes its easterly bend ; it then rises by successive steps and sloping plateaux to a summit altitude of about one thousand feet above Lake Winnipeg, or sixteen hundred feet above the sea.

These hill-ranges are known by the names of the Riding Mountain and the Duck Mountain.   On their eastern and south-eastern flanks they show an abrupt and broken escarpment, and within the space of five miles the country sinks from sixteen hundred to six hundred and eighty feet above the sea, or within eighty feet of the level of Lake Winnipeg.

At the foot of these hill-ranges, and east of them, lie the great Lakes Winnipego-sis and Manitobah, which are separated from Lake Winnipeg by a low, marshy, and nearly level tract, having an elevation rarely exceeding eighty feet above it.

The uniformity which obtains in the geographical distribution of the great lakes of the Winnipeg Basin is a beautiful illustration of the dependance of geographical features upon geological structure. It is equalled only by the relations of the great Canadian lakes, whose form and general features have been shown to be determined by the formations in which they are excavated.*

The outcrop of the different formations in the valley of Lake Winnipeg, as far as they are known, follows the general direction of the rim of the basin in which they are deposited with remarkable uniformity. Conforming to the direction of the Laurentian system exposed on the east side of Lake Winnipeg, and constituting the Laurentide Mountains, the Silurian series stretches from Pembina on the 49th parallel, to the Saskatchewan on the 54th, and thence towards the Arctic Sea. Following its outcrop, the Devonian series is symmetrically developed between the same distant boundaries; but the most singular feature of this region is, that the soft Cretaceous shales should also conform, with tolerable exactness, to the exposed edges of the unfossiliferous rim of the great

* On the Physical Structure of the Western District of Upper Canada, by Sir W. E. Logan, F.R.S.

basin in which they lie.   The present nucleus of the fossiliferous basin is occupied by the great lignite formation
of the Tertiaries on the Grand Coteau de Missouri; and
so symmetrical is the arrangement in this part of the
north-west, that a line drawn through any part of the
country from the Grand Coteau de Missouri, where it
strikes British territory, to any point between Pembina
and Cumberland on the Saskatchewan, would pass over
proportionally extensive areas of the Tertiary, Cretaceous,
Devonian, Silurian, and Laurentian series.

Besides the imposing Riding and Duck Mountains, the
Touchwood Hills may be enumerated as very important
and striking in a region whose marked characteristic is
that of a gently sloping plain.   These hills lie between
the head waters of the Assinniboine and the South Branch;
the elevation of the highest peak, the Heart Hill, probably does not exceed 700 feet above the general level of
the great plain.   The course of this range is from north-
east to south-west, and it forms the most prominent of
several ranges which lie parallel to one another.

South of the Assinniboine the Turtle Mountain is a
prominent and important feature.   It is cut by the 49th
parallel.   The Blue Hills of the Souris serve to destroy
the general sameness of the prairie level on the river after
which they are named; while the Blue Hills south of the
Assinniboine, and east of the little Souris River, offer
perhaps the wildest and most picturesque scenery in the
area here referred to.   The Porcupine Hill, Thunder
Mountain, and Pasquia Hill, were not included within
the area explored.   They are eminences which lie between
the Grand Rapids of the Saskatchewan and the head
waters of the Assinniboine, all of them probably forming,
in connection with the Riding and Duck Mountains, at a
former epoch, a continuation of a vast tableland, now

broken into detached mountain ranges by denudation, with bold abrupt escarpments on their eastern exposures, and gently sloping terraces, separated by steps, on their western aspects.

Prominent among the physical features of this region are the vast expanses of water which occupy the larger portion of its low eastern area. Lake Winnipeg is two hundred and eighty miles long, and in several parts more than fifty miles broad. Lakes Manitobah and Winnipego-sis together are nearly of the same length, and the broadest part of the first named is not less than twenty-six miles across. Nearly the whole country between Lake Winnipeg and its western rivals is occupied by smaller lakes, so that between the valley of the Assinniboine and the eastern shore of Lake Winnipeg fully one third is permanently under water. These lakes, both large and small, are shallow, and in the same water area show much uniformity in depth and coast line. Some of the smaller lakes are of dimensions which entitle them to notice. Such are St. Martin's Lake, with an area exceeding three hundred square miles: Water-hen Lake, Ebb and Flow Lake, and Dauphin Lake, each covering an area exceeding one hundred and fifty square miles.

West of the Assinniboine are the Qu'appelle Lakes, situated in the Qu'appelle valley, eight in number, and with an aggregate length of fifty-three miles. Besides these, the last Mountain Lake is probably forty miles long, and varies from three quarters of a mile to two miles in width. The Qu'appelle Lakes are very deep, eleven fathoms or sixty-six feet having been recorded.

North-east of the Touchwood Hills there are numerous large lakes, having areas varying from one hundred and twenty to one hundred and thirty square miles. Some of

these are strongly impregnated with saline ingredients, and
are the haunts of innumerable hosts of geese and other
aquatic birds. On the south-east flank of the same range,
and throughout the plain stretching towards the Assinniboine, lakes and ponds are everywhere distributed; the
same remark applies to the western flanks of the Riding
Mountain and Duck Mountains, as well as to a large area
south of the Assinniboine and east of the Little Souris.

Lake Winnipeg receives the waters of numerous rivers
which, in the aggregate, drain an area of about 400,000
square miles. The Saskatchewan (the river that runs
swift) is its most important tributary. The South Branch
of this magnificent river flows for fully two hundred miles
below the Elbow, at the foot of a continuation of the
Eyebrow Hill Range, a low offset of the Grand Coteau,
in a north-easterly direction, and its deep excavated valley
appears to lie at an average distance of twelve miles
from it. This range is cut by several narrow but deep
valleys, and from the small lakes or ponds which occupy
their summits, water flows during spring freshets to the
Saskatchewan and Assinniboine.

The valley of the Qu'appelle River is a remarkable
and important instance of this interlockage, but not the
only one which connects two different drainage slopes in
this region. Within fifty miles south-west of the Grand
Forks, and a short distance south of the Lumpy Hill,
there is another deep valley in the dividing ridge, from
whose summit lakelets water flows in the spring to the
South Branch, a distance of ten or twelve miles, and also
to the main Saskatchewan, which it reaches below Pine
Island Lake, a distance exceeding 160 miles. One other
interlockage between the South Branch and the valley of
the Assinniboine will be noticed in the proper place.

The South Branch, eighteen miles below the Elbow,

and 584 miles from its mouth, is 600 yards broad. The rate of the current is here $2\frac{3}{4}$ miles per hour, the greatest depth 10 feet in the main channel, the mean depth across being 4·6 feet. There are channels on both sides of the river, one being 6 and the other 10 feet deep. After passing the Moose Woods, about ninety miles from the Elbow, the river channel is much contracted, its current is uniform and swift, varying from $2\frac{3}{4}$ to $3\frac{1}{4}$ miles per hour : mud and sandbars disappear, and it flows between high banks of drift clay, with a treeless, arid prairie or plain on either hand. At the Moose Woods, where the river is very broad and sandbars numerous, the paddles of canoes have touched the bottom from one side to the other with the ordinary stroke of the voyageurs ; this occurred during a season of low water. In August 1858, Indians were crossing on horseback from the right to the left bank above the Elbow, the depth not exceeding four feet. Before joining the North Branch the current becomes very strong, often from $3\frac{1}{2}$ to 4 miles an hour. The river winds between high precipitous banks, with forests of oak, elm, ash, aspen and birch covering the low points, the opposite hill banks being clothed chiefly with birch and aspen. Groves of spruce show themselves on approaching the North Branch, but the soil on the prairie plateau maintains the most luxuriant growth of vetches, roses, and berry-bearing bushes of different kinds wherever the aspen forests have been burnt and open areas formed. From the Elbow to the Grand Forks the distance is 250 miles, and generally throughout the last fifty miles of its course the South Branch flows through a thinly wooded country, but possessing a soil of great depth and fertility.

The main Saskatchewan, opposite Fort à la Corne, is 320 yards broad, 20 feet deep in the channel, and flows at the rate of 3 miles an hour. The mean depth across the river here is 14 feet, but it is in the memory of those living at the fort, when the river was crossed on horseback during a very dry season.

About 158 miles below Fort à la Corne, near Tearing River, the main Saskatchewan is 330 yards broad, 22 feet deep in the channel, has a mean sectional depth of 20 feet, and flows at the rate of two miles an hour. Two hundred and ninety one miles below the Grand Forks the main Saskatchewan enters Cedar Lake, a dilatation of the river thirty miles long. Issuing from this large body of water, it expands into a small lake, but soon again contracting its channel, the Cross Lake Rapids come into view; these rapids have a fall of $5\frac{3}{4}$ feet. Hudson's Bay Company's boats of four or five tons are tracked up them with half cargo, but loaded boats, descending, run the rapids. The Saskatchewan then enters Cross Lake, and after issuing from this elongated expanse of water begins a rapid course to Lake Winnipeg, with a current often 3 and sometimes $3\frac{1}{2}$ miles an hour. The head of the Grand Rapid is about four miles from the mouth of the river. The length of the portage is one mile seven chains, and the rapids below the portage are about $1\frac{1}{2}$ mile long, so that the total length of the Grand Rapid exceeds $2\frac{1}{2}$ miles. The fall from the west to the east end of the portage, as ascertained by leveling, is $28\frac{1}{2}$ feet; the fall below the portage is estimated to be 15 feet; consequently the total fall is about 43 feet. The Grand Rapid is run by Hudson's Bay Company's loaded boats; in ascending from the foot of the Rapids to the east end of the portage, boats are tracked or towed up with half cargo; they are then run back again, and again tracked up with the other half of their freight. From east to west end of the portage boats are tracked up on the south side of the river, with a load of fifteen pieces (1350 lbs.), the remainder of the freight is carried over the portage. The distance from the Grand Forks to the mouth of the Saskatchewan is 342 miles; the distance from the Elbow of the South Branch to the mouth is 603 miles.

The Saskatchewan receives several affluents on its south side which are important only on account of the fertile tracts of country they drain.

About 110 miles in an air line south from the Grand Rapid, and 136 miles by the canoe route along the coast, Lake Winnipeg receives the Little Saskatchewan or Dauphin River, through which Lakes Manitobah and Winnipego-sis discharge themselves. During ordinary summer levels, the Dauphin River offers no impediment to small steamers of light draft; it thus forms a valuable and direct communication between the vast water areas which it links together. It flows through a flat and swampy country offering very few inducements, or indeed opportunities, for settlement. The mission of Fairford is situated on that part of this river which lies between St. Martin Lake and Lake Manitobah, having been removed to its present position from the lower part of Dauphin River in consequence of the occurrence of destructive floods, the surface of the country not being above eight feet over the summer level of the river. Dauphin Lake is connected with Lake Winnipego-sis by Moss River, navigable in high water by Red River freighters' boats. The tributaries received by Dauphin Lake scarcely require notice here, although they may become useful as affording means for transporting the valuable spruce of the Riding and Duck Mountains to Lake Manitobah; the most important of these tributaries is the Valley River, which separates the Duck from the Riding Mountain.

Lake Winnipego-sis receives the Red Deer River and Swan River, which open communication to an important tract of country east and north-east of the head waters of the Assinniboine. The south-western extremity of Lake Manitobah is distinguished by the extent and richness of the prairies which at a higher lake level it has assisted in forming. The White Mud River which meanders through them may be classed among the most valuable of the lesser tributaries of the great lakes of the Winnipeg basin.

The Red River of the north and the Assinniboine having been already described, require no further notice. Some of the affluents of the last-named river are sufficiently important to deserve a separate notice.

The Qu'appelle or Calling River falls into the Assinniboine about five miles below Fort Ellice. At its mouth this stream is 88 feet broad, 12 feet deep in the main channel, and shows a mean sectional depth of 8 feet; its current is at the rate of $1\frac{1}{2}$ miles an hour. The valley in which it flows inosculates with the South Branch of the Saskatchewan at the Elbow. It is 269 miles long, and seventy miles from the Assinniboine about one mile broad, and 310 feet below the prairie, which stretches north and south from its abrupt edges as far as the eye can reach. At the Qu'appelle Mission, 119 miles from the Assinniboine, the valley is one mile and a quarter broad, and 250 feet deep. The river here is 48 feet wide, 6 feet deep in the channel, with a mean sectional depth of 3 feet 6 inches, and a current of one mile an hour. The lakes at this point have a depth of 57 feet, so that the total excavation below the prairie on either hand is 307 feet.

Near the first or Qu'appelle Forks the valley is one mile and one third broad, and 220 feet deep. At the east end of Sand Hill Lake, 239 miles from the Assinniboine, and thirty-one miles from the South Branch, the valley is one mile and five chains broad, with a depth of 140 feet below the prairie. Eight miles from the west end of Sand Hill Lake, or fifteen miles from the Saskatchewan, the excavation is one mile and seventy chains broad and 150 feet deep. At the height of land where it has been invaded by sand dunes from the west and south-west, it is still nearly one mile broad and 110 feet deep, estimated from the well defined edge of the valley, where a low escarpment of rock still uncovered by the advancing sand of the dunes, serves to mark its limit and the power of the forces which excavated it. The level of the prairie studded with sand hills and dunes is some 30 feet above the edge of the rock noticed above.

The Little Souris, or Mouse River, joins the Assinniboine 140 miles from Fort Garry, by the windings of the river valley, and 116 by the buffalo hunters' trail. At its mouth the Little Souris is 121 feet broad, 3 feet 6 inches deep in the channel, with a mean sectional depth of 2 feet 4 inches, and a current of half a mile an hour. Its valley, at the Back-fat Creek, twenty-five miles from the Assinniboine, is one mile and a half broad (8,016 feet), and 225 feet deep, with a level prairie on either hand. Near Snake Hill, sixty-one miles from the outlet, the valley is only 110 yards broad, and 66 feet deep, with open prairie on both sides. The river here is 100 feet broad, and 4 feet deep in the channel. In its passages through the Blue Hills of the Souris, the river has excavated a profound valley between 400 and 500 feet deep, making a sudden turn from a due easterly course to one almost northerly, and avoiding what appears to be an ancient channel but slightly elevated above its present level. This old channel pursues a straight course to Pembina River, with which, on the authority of half-breeds familiar with the country, it is said to be connected. Entering the territory of the United States against the course of the stream, the Little Souris may be traced as far south as the 48th parallel of latitude, when suddenly turning to the north-west it pursues a course parallel to the flanks of the Grand Coteau de Missouri, re-entering British territory about the 102nd degree of longitude.

WOODED AND PRAIRIE LAND.

The western and south-western slopes of the Riding and Duck Mountains support heavy forests of white spruce, birch, aspen, and poplar. The trees are of a large size, and often exceed $1\frac{1}{2}$ and 2 feet in diameter, with an available length of 30 to 50 feet. On the summit plateau of the Riding Mountain the white spruce is the largest tree; here it attains dimensions, and is found in quantity sufficient to give to this region a great economic value. The wooded area over which timber consisting of the four kinds of trees enumerated, is found on the Riding and Duck Mountains, has a length of one hundred and twenty miles, with a breadth exceeding thirty miles. The affluents of the Assinniboine will serve during spring freshets, to bear these valuable forest productions to areas which will probably first attract settlement, and where they will be most required.

In the valley of the Assinniboine is an extensive and valuable forest of oak, elm, ash, maple, poplar, and aspen, with an average breadth of four miles; its length is about thirty miles. The flats and hill sides of the deep eroded valley, through which this river flows above Prairie Portage, sustain a fine forest, in which aspen, oak, birch, elm, and maple appear to prevail in numbers corresponding with the order in which they are enumerated; but this forest does not extend beyond the excavated valley of the river or its tributaries. All the affluents of the Assinniboine flow through deep ravines, which they have cut in the great plain they drain; these narrow valleys are well clothed with timber, consisting chiefly of aspen and balsam-poplar, but often varied with bottoms of oak, elm, ash, and the ash-leaved maple. On the west side of the main river, the valleys of the tributaries, such as the Little Souris and the Qu'appelle River, are timbered continuously for a distance of thirty to seventy miles from their outlets, and at intervals only, further up stream. On the Qu'appelle River good timber is found as far as the mission; but in progressing westward it is seen gradually to diminish in size, and finally to disappear altogether.

The Touchwood Hill Range, together with small parallel ranges, such as the Pheasant Mountain and the File Hill, averaging twenty miles in length by ten in breadth, are in great part covered with aspen forests, but the trees are generally small. At the Moose Woods, on the South Branch of the Saskatchewan, forests of aspen begin to appear; they continue, with occasional admixtures of birch and oak, more rarely of oak and elm, as far as the Grand Forks; here the spruce becomes common, and, with aspens, occupies the excavated valley of the main Saskatchewan for many miles. The hill-banks and the plateau on the south side of the river, for a distance of three or four miles south, sustain the Banksian pine, which disappears as the soil changes from a light sand to a rich and deep vegetable mould, supporting detached groves of aspen and clumps of willows.

On the Little Souris, especially in the neighbourhood

of the Blue Hills, the country is fertile and beautiful, but
the areas adapted for settlement lose much of the value
which would otherwise belong to them, from the absence
of wood.   West of the Souris is a boundless, treeless
prairie; so that in crossing from Red Deer's Head River
to Fort Ellice it was found necessary to carry wood for
fuel for a distance of sixty miles.   At Sand Hill Lake, on
the Qu'appelle, timber is so scarce in the river valley and
gullies leading to it, that bois de vache is the only available fuel.   The South Branch, from the Elbow to the
Moose Woods, flows through a treeless region, as far as
relates to the prairie on either side; but in the ravines
leading to the river detached groves of small timber occur.
The boundary of the prairie country, properly so called,
may be roughly shown by a line drawn from the great
bend of the Little Souris, or Mouse River, to the Qu'appelle mission, and from the Mission to the Moose Woods,
on the South Branch.

AREAS FIT FOR SETTLEMENT.

Valley of the Assinniboine.—Issuing from the Duck
Mountain are numerous streams which meander through
a beautiful and fertile country.   This area may be said to
commence at the Two Creeks, ten miles from Fort Pelly,
thence on to Pine Creek, fifteen miles further.   The vegetation is everywhere luxuriant and beautiful, from the great
abundance of rose-bushes, vetches, and gaudy wild flowers
of many species.   After passing Pine Creek the trail to
Shell River pursues a circuitous route through a country
of equal richness and fertility.   Shell River is forty-two
miles from Pine Creek, and in its valley small oak appear,
with balsam-poplar and aspen, covering a thick undergrowth of raspberry, currant, roses, and dogwood.
Between Shell River and Birdstail River, a distance of

thirty-nine miles, the country is level and often marshy, with numerous ponds and small lakes, but where the soil is dry the herbage is very luxuriant, and groves of aspen thirty feet high vary the monotony of the plain.

Between the trail and the Assinniboine the soil is light, and almost invariably as the river is approached it partakes of a sandy and gravelly nature, with boulders strewn over its surface.

The flanks of the Riding Mountain are covered with a dense growth of aspen and poplar, and cut by numerous small rivulets. From Birdstail River to the Little Saskatchewan, or Rapid River, a distance of thirty-three miles, the same kinds of soil, timber, and vegetation prevail. About one hundred miles from its mouth the Rapid River issues from the densely wooded flanks of the Riding Mountain through a narrow excavated valley filled with balsam-poplar, and an undergrowth of cherry and dogwood, with roses, convolvuli, vetches, and various creepers. The slopes are covered with poplar eighteen inches in diameter. Descending the river, groves of poplar and spruce show themselves, with thick forests of aspen and balsam-poplar covering the terrace on either hand. The river is here forty feet wide, with a very rapid current. Before it makes its easterly bend the ash-leaved maple shows itself in groves, and on both sides is an open undulating country, attractive and fertile, with detached clumps of young trees springing up in all directions. The region drained by the Rapid River continues beautiful and rich until within twenty-five miles of the Assinniboine, so that it may with propriety be stated, that for a distance of seventy-five miles this river meanders through a country admirably adapted for settlement. Ponds and lakes are numerous, wild fowl in

great numbers breed on their borders, and the waters of
the Rapid River abound in fish. It will probably become
important as a means of conveying to the settlements on
the Assinniboine and Red River supplies of lumber from
its valley and the Riding Mountain.

From the Rapid River to White Mud River the
distance is thirty-three miles, and the country continues
to preserve the same general character with respect to
fertility and fitness for settlement which has now been
traced out for a space of 164 miles. White Mud River
flows into Lake Manitobah at its south-western extremity. This river drains an extensive area of the richest
prairie land, similar in all respects to the White Horse
Plain on the Assinniboine, or the rich wastes on Red
River. White Mud River is connected with Prairie
Portage by an excellent dry road, the crossing place
being about eighteen miles from the Portage. The river
banks are well timbered with oak, elm, ash, maple, aspen,
and balsam-poplar. It possesses valuable fisheries, and
communicates by an uninterrupted canoe navigation with
Lake Manitobah for a length of thirty miles. The soil
on its banks and far on either side is of the finest quality.
At the mouth of the river a fishing establishment has
been maintained by the people of the Portage for several
years.

The valley of La Rivière Sale has a general direction
parallel to that of the Assinniboine, and about sixteen
miles south of it. The country between the two rivers
is wet and marshy, with large areas covered with willow thickets and clumps of small aspen. South of the
valley of the first-named river the prairie is magnificent, and not surpassed by any area of equal extent on
Red River.

The area of the region well adapted for settlement on

the east and north of the Assinniboine, and in the valley of La Rivière Sale may be assumed to occupy 3,500,000 acres. In the valleys of Mouse River, the Qu'appelle River and White Sand River, the area of land likely to invite settlement does not exceed one million acres. The lakes in the valley of the Qu'appelle River are important, they abound in fish, among which white fish are numerous, large in size, and of excellent quality; the gray and red suckers, pike and pickerel are also abundant.

Valley of the Saskatchewan.—1. The country between the Lumpy Hill of the Woods and Fort à la Corne, or the Nepowewin Mission, including the valley of Long Creek and the region west of it, bounded by the South Branch and the Main Saskatchewan. This area may contain about 600,000 acres of land of the first quality.

2. The valley of Carrot River and the country included between it and the Main Saskatchewan, bounded on the south by the Birch Hill range. There is a narrow strip on the great river, about five miles broad, where the soil is light and of an indifferent quality. The area of available land probably does not exceed 3,000,000 acres.

3. The country about the Moose Woods on the South Branch.

4. The Touchwood Hill range.

5. The Pheasant Hill and the File Hill.

The aggregate area of these fertile districts may be stated to extend over 500,000 acres.

Assuming that the prairies of Red River and the Assinniboine east of Prairie Portage contain an available area of 1,500,000 acres of fertile soil, the total quantity of arable land included between Red River and the

Moose Woods on the South Branch of the Saskatchewan will be as follows :—

	ACRES.
Red River and the Assinniboine Prairies east of Prairie Portage	1,500,000
Eastern watershed of the Assinniboine and La Rivière Sale .	3,500,000
Long Creek and the Forks of the Saskatchewan . . .	600,000
Between Carrot River and the Main Saskatchewan . . .	3,000,000
The Touchwood Hill range, the Moose Woods, &c., &c. . .	500,000
Mouse River, Qu'appelle River, White Sand River . . .	1,000,000
The region about the head waters of the Assinniboine, including the valley of Swan River	1,000,000
Total area of arable land of first quality	11,100,000

Of land fit for grazing purposes, the area is much more considerable, and may be assumed equal in extent to the above estimate of arable land.

East of the Riding and Duck Mountains.—The region east of the Riding and Duck Mountains, when viewed as a whole, will furnish a very insignificant field for settlement. Where permanently dry land exists, the limestone rock generally approaches so near to the surface, as to be exposed whenever small trees are blown down, or the soil is penetrated to the depth of six or eight inches. The greater portion of the area on the shores of Lake Winnipeg, Lake Manitobah, the Little Saskatchewan, Moss River, Dauphin Lake, and St. Martin Lake, together with the region between Lakes Winnipeg and Manitobah, always excepting the southern shore of the latter lake, is not generally fitted in its present condition for the habitation of civilized man.

CHAP. XII.

THE JOURNEY TO CANADA VIÂ ST. PAUL.

Isolation of the Valley of Lake Winnipeg.—The Country drained by the Saskatchewan.—Routes to the Valley of Lake Winnipeg.—The Northern Route.—The Southern Route.—Pembina.—St. Joseph.—Deux Rivières. —Pine River.— The Mail.— The Red River Post Office. — Red Lake River.—War Path of the Sioux and Ojibways.—Turtle Creek.—Burning Prairies.— Height of Land Hills.—Caravans.—The Southern Slope.— Leaf River.—Crow Wing River.—Table of estimated Distances between Camps.—Crow Wing.—St. Paul.—Toronto.

THE valley of Lake Winnipeg is separated from the valleys of the Mississippi and St. Lawrence by extensive barriers, which have hitherto been instrumental in preserving it from the approach and intrusion of civilized races. The time has now arrived when this secluded region is likely to attract a wide-spread attention, and inquiry will naturally be turned not only to its resources, but also to its relations in point of geographical position and means of communication with the commercial world, as well as the opportunities it may supply for establishing a direct line of communication across the continent of America between the Pacific and Atlantic Oceans.*

* The Missouri is navigable as far as Fort Benton in long. 110° 30′, lat. 47°40′; 3,120 miles from its junction with the Mississipi. The flat-bottomed steamer "Chippewa" left St. Louis on the 1st June, 1857, and arrived at Fort Benton on the 17th July. Returning she reached St. Louis on the 19th August, performing a distance of 6,240 miles in 80 days, or 78 miles a day. At Dauphin's Rapids, above the Yellow Stone, the steamer was hauled by line, with this exception no difficulty in navigating the Missouri was experienced. This pioneer steam vessel carried 130 packages for the American Fur Company.

The Saskatchewan together with the Red River of the North drain an area exceeding 350,000 square miles, and the part of the valley they drain included within British territory, lies between the 49th and 55th parallels of latitude, and the 93rd and the 118th degrees of longitude west of Greenwich.* A European area, similarly situated east of the 10th degree of longitude, would comprehend very nearly the whole of England and Ireland, part of the German Ocean, the English Channel, the north-eastern corner of France, the whole of Belgium and Holland, and the greater part of the valley of the Rhine, together with the kingdom of Hanover.

The routes by which access is obtained to this great valley lie in the courses of three different watersheds. First, the present Hudson's Bay Company's route from York Factory, Hudson's Bay, _viâ_ Hayes River, &c. to Lake Winnipeg ; second, the Lake Superior route, _viâ_ Dog Lake and Rainy Lake ; third, the Mississippi Valley route from St. Paul to Red River.

THE NORTHERN ROUTE VIÂ YORK FACTORY.

As it is not in the least degree probable that the Northern or Hudson's Bay route will ever be selected as a permanent line of communication between Canada and the country drained by the Saskatchewan, a very brief notice of its most prominent features will be sufficient.

This route begins at York Factory, Hudson's Bay, and goods are forwarded in boats of the same size and build as those employed by the Red River freighters, of which a description is given in the second volume. York Factory is situated on the left bank of Hayes River, up which the boats are tracked, poled, or sail as far as Rock

* A description of the boundaries of the valley of Lake Winnipeg is given in the Second Volume.

Portage, a distance of 124 miles.* From Rock Portage to Lake Winnipeg the route lies through a low region, involving numerous portages, and occupied by the primary unfossiliferous rocks. The distance between Hudson's Bay and Lake Winnipeg is about 400 miles, the difference in level being 628 feet. Vessels cannot remain with safety longer than six weeks at York Factory, in consequence of the probability of ice arresting their passage to the Atlantic. Nelson River, through which the drainage waters of Lake Winnipeg find their way to the sea, is characterized by heavy falls and rapids which effectually oppose communication even by canoes or boats.

THE SOUTHERN OR ST. PAUL ROUTE.

On the 8th October, 1857, I despatched those members of the expedition who were to return to Canada by the road on the west side of Red River to Pembina, remaining behind myself in order to see Mr. Dawson, who, having partially recovered his health, was coming up the river with the Rev. J. Macdonald from Islington Mission, where, it will be recollected, he had remained in consequence of illness. Mr. Gladman had returned to Lake Superior in canoe early in September, Mr. Napier and his assistants were comfortably lodged in the Middle Settlement, Mr. Dawson had secured excellent accommodation within half a mile from Fort Garry, and nearly opposite to the Roman Catholic Cathedral of St. Boniface, where he could with ease enjoy the opportunity of attending the religious services of that communion, of which he is a warm and zealous member. Their winter's work before them was to explore the country between Red

* See Report by Lieut. T. Blakiston, R.A.

River and the Lake of the Woods, write reports on the route from Lake Superior, and collect information respecting the resources of the country and the condition of its inhabitants.

On the 9th I overtook my party about six miles from Fort Garry. All told, we were five gentlemen, five half-breeds, six saddle-horses, and five carts, to which were respectively attached four poor horses and one refractory mule. During the afternoon we met the Bishop of Rupert's Land and Miss Anderson, who were returning to Red River from England.

On the 11th we camped at Pembina, near the mouth of the river of the same name. Mr. Murray, the gentleman in charge of the Hudson's Bay Company's post, two miles north of the boundary line, gave us an excellent dinner, thus maintaining to the last the reputation for hospitality which the officers in charge of the posts of the company have justly earned. Whatever may have been the former condition of the village of Pembina, it is now only a small and scattered collection of log-houses, situated on the right bank of the Red River, in the new territory of Dakotah. The ruins of several good houses, formerly occupied by the Roman Catholic mission, are still to be seen, but in all other respects the town and port of Pembina exist only on paper. The few log-houses which have given it a name and a certain reputation, derived probably from its being formerly a frontier post of far more pretensions than at the present time, still serve for an excuse to attract public attention to the fancied progress of the Americans in this part of the Red River valley. In the late returns for the election of officers in the new state of Minnesota (October, 1857), the names of many resident voters are recorded, but it would be a matter of great difficulty to discover their abode now. The pre-

sence of some United States dragoons, forming part of an
exploring party which camped near Pembina two years
ago, gave rise to a report which has often appeared in print
and on maps, that Pembina is a post garrisoned by United
States troops, instead of being a small village containing
about a dozen scattered log-houses. About a day's journey west of Pembina the village of St. Joseph is situated,
in the territory of Dakotah, seven miles south of the boundary line. It was founded by the Red River half-breeds,
who were induced to settle there in order to escape the
floods of Red River, from which they had suffered or an-
ticipated severe losses. The village has already acquired
considerable importance as a depôt for articles of trade
which are brought by the citizens of the United States
from St. Paul.

The village of Pembina, Dakotah Territory.

The country about Pembina is remarkably fertile, and
so free from undulations, that towards the west the flanks
of Pembina Mountain, already briefly described, may be
seen in clear weather thirty miles distant.

Our course lay on the east side of the river, through a

beautiful level prairie studded with willow bushes, for about nineteen miles in a south-east direction, when we struck the first of the "Deux Rivières" at sunset, crossed the river and camped, having traveled twenty-two miles. A very perceptible change was observed in the prairie the next day on approaching Pine or Tamarac River; the soil consisted of a light vegetable mould, and wherever rain had fallen and collected in little hollows, sand showed itself. Hummocks of aspen and willow relieved the sameness of the scenery, and a distinct rise by ridges, at the base of which the river flowed, was easily recognized. Pine River at the crossing-place is about 25 feet broad with a rapid current. Between Pine River and Middle River the soil preserves its light character, the trail running for many miles on ancient lake ridges or beaches which are similar in every respect to those observed between the Roseau and Fort Garry. The night of the 13th October was cold and fine; a few grasshoppers still lingered on the prairies, and their eggs in many places lay in vast numbers on the surface of the ground. The following day was clear, beautiful and warm, but as night approached, the sky in the north-west began to assume a ruddy tinge, and finally a lurid red, produced by the fires in the rich prairies beyond the Assinniboine, at least ninety miles in an air line from Pine River.

During the morning of this day we traveled along an ancient lake ridge, doubtless a continuation of one of those which appear some miles east of the settlements on Red River. Prairie hens were seen here in great abundance, and numerous flocks of wild geese passed overhead. Near Pine River we met "*The Mail*," borne on the back of a half-breed, who was accompanied by a boy fifteen or sixteen years old carrying the blankets and cooking utensils. The mail bearer was ill, and had not eaten food for two days, having been longer on his journey than he expected. He had no means of killing the prairie hens which were so abundant on the trail, and which might have provided him with food. He carried the mail in a large leather bag, by means of a strap passing round his head; he was poorly clothed, wet, and miserable, and had been fifteen days coming from Crow Wing. We gave him some buffalo meat and pemmican, on the strength of which he hoped to reach Pembina in two days.* Serpent or Snake River, where we arrived during the afternoon, flows between steep sand-banks and hills; the soil is very light, and after passing Serpent River is scarcely fitted for arable farms, but might fur-

---

* The subjoined notice of the "Red River Post Office" is from the Nor'wester of the 28th January, 1860 :—

THE POST OFFICE.

"The year 1858 witnessed a new feature in the postal arrangements of the country. In that year, the Canadian Government authorised the conveyance of mails to and from the settlement, *viâ* Fort William. Since that time, therefore, and up to the beginning of the present winter, we have had two lines of mail communication—one through American and one through British territory. The former has hitherto on the whole given great satisfaction, and we doubt not the latter will also give satisfaction, after some more experience of the route and its requirements. In June, 1859, the two lines together brought in 713 papers and 400 letters, besides a number of magazines and reviews. The last mail, which arrived on the evening of the 19th instant, brought in 880 newspapers and 210 letters. This is the largest number of papers ever brought in by a single mail. The number of letters is smaller than by previous mails; but there will of course be a fluctuation. Our mails are only monthly. We may, it is true, send to Pembina twice a month; but as there is but a monthly mail from there, the arrangement is but unsatisfactory. Let us have a genuine fortnightly mail. There was one from July, 1858, to July, 1859: why was it given up?

"The outgoing mail on the 28th ultimo conveyed 350 letters and a large numbers of newspapers. As that was the first instance of newspapers being sent abroad from this settlement, it will mark an important era in the history of the Red River Post Office."

nish very extensive and excellent sheep pasturage. The prairies here are altogether denuded of timber, so that this day we were compelled to carry our fuel for cooking purposes from Serpent River to the middle of the plain where we encamped for the night.

In the morning ice was found in the kettles, but the coolness of the night was not unpleasant. The trail continued for many miles on a level and rounded Lake Ridge, and then descended into a low, rich, wet prairie, towards Red Lake River, 186 miles from Fort Garry by our estimate. Across this fine stream the baggage was passed in two small canoes, the horses swam across, and the carts were hauled with ropes. The valley of Red Lake River is heavily timbered, and will probably become an important stream as the settlements from the south begin to descend the valley of Red River. Already a trading-house has been built at the crossing-place on Red Lake River, and on some of the American Government maps it figures as Douglas, the name of the town which it is proposed to build there. The valley of Red Lake River is the war path of the Sioux and Ojibways, and our half-breeds asked us not unnecessarily to fire off any guns or pistols as long as we were within ten or twelve miles of Red Lake River, that we might not attract the attention of any stray parties of Sioux who might possibly be within hearing.

On the 16th we passed over a high prairie, rising in steps at long intervals apart and level at its marshy summit. The breadth of this prairie is about twenty-three miles, Turtle Brook forming its southern termination. No trees are visible, the soil is generally light, and the higher portions gravelly, but in depressions the soil is of the first quality. Boulders of the primary unfossiliferous rocks were observed in great numbers on the north flanks of the ancient lake ridges. On this desolate prairie we met a caravan of nine Red River carts containing merchandise, which the owners had purchased at St. Paul; they had been twenty-one days coming a distance of 320 miles. In the afternoon we arrived at a part of the prairie where the fire had run; as far as the eye could see westward the country looked brown, or black, and desolate. The strong north-westerly wind which had been blowing during the day drove the smoke from the burning prairies beyond Red River, in the form of a massive wall towards us; a sight more marvellously grand, and at the same time gloomy and imposing, could scarcely be conceived than that approaching wall of smoke over the burnt expanse of prairie stretching far away to the west. The upper edge was fringed with rose-colour by the rays of the sun it had just obscured; and as it swept slowly on, the rich rose-tints faded into a burnt sienna hue, which gradually died away as the obscuration became more complete, until, though early in the afternoon, with a bright cloudless sky towards the east, a twilight gloom began to settle around us, and the rolling folds of smoke sweeping over the prairie, rapidly enveloped all things in a thin but impenetrable haze. When the sun was still some degrees above the horizon the light was that of dim twilight; the prairie hens flew wildly across the trail, and without, as is usual with them, any determined and uniform direction; our horses appeared to be uneasy or alarmed, and the whole scene began to wear an aspect of singular solemnity and gloom. Night came on suddenly, and with a darkness which might be " felt," as we reached the valley of Sand Hill River; here, trusting to the sagacity of our horses, we let them find their way to the stream, on the banks of which we encamped. During the night the horses were very restless, often galloping suddenly among the carts and tents, and at no time appearing to venture far from the camp.

The wind changed during the night, and the morning of the 17th brought a bright and brilliant sky, with a glistening hoar frost on the prairie. Ice was observed in the ponds, and at our camp it was found about a quarter of an inch thick in the kettles which were exposed. Numerous pelicans and geese were seen flying south, and all the customary indications of approaching winter were observed from time to time. The trail this day lay through a fertile rolling prairie intersected by sandy ridges, the slopes were very rich, but the valleys wet. Here we saw the Height of Land Hills, apparently about twenty-five miles off, and having arrived at Marsh River we encamped on a hill near it, on the west side of the dividing ridge.

During the day we met a caravan of six carts, nineteen days from St. Paul; they belonged to private Red River speculators, and were laden with ploughs, whisky, stoves, scythes, &c.

Rising half an hour before daybreak we found ice in the kettles, a strong wind from the north, and a snow storm approaching, which just whitened the ground at nine A.M. but soon passed away. From Wild Rice River, we crossed an undulating prairie about twenty miles broad, to the foot of a low range of hills constituting a spur of the Height of Land, and camped on the north side of the undulating plateau which forms the dividing ridge. A heavy snow storm occurred during the night, and on the morning of the 19th we found the wind strong and very cold; ice half an inch thick having formed in the kettles two yards from the fire. The trail continued through a very beautiful rolling plateau, with clumps of wood here and there, and lakelets between the hills. Camped at noon near the edge of the southern slope, which however does not send water into the Gulf of Mexico except during floods, when a communication is established between Otter Tail Lake and Leaf Lake. Our course lay in the valley of Otter Tail River, a tributary of Otter Tail Lake from which Red River takes its rise. The direction of this stream is due south, and although our route descended rapidly towards the Mississippi, yet we were still in the valley of a river flowing into Hudson's Bay. The distribution of the drift on this dividing ridge frequently determines the course of a stream in a direction diametrically opposite to the general trend of the surface. Hauteur de Terre River, or the upper part of Otter Tail River and Red Eye River, for instance, flow for a distance of forty miles nearly parallel to one another, in a general direction to the S.S.E., but one flows into Otter Tail Lake and thence into Hudson's Bay, the other is an affluent of Crow Wing River, which joins the Mississippi near Fort Ripley.

Even on the summit of the southern slope the aspect of the country begins to change, and prettily wooded lakes become numerous, affording in summer a most delightful variety of scenery. The soil, however, is light, and not favourable for cultivation. Camped at Forty-fourth Lake, about 110 miles from Crow Wing.

The country passed through on the 20th was extremely beautiful, the soil good, timber and prairie land being distributed in about equal quantities. The grackle in countless numbers were seen passing south ; the lakes were alive with ducks, geese, and several other kinds of water fowl, recalling to mind the appearance of the ponds in Red River and the Assinniboine valleys. In the woods we met sixteen carts from St. Paul, bound to St. Joseph's, and laden with tea, sugar, powder, and dry goods. We descended the successive steps of the southern slope rapidly, and soon reached a warmer climate; passed little Red River at noon, and camped in the middle of the prairie, hearing, not without pleasurable feelings, the barking of dogs during the night, which, besides indicating our approach to settlements, was suggestive of pleasant thoughts and delightful anticipations. The prevailing character of the soil, hitherto, is light, but the country is truly beautiful.

On Wednesday, 21st, we arrived at a house near Leaf River, called by its occupants Leaf City, and so represented on the official map of Minnesota; it is within a few miles of Otter Tail city, on Otter Tail Lake. Otter Tail city contains half a dozen log-houses, and is intended by its present proprietors to become a town of importance. Leaf River connects the waters which flow into Red River with those which seek the Mississippi basin, and during floods a canoe can pass from one watershed to the other without difficulty. South of Leaf River the country becomes rolling with deep valleys and extensive swamps between the hills. Leaf River is fringed with a magnificent forest, in which we recognised many trees not seen since leaving Canada.

Camped on the 22nd seven miles from Crow Wing River. During the day we met the Roman Catholic bishop and two priests, together with some French Canadian emigrants bound to Red River. Early the next morning, after passing through a poor country, we arrived at Crow Wing River, where we found a new store well-stocked with goods, which the enterprising owner said he had built and furnished for the benefit of the Red River people; he thinks he will be able to drive a very profitable trade with them. Our road now lay through pine woods and bad swamps, which continue for eight miles, until within twenty-five miles of Crow Wing. The communication through these swamps is wretched, but there is every prospect of the State constructing a new road next year.* Reached Crow Wing at sunset, Saturday, the 24th October, having been sixteen days out from Fort Garry.

The subjoined table of distances affords a close approximation to each day's journey:—

TABLE OF ESTIMATED DISTANCES BETWEEN CAMPS.

Fort Garry	0	Marsh River . . .	242
Stinking River . . .	9½	Wild Rice River . . .	247
Scratching River . . .	37½	Spur of dividing ridge . .	270
Plum River	51	Buffalo Creek . . .	279
Pembina	70	Forty-fourth Lake . .	310
First of the Two Rivers to		Little Red River . . .	320
the upper crossing . .	95	109th milestone from Fort	
Little Bridge Creek . .	104	Ripley	329
Middle River . . .	110	Rush Lake	338
Second of the Two Rivers .	114	Seventy-seventh-mile Lake .	361
Pine or Tamarac River .	136	Seven miles east of Leaf	
Rock River	142	River, 62½ miles from Crow	
Serpent or Snake River .	147	Wing	376
Middle of Prairie . . .	160	Twenty-four miles from Crow	
Red Lake River . . .	186	Wing	403
Turtle Brook . . .	212	Crow Wing	428
Sand Hill River . . .	216	St. Paul	558

Crow Wing is a small, new town, depending chiefly upon the pineries in its neighbourhood for support, as

---

* A new road has been completed, and all the rivers bridged as far as Graham's Point on Red River, opposite Fort Abercrombie. (1859.)

well as upon the prospect of a road between it and
Superior City.   Its position in relation to Lake Superior
and the valley of Red River, is thought to be very
favourable, and the people say that a plank road from
Superior City to Crow Wing, which need not exceed one
hundred miles in length, would secure the trade of the
valley of Lake Winnipeg.

From Crow Wing we travelled by stage to St. Paul,
thence by steamer down the Mississippi to Prairie du
Chien.   From this far western town there is direct communication, by rail, with Toronto, where we arrived on
the 4th of November, three months and a half from the
day of our departure on this exploration.

THE

ASSINNIBOINE & SASKATCHEWAN EXPLORING EXPEDITION

OF 1858

INTRODUCTION.

The following instructions will explain the object of the Assinniboine and Saskatchewan exploring expedition.

No. I.

Secretary's Office, Toronto, 14th April, 1858.

Sir,—During the last week I communicated to you, verbally, instructions in reference to the proposed expedition to the neighbourhood of the Red River during the present year.

2. It has been decided, as you are aware, with a view to keep down as much as possible the expenditure this year, to dispense with the services of Mr. Gladman as its general manager.

3. The exploration party this year will consist of two divisions, one to be placed under your direction and control, and the other under the direction of Mr. Dawson.

4. His Excellency in Council has been pleased to place under your charge the topographical and geological portion of the exploration, respecting which full instructions will be given in another letter, while Mr. Dawson will continue to perform the same duties as last year, viz., those of surveyor, &c.

5. The estimate of the probable expenditure of the expedition, submitted by you on the 6th instant, was laid before his Excellency in Council, and has been approved of by them, and I have accordingly now to direct you to be guided as much as possible by that estimate in engaging your assistants, hiring your

men, as well as in the other necessary expenditures of the expedition.

6. It is hardly necessary to say that his Excellency relies upon your exercising a due economy in all matters connected with the expedition.

7. As soon as you have completed your contemplated party, you will furnish me with a schedule, giving the names of all the persons composing it, and stating their rates of pay, and the dates from which their pay is to commence. Such a schedule will be necessary to supply the auditor with the means of auditing your accounts.

8. Having organised your party, you will lose no time in repairing with them to Red River, taking with you the supplies (referred to in the estimate) required for Mr. Dawson.

9. On your way to the Red River, you will take possession of the canoes, provisions, and other articles belonging to the Government, either at Collingwood or Sault Ste. Marie. These, with the men intended for Mr. Dawson, you will deliver over to that gentleman when you meet him, either at Red River or on his way back.

10. You are to consider all the articles and materials of any description belonging to the Canadian Government, connected with the late expedition, as available for the purposes of the present expedition, and you and Mr. Dawson may therefore divide them between you in whatever way you may think most advantageous. Such articles, if any, as may not be required by either of you, should be left in the custody of some trustworthy person to await the orders of the Government.

11. As soon as you shall have put Mr. Dawson in possession of the men and canoes intended for him, each of you will be held separately responsible for the expenses of his own party. You will therefore be careful to keep an accurate account of your expenditure.

12. The auditor-general of public accounts will give you any information you may require as to the most convenient mode of making out and furnishing your accounts, &c.

13. On your return from Montreal I shall be prepared to give you your instructions with reference to the localities in which

your explorations are to be conducted, and as to the objects to which your attention is to be more especially directed.

I have the honour to be, Sir,

Your obedient servant,

T. J. J. LORANGER,

Secretary.

Henry Y. Hind, Esq., Toronto.

No. II.

Secretary's Office, Toronto, 27th April, 1858.

SIR,—I have the honour to communicate to you the instructions promised in the last paragraph of my letter to you of the 14th instant, for your guidance in connection with the branch of the expedition to the west of Red River, which has been committed to your charge.

2. The instructions contained in that letter will suffice for your guidance up to the time of your arrival at the Red River settlement, and the present instructions therefore have reference merely to your operations after having left that settlement.

3. The region of country to which your explorations are to be then directed is that lying to the west of Lake Winnipeg and Red River, and embraced (or nearly so) between the river Saskatchewan and Assinniboine, as far west as "South Branch House," on the former river, which latter place will be the most westerly point of your exploration.

4. It will be your endeavour to procure all the information in your power respecting the Geology, Natural History, Topography, and Meteorology of the region above indicated.

5. As to the general character of the geological portion of your labours, it is unnecessary to add anything to the instructions

communicated to you last year, and which, so far as this point is concerned, will serve for your guidance for the present season.

6. There are, however, two matters to which I am to request you to direct your particular attention, namely, the salt region in the neighbourhood of Lake Manitobah, adverted to in your report for last year, and the deposit of Tertiary coal or lignite, reported to exist in the valley of Mouse River.

7. It is most important that you should ascertain, by actual examination, as far as possible, the existence, extent, and character of these deposits.

8. In ascending or descending the different rivers you may have occasion to explore, it is advisable that you should note with care, their breadth, depth, rate of current, and the probable quantity of water discharged by them at different points, and at different seasons of the year; their facilities for navigation by boats or steamers, and whether they overflow their banks to any extent at any season of the year.

9. The general aspect of the whole region should be carefully described. The character of the timber and soil observed, and the general fitness of the latter for agricultural purposes ascertained as far as may be from observation and inquiry.

10. It is desirable that your meteorological observations should be made with the maximum and minimum thermometer, and with the wet and dry bulb. The temperature of the rivers, lakes, and springs should also be recorded, and the rainfall observed.

Any reliable information you can obtain as to the quantity of snow precipitated during the winter, would also be of interest.

11. Your topographical explorations should be made with reference to the construction of a map (as complete as possible) of the region explored, on a scale of two miles to one inch—and your operations should be conducted in view of a possible extension, at some future time, of the exploration, so as to embrace the entire valley of Lake Winnipeg and its feeders.

12. With a view to illustrate the natural history of the country, you will avail yourself of such opportunities as may present themselves to collect any objects that may be useful for that purpose.

13. Any geological or natural history specimens which you

may have collected during your explorations, may be left by you at Red River, on your return, with the other property of the Government belonging to the expedition, to await the orders of the Government, with the other articles referred to in the tenth paragraph of my letter of the 14th inst.

14. I am to add that his Excellency, having every confidence in your judgment and discretion, does not wish to trammel you with more detailed instructions, and that you are left at liberty to make any other exploration, in addition to those particularly named therein, should you, upon information obtained in the locality, deem it desirable for the general purposes of the expedition.

15. It is hardly necessary to state that you will be held responsible for the conduct, diligence, and fidelity of the party under your charge.

16. With a view to distinguish your branch of the expedition for the present year, it will be convenient to designate it as the "Assinniboine and Saskatchewan Exploring Expedition;" by this title therefore you will describe it in your reports.

I have the honour to be, Sir,

Your obedient servant,

T. J. J. LORANGER,
Secretary.

Henry Y. Hind, Esq., Toronto.

I am indebted to Mr. F. B. Meek, of Washington, U.S., for the description of the Cretaceous fossils collected during the exploration. Mr. Meek's remarks and descriptions are contained in Chapter XIX. of my Report on the expedition to the Canadian Government.

The Silurian and Devonian fossils were examined and described by Mr. Billings, the palæontologist to the Canadian Geological Survey. The descriptions and remarks of Mr. Billings are contained in Chapter XX. of the Report referred to in the preceding paragraph. I am glad to have another opportunity of expressing my obligations to these gentlemen.

I have also much pleasure in tendering my warmest thanks to Sir George Simpson, not only for the letters of introduction with which he favoured me to the officers of the Hon. Hudson's Bay Company's service in Rupert's Land, but also for his personal efforts when at Fort Garry, to facilitate the progress of the expedition by every means in his power. The assistance rendered by Sir George Simpson was of the greatest use to me, and the courteous manner in which it was granted increases my indebtedness to him.

From the officers of the Hon. Hudson's Bay Company's service in charge of the different posts, I received without any exception kind attention and valuable assistance; in the following pages many friendly actions are faithfully recorded.

CHAP. XIII.

THE gentlemen whose names are subjoined composed
the expedition which started from Toronto for Red River
on the 29th April, 1858.

> HENRY YOULE HIND, M.A., in charge of the Expedition.
> JAMES AUSTEN DICKINSON, B.A., Surveyor and Engineer.
> JOHN FLEMING, Assistant Surveyor and Draughtsman.
> HUMPHREY LLOYD HIME, Photographer.

The following Iroquois Indians were engaged at Caughnawaga near Montreal to man the canoes :—

Charles Skanasati, guide.
Martin Takatsitsienseré.
Louis Pekageiaien.
Ignace Tior-ateken.
Lazare Aneratentka.
Mathias Shaitikarenes.
Thomas Orite.
Louis Atioksisaks.
Thomas Shakashetstha.
Mathias Asinrathon.
Ignace Taseraren.
Thomas Tekarenhonte.
Pierre Aronhiakenra.

Fortunately the steamer "Illinois" from Detroit to Lake Superior Ports had some freight for Grand Portage, the Lake Superior termination of the Pigeon River route to Red River. By this long-neglected communication I determined to go. Having secured a passage to Grand Portage direct, and embarked our baggage and stores, which weighed nearly six thousand pounds, we left Detroit on the 30th April, and reached the Sault Ste. Marie on the 2nd of May.

A week before leaving Toronto I wrote to Mr. Simpson, the officer in charge of the Hudson's Bay Company's post at Sault Ste. Marie, enclosing a copy of a letter from Sir George Simpson, and requesting him to send two new North canoes, fully appointed, to the United States side of the St. Mary's river, in order that their embarkation might occasion as little delay as possible.

The canoes were lying in readiness on the wharf at the Sault together with a supply of gum, watap *, cod-lines, and other indispensables to canoe navigation.

---

* The root of the tamarac, used for sewing together the pieces of birch bark of which the canoes are constructed.

We arrived at the Grand Portage on the 5th of May and immediately made preparations for crossing the Portage. I found it necessary to engage three more voyageurs to man a third canoe which the large quantity of baggage rendered necessary; these were Wigwam, an Ojibway half-breed; Charles Louis, a French Canadian, and François Chabot, also of French Canadian origin.

We started from the west end of the Grand Portage on the 9th May, reached Fort Frances on the 24th, and the mouth of the Winnipeg on the 29th.*

We camped off the mouth of Red River seven days after leaving Fort Frances, and might easily have reached the settlements on the first day of June, but in view of our rapid voyage from Rainy Lake, and being in advance of Sir George Simpson, I did not think it necessary to press the guide, we therefore waited for a few hours at Fort Alexander, and enjoyed the hospitality of Mr. Sinclair, the gentleman in charge.

The exact time the expedition spent in canoes between Lake Superior and Red River, after deducting the delays at the forts before mentioned, was twenty-one days and six hours, as opposed to twenty-seven days and six hours by the Kaministiquia route the year before. The average daily progress was twenty-eight and a half miles against twenty-five miles in 1857.

On my arrival at the Middle Settlement, where Mr. Dawson and his party had their quarters, I found Mr. Russell in charge of the house and effects, Mr. Dawson with other members of his party having started some days previously for the Saskatchewan, whence they were not expected to return until the end of June. I therefore placed Mr. Russell in possession of the canoes and men intended for

* For an estimate of the cost of opening the Pigeon River Route, see Appendix, Vol. II.

Mr. Dawson, and immediately commenced to organize a party to explore the country drained by the Assinniboine and Little Souris.

On the morning of the 14th June, the half-breeds engaged for the expedition into the prairie country west of Red River, assembled at our temporary quarters and began to load five Red River carts and a waggon of American manufacture, with two canoes, camp equipage, instruments, and provisions for a three months' journey. At noon the start was made, and the train proceeded to Fort Garry*, a distance of eight miles, to take in a supply of flour and pemmican. We camped about half a mile from the fort and took an inventory of our baggage, and made such regulations and arrangements as are considered necessary at the commencement of a long journey through a country partly inhabited by hostile tribes of Indians, and not always affording a supply of food even to skilled hunters.

The party was composed of the gentlemen already named, six Cree half-breeds, a native of Red River of Scotch descent, one Blackfoot half-breed, one Ojibway half-breed, and one French Canadian. Our provisions consisted of one thousand pounds of flour, four hundred pounds of pemmican, one thousand rations of Crimean vegetables, a sheep, three hams, and a supply of tea for three months, with a few luxuries, such as pickles, chocolate, a gallon of port wine, and a gallon of brandy. Each cart was loaded with about 450 lbs. and the waggon with double that amount. The birch bark canoes were 18 feet

---

* "The mean of five observations at Upper Fort Garry, at the mouth of the Assinniboine, for latitude, three meridian by altitude of the sun and two by Polaris, gave for the latitude 49° 53′ 24″. Mr. Calhoun, who was attached to Major Long's expedition in 1823, made it 49° 53′ 35″, but according to a record in the possession of one of the officers of the fort, Lefroy placed it in latitude 49° 58′." Owen's "Geological Survey of Wisconsin, Iowa, and Minnesota," p. 180. Capt. Palliser places Fort Garry in latitude 49° 52′ 6″. Longitude, 96° 52′ 27″.

long, and weighed 150 lbs. each. At the White Horse Plain, twenty-two miles from Fort Garry, I procured an additional cart, and purchased an ox to serve as a *dernier ressort* in case we should not meet with buffalo ; and at Prairie Portage, the last settlement on the Assinniboine, I engaged the services of an old hunter of Cree origin, who had been from his youth familiar with Indian habits and stratagems. This addition increased the party and equipment, before we left the last settlement, to fifteen men, fifteen horses, six Red River carts, one waggon, and one ox.

Leaving our camp early on the morning of the fifteenth, we ascertained by leveling the altitude of an ancient lake ridge, near to St. James's Church*, to be eleven feet above the prairie at Fort Garry, and about two miles from it. These ridges are common in the prairies of Red River, and do not necessarily point to an ancient lake margin, as it is probable that most of them were formed under water. They may be traced for many miles, but are sometimes lost in the general rise of the prairie, or blend with higher ridges.

On arriving at St. James's Church we separated into two divisions, Mr. Fleming and Mr. Hime, with the carts and waggon, proceeding to Lane's Post on the Assinniboine, twenty-four miles from Fort Garry, while Mr. Dickinson and myself with two half-breeds, struck in a north-westerly direction across the prairie to Stony Mountain, and thence to the Big Ridge, having arranged to meet at Prairie Portage.

In a wheat field opposite St. James's Church were several pigeon traps, constructed of nets 20 feet long by 15 broad, stretched upon a frame ; one side was propped up

---

* The Rev. W. H. Taylor's Mission. The income of this Mission is derived from the following sources : £100 from the Society for the Propagation of the Gospel, £100 from the Bishop of Rupert's Land.

T 3

by a pole 8 feet long, so that when the birds passed under the net to pick up the grain strewed beneath, a man or a boy concealed by a fence or bush, withdrew the prop by means of a string attached to it, and sometimes succeeded in entrapping a score or more of pigeons at one fall. Near the net some dead trees are placed for the pigeons to perch on, and sometimes stuffed birds are used as decoys to attract passing flocks.

In pursuing our course to Stony Mountain we endeavoured to follow the ridge before alluded to, but after tracing it for several miles it became imperceptibly blended with the level prairie. Several ridges were crossed after we lost the first, but in all instances they died away after having preserved their rounded form for two or three miles. Stony Mountain is a limestone island of Silurian age,

Stony Mountain.

having escaped the denuding forces which excavated the Red River valley. It is about four miles in circumference, its highest point is sixty feet above the prairie level. Horizontal layers of limestone, holding very few and obscure fossils, project on its western cliff-like sides. Its eastern exposure is gently sloping, and some ten feet from the summit, the remains of an ancient lake beach are well preserved. Viewed from a distance, Stony Mountain requires little effort of the imagination to recall the time when the shallow waters of a former extension of Lake Winnipeg, washed the beach on its flank, or threw up as they gradually receded, ridge after ridge, over the level floor of the lake, where now are to be found wide and beautiful prairies covered with a rich profusion of long grass.

Leaving Stony Mountain our course lay westerly through a wet prairie to the Big Ridge. Grey cranes, ducks, and plover were numerous on the marshy areas, and in every little bluff * of aspen or willow, the beautiful rice birds (*Dolichonyx oryzivora*) were seen or heard. Where we camped on the edge of a lake near the foot of the Big Ridge, bittern, grackle, and several varieties of duck flew to and fro in alarm at our invasion of their retreats. On the flank of the Big Ridge the cinnamon thrush (*Turdus rufus*) was noticed, but most common of all was the tyrant flycatcher (*Muscicapa tyrannus*), who endeavoured to hold undisputed sway over the bluff he had selected as his home. Near and west of Stony Mountain, many small barren areas occur, covered with a saline efflorescence ; they may be traced to the Assinniboine, and beyond that river in a direction nearly due south to La Rivière Sale, and the 49th parallel. These saline deposits are important, as they in all probability serve, as will be shown hereafter, to denote the presence of salt bearing rocks beneath them, similar to those from which the salt springs of Swan River, Manitobah Lake, and La Rivière Sale issue.

Early on the morning of the 17th we ascended the Big Ridge. Its elevation above the prairie is about 60 feet ; on its south side it slopes gently to the prairie level, to the north is a plateau, well wooded with aspens, stretching

---

* The half-breeds call little groves of aspens or willows in the prairies "bluffs."

towards Lake Manitobah; the view from its summit extends far and wide over the Assinniboine prairies; and skirting its base on the south flank are groves of aspen and balsam poplar, with scattered oak trees and willow bushes. The pasturage in the open glades is of excellent quality and very abundant. The ridge is quite level and from 80 to 100 feet broad, devoid of trees, slightly arched and composed of gravel. Here and there it is cut by rivulets draining the marshes in the plateau on its northern side. As it approaches Prairie Portage its apparent elevation diminishes, until at the Portage River it is no longer discernible. We traced it for a distance of seventy miles, and it will be mentioned further on that a similar ridge, but one formed at an earlier period and at a higher level, is seen west of Manitobah Lake, near the Hudson's Bay Company's post, Manitobah. House. The older ridge preserves there the same characters of horizontality, uniform outline, gravelly formation, and admirable adaptation to the purposes of a road, which have been already noticed in connection with the Big Ridge north of the Assinniboine and east of Red River. For many miles, ties for a railway might be laid upon both without a pebble being removed, and the only breaks in their continuity occur where streams from the plateau and higher grounds in the rear have forced a passage through them. The older ridge, however, follows the western contour of Lakes Winnipeg and Manitobah, and passes through a country not likely to be first selected by a large body of settlers. The Big Ridge is important in so far that it forms the boundary of land of the first quality, which occupies the low prairie valleys of the Assinniboine and Red River. Soundings subsequently made in Lake Manitobah showed a uniform depth of eighteen feet for a distance exceeding sixty miles along its south-eastern coast, so that if its beds were exposed, it is pro-

bable that in process of time it would also become a rich
and extensive prairie country, with its present beach distinctly visible as its old boundary.   Indeed, the surface
of the country between the Big Ridge, the Assinniboine,
and Red River, is similar in gently undulating outline to
the succession of undrained marshes, ridges, and bogs
which exist between the west coast of Lake Manitobah
and the older ridge, pointing to a very gradual but
constant drainage of this region after a long period of
submergence.

'We reached Prairie Portage in the evening, where we
joined the main party.   The Assinniboine at Lane's Post
(June 16th) is about 120 feet broad, and its turbid
water flows at the rate of 1½ miles per hour.   A few
miles west of Lane's Post, the saline efflorescence, before
noticed as occurring in patches on the prairies and forming small barren areas, is no more to be seen; it consists
of chloride of sodium and sulphate of magnesia, with a
little chloride of calcium.   The first grasshoppers were
observed this year at Lane's Post; they were a brood
from the eggs deposited by a swarm which alighted on
the White Horse Plain in September last.

At Prairie Portage we found an Ojibway encampment
in which were some of the refractory personages who
had hitherto resisted the humane and unceasing efforts
of Archdeacon Cochrane to christianize them.   Among
the various methods tried by the archdeacon to induce
these wanderers to settle and farm,— the first preliminary
to the progress of Christianity among wild Indians,—that
of presenting the most docile with an ox and plough and
teaching them to use it, was the least successful.   At the
first good opportunity, or during a time of scarcity, the
ox and plough would be sold to the highest bidder for
very much less than it cost.   A promise to add another

ox at the end of a year if the first gift was faithfully preserved was of no avail, the charms of the buffalo plains
were too tempting, or the seduction of gambling too
powerful to be withstood, notwithstanding the most
solemn heathen promises. The school, however, gives
better hope, and no doubt the rising generation of Indian
and half-breed origin at Prairie Portage, will form a
thriving, industrious, and Christian community.

Prairie Portage is very delightfully situated sixty miles
west of Fort Garry on the banks of the Assinniboine.
The prairie here is of the richest description, and towards
the north and east boundless to the eye. The river bank is
fringed with a narrow belt of fine oak, elm, ash, and ash-
leaved maple, but on the south side a forest extends from
two to four miles in depth, and then passes into aspen
groves; the river abounds in sturgeon and gold eyes, and
within eighteen miles there is an excellent fishing station
on the coast of Lake Manitobah, where the Portage people
take vast numbers of white fish every fall. The old water
course of the Assinniboine, near Prairie Portage, is now a
long narrow lake, fringed with tall reeds, a favourite
haunt of wild fowl and grackle, among which we observed the showy yellow-headed blackbird (*Agelaius
xanthocephalus*).

Prairie Portage will eventually become an important
settlement, not only on account of the vast extent of
fertile country which surrounds it, but because it lies
in the track of the buffalo hunters proceeding to the
Grand Coteau and the South Branch by way of the
Souris River. It is also near to the fertile region
drained by White Mud River, and the road to the south-
western flanks of the Riding Mountain, passes by the
Portage. The current of the Assinniboine is very uniform here, careful leveling showed that it fell one foot
two inches in a mile, with a velocity of two miles an

hour where the trial was made, but in the middle of the stream the velocity must be considerably greater. The cliff swallow (*Hirundo fulva*) had built its nests in great numbers on the banks of the river, which are about sixteen feet above the level of the water. I counted no less than thirteen groups of nests within a distance of five miles, when drifting down in a canoe. The cliff swallow was afterwards seen in great multitudes on the Little Souris, the South Branch of the Saskatchewan, and the Qu'appelle River.

The first of a series of thunder storms which lasted for some weeks visited us on the 17th; the warm rain fell in torrents and thoroughly wetted all who were exposed. Pigeons were flying in vast numbers across the Assinniboine, and the black tern (*Sterna nigra*) was numerous in the prairies near the settlement. In descending the river for a few miles to inspect its banks, we had occasion to pass by a fish weir, where several Ojibways from the camp near the Portage were watching for sturgeon with spears in their hands. They took no notice of us as we passed, being too busily engaged, but on our return to the encampment we found them waiting with fish to barter for tobacco and tea. We made them a few trifling presents, and, by way of recompense, sustained during the night the loss of a cheese, which after curiously eyeing during supper, they modestly asked for permission to taste. They pronounced it excellent, and in the dead of night (when our half-breeds were sound asleep) they opened the basket in which it had been placed, and quietly abstracted it. In future, when Indians were around, all eatables and articles they might covet were properly secured, and the cheese proved to be our only loss during the exploration.

Leaving Prairie Portage on the morning of the 19th, we took the trail leading to the Bad Woods, a name given to a wooded district about thirty miles long, by the buffalo hunters in 1852, who, in consequence of the floods of that year, could not pass to their crossing place at the Grand Rapids of the Assinniboine by the Plain or Prairie Road. There were four hundred carts in the band, and the hunters were compelled to cut a road through the forest of small aspens which forms the Bad Woods, to enable them to reach the high prairies. This labour occupied them several days, and will be long remembered in the settlements in consequence of the misery entailed by the delay on the children and women.

The trail continued for three miles through a continuation of the low prairies of the Assinniboine, when a sudden ascent of 20 to 25 feet introduced us to a different kind of country, the plateau beyond the Big Ridge, which here crosses the river, and forms the lowest or first step of the Pembina Mountain. The physical features of this boundary to a great table-land will be noticed at length in the sequel. The soil continues poor and sandy for several miles, supporting clumps of aspen interspersed with a few oaks in low places. The view across the Assinniboine reveals in the distance the Blue Hills, and between them and the river is a vast forest, which a subsequent exploration in the autumn showed to consist of oak, elm, ash and aspens, for two to three miles nearest to the Assinniboine; but beyond this limit the forest is almost entirely composed of aspens of small growth.

Grasshoppers were now seen in great numbers, and the first humming-bird was noticed here. The banks of the river showed recent watermarks twelve feet above its present level, willow and other trees overhanging the stream being barked by the action of ice during spring freshets at that elevation. Everywhere rabbits were numerous, and considerable areas occur covered with dead

HALF-WAY BANK, ASSINNIBOINE RIVER.

Printed by Spottiswoode and Co.

[New-street Square, London.

willows and aspens, barked by these animals in the winter, about two feet six inches above the ground.   The height of the bank twenty-two miles from Prairie Portage is about 80 feet above the river, denoting a rapid rise in the general level of the country.

On the morning of the 20th we entered the Bad Woods and followed the road cut by the hunters in 1852.   The aspens were much disfigured by countless numbers of caterpillars resembling those of the destructive Palmer worm.   In the afternoon we arrived at the Sandy Hills; they consist of rounded knolls covered with scrub oak and aspens.   Our latitude to-day at the Half Way Bank was ascertained to be 49° 46′ 19″, the height of the prairie 150 feet above the river, the breadth of the valley in which the river flowed 5,680 feet, and the variation of the compass 13° E.   After passing the point where the foregoing observations were made, the trail again enters the Bad Woods and continues through them until it strikes the Sandy Hills again. These rounded eminences have all the appearance of sand dunes covered with short grass and very stunted vegetation.

As we emerged from the Bad Woods a noble elk trotted to the top of a hillock and surveyed the surrounding country; a slight breath soon carried our wind as the hunter was endeavouring to approach him, he raised his head, snuffed the air and bounded off.   Another terrible thunderstorm came on at sunset, with heavy rain and boisterous wind.   The aspect of the country for many miles is that of a plain sloping gently to the east, and studded with innumerable mounds or hillocks of sand, thinly covered with a poor and scanty growth of grass; here and there small lakes or ponds occur fringed with rich verdure, but its general character is that of sterility. From the summit of an imposing sand-hill, formerly a drifting dune, which we ascended on the 21st, the country lay mapped at our feet; as far as the eye could reach, sand-hills, north, east, and west, sometimes bare and ripple-marked, but generally covered with short grass, were exposed to view.

On the afternoon of this day a hail storm of unusual violence caused us to halt. The stones penetrated the bark of our canoes and broke off the gum. The grass-hoppers, which were very numerous just before the storm began, suddenly disappeared, but they might be found quietly clinging to the leaves of grass in anticipation of the storm. After it had passed they re-appeared, apparently in undiminished numbers, although every member of the party, crouching for shelter under the carts and waggon, fully expected the complete annihilation of these destructive and troublesome insects. A wonderful instinct enables them to seek and find refuge from a pitiless hailstorm or a drenching rain. The same evening a thunderstorm of short duration again arrested our progress, but the sun set in gorgeous magnificence, with a brilliant rainbow and vivid flashes of lightning in the east. The cinnamon thrush is not uncommon among the sandy hills; we saw several during the day. The following morning we reached the " pines," for which we had been anxiously looking, but to our disappointment they proved to be nothing more than balsam spruce in scattered clumps.

The heat of the weather began now to be very oppressive, and joined to the incessant attacks of mosquitoes and " bulldogs" proved very exhausting to the horses. At each camping-place we were compelled to make "smokes" to drive away these tormentors, otherwise the persecuted animals would endeavour to approach our camp fires, creating no small confusion in our cooking arrangements. On the 23rd, we traversed a region of sand-hills and ridges, until we arrived at Pine Creek, a distance of eight miles from the preceding camp. Here the sand-hills are absolutely bare, and in fact drifting dunes. Sending the main party in advance, Mr. Dickinson and I set out to examine the valley of the Assinniboine where Pine Creek disembogues. The sand dunes were seen reposing on the prairie level about 170 feet above the river. In crossing the country to regain the carts, our course lay across a broad area of drifting sand beautifully ripple-marked, with here and there numbers of the bleached bones of buffalo protruding from the west side of the dunes, memorials probably of former scenes of slaughter in buffalo pounds, similar to those which we witnessed some weeks afterwards at the Sandy Hills on the South Branch of the Saskatchewan. The progress of the dunes is very marked ; old hillocks partially covered with herbage are gradually drifted by the prevailing westerly wind to form new ones. Sometimes the area of pure sand was a mile across, but generally not more than half that distance. The largest expanse we saw was near the mouth of Pine Creek, it is called by the Indians "the Devil's Hills," and a more dreary, parched-looking region could scarcely be imagined.

Opposite the Souris the subjoined section of the north bank of the Assinniboine was measured :—

1. 22 feet pale yellow sand.
2. 1 foot dark blue clay.
3. 12 feet ferruginous gravel and sand.
4. 4 inches dark green hard gravelly clay.
5. 15 feet soft unctuous blue clay to the water's edge.

It is very probable that, of this section, the lowest number represents the weathered shales which were afterwards found exposed a few miles up the Little Souris valley.

We reached the mouth of the Little Souris River on the 24th, and made preparations to cross the Assinniboine at this point.   The distance traveled through the Sandy Hills was about forty-eight miles, their breadth does not exceed ten miles.   At the mouth of the Souris the grasshoppers were in countless numbers, and so voracious as to attack and destroy every article of clothing left for a few minutes on the grass.   Saddles, girths, leather bags, and clothing of every description were devoured without distinction.   Ten minutes sufficed them, as our half-breeds found to their cost, to destroy three pairs of woollen trowsers which had been carelessly thrown on the grass. The only way to protect our property from the depredators was to pile it on the waggon and carts out of reach. There were two distinct broods of grasshoppers, one with wings not yet formed, which had been hatched on the spot, the other full-grown, invaders from the prairies south of the Assinniboine.   We saw here one of the vast flights of these insects which afterwards were witnessed on a scale of alarming magnitude, giving rise in their passage through the air to optical phenomena of very rare and beautiful descriptions.   As we cautiously approached the bank of the river opposite the mouth of the Little Souris, on the look-out for Sioux Indians, some jumping deer and a female elk were observed gamboling in the river.   A shot from a Minie rifle dispersed them and started from their lair two wolves who were watching the deer, patiently waiting for an opportunity to surprise them.

The volume of water in both rivers was carefully measured at their point of junction.   The Assinniboine was found to be 230 feet broad, with a mean depth of 6 feet, and a current of one mile and a quarter per hour.   The Little Souris was 121 feet broad, 2 feet 4

CONFLUENCE OF THE LITTLE SOURIS AND THE ASSINNIBOINE.

Printed by Spottiswoode and Co.]

[New-street Square, London.

inches in mean depth, and flowing at the rate of half a mile an hour. Observing numbers of fish rising at grasshoppers in the Souris, we stretched a gill net across the mouth of the river, and succeeded in taking pickerel, gold-eyes, and suckers, the grey and the red. In a second attempt we caught a tartar ; a huge sturgeon got entangled in the meshes of the gill net, and before we could land him he succeeded in breaking away and carrying a portion of the net with him.

Signs of Sioux Indians in the neighbourhood led to our keeping watch during the night ; and on the morning of the 25th we proceeded cautiously up the valley of the Souris, keeping a sharp look out. On the left bank the Blue Hills of the Souris are visible ten miles from the mouth of the stream, and towards the west, the Moose Head Mountain is seen to approach the Grand Rapids of the Assinniboine. The first rock exposure in the valley was observed about fifteen miles from the mouth of the Souris. It consisted of a very fissile, dark-blue argillaceous shale, holding numerous concretions containing a large percentage of iron, partly in the state of carbonate and partly as the peroxide. Some very obscure fossils were found, with fragments of a large inoceramus. The shale weathers ash white. It is exposed in a cliff about ninety feet high ; the upper portion of the cliff consists of yellow sand, superimposed by sandy loam holding limestone boulders and pebbles; the exposure of shale is seventy feet thick, in horizontal layers. The country west of the Souris, so far, is an open, treeless, and undulating prairie. On the east side, the Blue Hills are very picturesque, their flanks and summits are wooded with aspen. Rain as usual, the day closing with a thunderstorm.

On the 26th we arrived at the westerly bend of the

Souris in the midst of a very lovely, undulating country; the river is here fifty feet broad, and in its passage through the Blue Hills it has excavated a valley fully four hundred and fifty feet deep.   Rock exposures are of frequent occurrence, the dip being 3° south.   Fragments and perfect forms, but very fragile, of a large inoceramus are common.   The ferruginous concretions continue disposed in regular layers, and appear to be a characteristic

Valley of the Souris, looking towards the Blue Hills of the Souris.

feature of the cretaceous rocks in this valley.   A continuation of the valley of the Souris extends in a direction nearly south-east towards Pembina River, with which it is said by the half-breeds to interlock.   Three lakes visible from our camp were stated to be the sources of Pembina River; a little stream issuing from the most westerly of these is called Backfat rivulet, it flows into the Souris.   Southwards, Turtle Mountain shows with a faint blue outline on the horizon.

Deer are very numerous at this beautiful bend of the

river; it appears to be a favourite watering-place. The half-breeds of St. Joseph often make it their crossing-place when on a hunting expedition to the Grand Coteau. It is not improbable that it will become a point of importance if ever an emigrant route should be established from Minnesota to the Pacific, *viâ* the south branch of the Saskatchewan; and from the great distance saved by going through St. Joseph instead of Fort Garry, it is not improbable that this may yet be the case.

Valley of the Souris, opposite the Valley of the Backfat Lakes.

On the 27th we succeeded in passing the Blue Hills, and enjoyed on the evening of the same day one of the most sublime and grand spectacles of its kind which it is possible to witness. Before leaving the last ridge of the Blue Hills, we came suddenly upon the borders of a boundless level prairie on the opposite side of the river, one hundred and fifty feet below us, of a rich, dark-green colour, without a tree or shrub to vary its uniform level and yet with one conical hill apparently in its centre. Here we expected to find buffalo, but not a sign of any living creature could be detected with the aid of a good glass The prairie had been burnt last autumn, and the Buffalo had not arrived from the south or west to people this beautiful level waste. What a magnificent spectacle this vast prairie must have furnished when the fire ran over it before the strong west wind!

From beyond the South Branch of the Saskatchewan to Red River all the prairies were burned last autumn, a vast conflagration extended for one thousand miles in length and several hundreds in breadth. The dry season had so withered the grass that the whole country of the Saskatchewan was in flames. The Rev. Henry Budd, a native missionary at the Nepowewin, on the North Branch of the Saskatchewan, told me that in whatever direction be turned in September last, the country seemed to be in a blaze; we traced the fire from the 49th parallel to the 53rd, and from the 98th to the 108th degree of longitude. It extended, no doubt, to the Rocky Mountains.

A few miles west of the Blue Hills, being anxious to ascertain the dip of a very remarkable exposure of shale with bands of ferruginous concretions, Mr. Dickinson leveled with the utmost care an exposure facing the south, and found it to be horizontal. At the base of the exposure, and on a level with the water's edge we succeeded in finding a layer of rock full of gigantic inoceramus. One specimen measured $8\frac{1}{2}$ inches in diameter, it was very fragile but the peculiar prismatic structure was remarkably well preserved. On attempting to raise it, it separated into thousands of minute prisms so characteristic of this shell.

Vast numbers of pigeons were flying in a north-westerly direction, and our friends the grasshoppers were everywhere abundant. From the Blue Hills to the south bend of the river, rock exposures possessing the characteristics already noticed, occurred at every bend. The first speci-

VALLEY OF THE SOURIS.

Printed by Spottiswoode and Co.]

[New-street Square, London.

men of lignite was seen near the mouth of Plum Creek,
where we camped on the 29th; it was a water-worn
rounded boulder.

After leaving the Blue Hills no trees or shrubs of any
description were seen until we arrived at Plum Creek.
On low points in the Souris valley some fine oak, elm,
balsam poplar, and aspen are found for the first twenty
miles. The guelder rose is common on the ravines, wild
prairie roses abundant, snowberry and two varieties of
cherry of frequent occurrence, as well as woodbine, wild
convolvulus, and hop, but for a distance of twelve miles
west of the Blue Hills the country is treeless on both sides
of the river, and the drift of small depth.

A little beyond Plum or Snake Creek we found numerous pebbles and boulders of lignite; and with a view to
ascertain whether the lignite existed *in situ*, an excavation
was made in the bank of the river and the stratification
for a depth of twenty-five feet exposed. The last outcrop
of the cretaceous shales was observed about three miles
east of the bank where this trial was made. A few hours'
labour revealed five old beaches, probably of a former
lake. These beaches were composed of sand and boulders
of lignite from the size of a hen's egg to one foot in
diameter. No fragment of lignite was found which did
not possess a rounded or spheroidal form and a roughly
polished or worn surface. An abundant supply was easily
obtained for a fire which was soon made on the bank; a
strong sulphurous odour was emitted from the iron pyrites
in the lignite, and some boulders when broken open exhibited streaks and small particles of a resinous substance
like amber.

The excavation exposed the section which is shown
in the woodcut on the next page.

The low hills about Plum Creek are sand dunes, and on

their sides an opuntia is very common. The prairie on the west of the Souris as well as on the east is treeless, but the banks of Plum Creek support a thin belt of small forest trees, such as oak and ash, with a few ash-leaved maple.

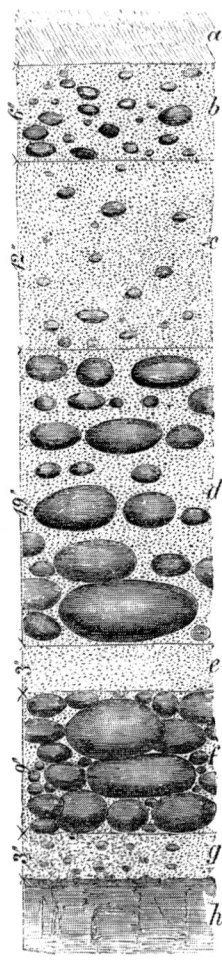

Section on theLittle Souris, showing ancient Beaches (*b, c, d, f, g*) with Lignite Boulders.

The annual fires prevent the willows and aspens from covering the country, which they would undoubtedly do until replaced by other species, if not destroyed to within a few inches of the root every time the fire sweeps over them. The banks of the Souris here are not more than 40 feet high, with level prairies on either hand, a few miles beyond the Snake Hills. Within four miles of the mouth of Plum Creek, Oak Lake, several miles in diameter, attracted the hunting portion of our party; they brought back some pelicans and a score of duck. Thunder storms as usual to day and yesterday.

On the 1st July we arrived at the Souris Sand Hills, and made a section of the river bank where a land slip had produced a fine exposure to the water's edge. The formation consisted of five feet of blue clay above the level of the river, supporting four feet of ferruginous sand and gravel, on which reposed twelve feet of sandy loam and sand to the prairie level. The blue clay, capped by the ferruginous sand, was traced for a distance of $2\frac{1}{2}$ miles, and showed a dip to the south of two feet in the mile, the clay disappearing beneath the water. No organic remains of any description were found, although a careful search was made, but boulders of lignite from 6 to 9 inches in diameter, were frequently seen in the bed of the river. The eggs of the nighthawk (*Chordeiles Virginianus*) were several times found on the bare ground, among the Sand Hills, with no approach to a nest for the helpless young. The parent birds endeavoured to draw us away from their eggs, fluttering as if wounded a short distance from them and uttering cries of distress. The Hudson's Bay Company have a post on the river among the Sand Hills, which is maintained only in winter, during the absence of the Sioux: these savage barbarians being altogether opposed to the approach of civilization near to their hunting-grounds, and entertaining besides a feeling of deadly hostility to the Red River half-breeds.

Near the Company's house we found on the river bank an extensive deposit of bog iron ore, capped by shell marl, and above the marl drifted sand. The banks of the river are here not more than 25 feet high, and on the east side there is a narrow fringe of fine timber. The Bois de Vache (dried buffalo dung) is distributed very abundantly in the prairie and through the Sand Hills and near to the company's post. In fact the buffalo were very numerous during the whole of the winter of 1856 and spring of 1857 on the banks of the Souris, but the great fires during the autumn of last year, have driven them south and north-west, and between the two branches of the Saskatchewan. The country becomes very low after passing the last sand-hills, and over a large extent of prairie south of them, drift timber is distributed, showing the extraordinary rise in the waters of the river during the floods of 1852.

u 4

On the low banks which are constantly wearing away, and revealing fresh surfaces at points or bends, some well stratified layers of a brown-coloured deposit, two feet thick, appear about four feet from the present prairie level. They resemble stratified bands of certain varieties of bog iron ore, but a very slight examination is sufficient to show that they are of vegetable origin. They consist of a series of accumulations of buffalo dung, collected doubtless by rains and floods from a considerable area, and deposited in the low part of the Souris valley, which lies to the south of the Sand Hills. A geological formation composed of bois de vache is a novelty not likely to be met with anywhere but on the continent of North America.

On the 2nd of July we observed the grasshoppers in full flight towards the north, the air as far as the eye could penetrate appeared to be filled with them. They commenced their flight about nine in the morning, and continued until half-past three or four o clock in the afternoon. About that hour they settled around us in countless multitudes, and immediately clung to the leaves of grass and rested after their journey. On subsequent days, when crossing the great prairie from Red Deer's Head River to Fort Ellice, the hosts of grasshoppers were beyond all calculation; they appeared to be infinite in number. Early in the morning they fed upon the prairie grass, being always found most numerous in low, wet places where the grass was long. As soon as the sun had evaporated the dew, they took short flights, and as the hour of nine approached, cloud after cloud would rise from the prairie and pursue their flight in the direction of the wind, which was generally S.S.W. The number in the air seemed to be greatest about noon, and at times they appeared in such infinite swarms as to lessen perceptibly the light of the sun. The whole horizon wore an unearthly ashen hue from the light reflected by their transparent wings. The air was filled as with flakes of snow, and time after time clouds of these insects forming a dense body casting a glimmering silvery light, flew swiftly towards the north-north-east, at altitudes varying from 500 to 1000 feet and upwards.

Some idea of the height of the flight of these insects may be gathered from the opportunity enjoyed by Mr. E. James, who, when standing upon the summit of a peak of the Rocky Mountains, 8500 feet above the level of the Plains in Nebraska territory (14,500 above the sea), saw them above his head as far as their size would render them visible."*

Lying on my back and looking upwards as near to the sun as the light would permit, I saw the sky continually changing colour from blue to silver white, ash grey and lead colour, according to the numbers in the passing clouds of insects. Opposite to the sun the prevailing hue was a silver white, perceptibly flashing. On one occasion the whole heavens towards the south-east and west appeared to radiate a soft grey tinted light with a quivering motion, and the day being calm, the hum produced by the vibration of so many millions of wings was quite indescribable, and more resembled the noise popularly termed " a ringing in one's ears," than any other sound. The aspect of the heavens during the greatest flight we observed was singularly striking. It produced a feeling of uneasiness, amazement, and awe in our minds, as if some terrible, unforeseen calamity were about to happen. It recalled more vividly than words could express the

* Explorations in Nebraska and Dakotah. Preliminary Report of Lieut. G. K. Warren, Top. Eng. U. S. Army, 1858.

devastating ravages of the Egyptian scourges, as it
seemed to bring us face to face with one of the most
striking and wonderful exhibitions of Almighty power
in the creation and sustenance of this infinite army of
insects.

In the evening, when the grasshoppers were resting
from their long journeys, or in the morning, when feeding on the grass leaves, they rose in clouds around us as
we marched through the prairie; if a strong wind blew
they became very troublesome, flying with force against
our faces, in the nostrils and eyes of the horses, and
filling every crevice in the carts. But fortunately, comparatively few flew on a windy day, otherwise it would
have been almost impossible to have made headway
against such an infinite host in rapid motion before the
wind, although composed individually of such insignificant members.

Those portions of the prairie which had been visited
by the grasshoppers wore a curious appearance; the
grass was cut uniformly to one inch from the ground,
and the whole surface was covered with the small, round,
green exuviæ of these destructive invaders.

The valley of the Souris, along which we travelled
during the day, varies from one quarter to one mile
broad; the river is not more than twenty-five feet across,
and very shallow. It flows through a rich open meadow,
twenty to thirty-five feet below the general level of the
prairie, which on either hand is undulating, treeless,
covered with short stunted grass, and showing abundance
of last year's bois de vache. The first fresh buffalo
tracks were seen to-day, and while taking observations
for latitude, tracks of a different character and greater
significance were discovered by one of the half-breeds;
the fresh print of horses' feet, pronounced to be a few

hours old, denoting the presence of Sioux or Assinniboines in our neighbourhood.

Before reaching the 49th parallel, the Souris meanders for several miles through a treeless valley, about a mile broad and sixty feet below the prairie level. Turtle Mountain on the east rises nobly from the great plain, the boundary line between British and American territory cutting it. The country west of the Souris is a treeless desert, in dry seasons destitute of water, and without a shrub or bush thicker than a willow twig. We ascertained the breadth of this arid, woodless tract to be at least sixty miles north of the Red Deer's Head River on the 49th parallel. Near the boundary line the Souris expands into a series of large ponds and marshes which are called the Souris Lakes. During periods of high water they form a continuous lake of imposing magnitude, extending many miles south of the 49th parallel, consequently far within the United States territory.

A vast number of gneissoid and limestone boulders are strewed over the hill bank of the Souris, near the 49th parallel, and on a point between a small brook and the river we found a number of conical mounds, and the remains of an intrenchment. Our half-breeds said it was an old Mandan village; the Indians of that tribe having formerly hunted and lived in this part of the Great Prairies. We endeavoured to make an opening into one of the mounds, and penetrated six feet without finding anything to indicate that the mounds were the remains of Mandan lodges. There is a Mandan village near Fort Clark on the Missouri, and in the country drained by the Yellowstone the remains of this once numerous and powerful tribe are now to be found.

The mouth of Red Deer's Head River is within a few

yards of the 49th parallel, Mr. Hime took a photograph of the valley while others of the party made an excursion to the Souris Lakes, within the United States territory, in the hope of finding buffalo to replenish our stores ; but although fresh tracks were seen, and skulls and bones in large numbers, the remains of last year's "run," yet no living animal but a "cabri," or prong-horned antelope (*Antilocapra Americana*), was visible.

Turkey buzzards (*Cathartes aura*) were observed hovering at a great height above us, and two young birds were shot in the valley of the Souris. The great scarcity of animal life near the Souris Lakes appeared remarkable, but it might be caused by the desert character of the surrounding country, which was so barren and arid as to be incapable of supporting a scanty growth of herbage on the sandy soil of the prairie.

Having reached the 49th parallel and traced the Souris in search of lignite in position for a distance of a hundred miles, we altered our course to a good camping-ground on Red Deer's Head River, and made preparations for crossing a treeless prairie, at least sixty miles broad, in a direction nearly due north.

The Little Souris or Mouse River rises in British territory, on the flanks of the Grand Côteau de Missouri, near the 105th meridian.*   Its valley was reported to us by the Crees of the Sandy Hills on the Qu'appelle to inosculate with Elbow Bone Creek, or the Souris Forks, as this stream is also termed, which flows into the Qu'appelle a few miles west of Long Lake. Captain Palliser indicates † its connexion with the Moose Jaws Forks, also an affluent of the Qu'appelle. It is not in the least degree

---

* See Capt. Palliser's Map, published in the Blue Book, 1859.
† Ibid.

improbable that connexions exist between both rivers. The Souris crosses the boundary line near the 102nd meridian, and flows at the base of the Grand Coteau, nearly as far south as the 48th parallel, when suddenly turning northward, it re-enters British territory near the 101st meridian.

South of the Souris Lakes it flows in a valley 200 feet below the level of the prairie, with a wooded bottom from one half to two miles wide. The nearest serviceable timber east of the Souris, in the direction of the proposed Pacific railroad near the 47th parallel, is in the valley of Red River, 200 miles distant, and with the exception of cotton wood there is no timber west of the Souris for 400 miles.*

* Governor Stevens's Explorations and Surveys.

CHAP. XIV.

FROM THE BOUNDARY LINE TO THE QU'APPELLE LAKES *viâ*
FORT ELLICE.

Indian Signs —Smell of Fire.—The Sioux.—Precautions.—"Something."—
Red Deer's Head River.—The Great Prairie, Character of.—Mirage.—Birds.
—Grasshoppers.—Limit of Burnt Land.— Pipestone Creek.— Standing
Stone.—Country changed. —Forest disappeared.—Approach to the Assin-
niboine.— Cretaceous Rocks.— Buffalo Bull.— Fort Ellice.— McKay.—
Crees.—Hunters.—Provision Trading Posts.—Pemmican.— Dried Meat.—
Thunder Storms.—Mammoth Bones.— Ojibway Hunter.— Half-breeds.—
En Route for the Qu'appelle Mission.—Grasshoppers.—Thunder Storm.
—Trail.— Weed Ridge. — Kinni-Kinnik. — Mode of Manufacturing. —
Boulders.—White Crane.— Magpies.— Birds.— Dew.— Aridity, of Great
Prairie.— Charles Pratt.— Chalk Hills. — Indian Turnip. — Qu'appelle
Lakes.— Fresh Arrangements.—Descent and Ascent of the Qu'appelle.—
Qu'appelle Mission. — Dimensions of Valley. — Character of Lakes.—
White Fish.—Rev. James Settee.—Garden of Mission.—Grasshoppers.
—Christian Worship. — Baptism. — "Praying Father" and "Praying
Man." —Rum.— Indian Wishes.— Objection to Native Missionaries.—
Difficulties arising from the Prejudice of Tribes.—Plain Crees passing
away.—En Route.

WHILE engaged in taking observations for latitude at
the mouth of Red Deer's Head River, on the night of the
2nd July, John McKay, a Scotch half-breed, observed
what he thought to be a wolf, approach the brow of a hill
about two hundred yards from us, and after apparently gaz-
ing at the encampment for a few minutes it retired beyond
view. The night was clear, and our tents being placed in
the valley of the river close to its junction with that of the
Souris, surrounded by steep hills about one hundred and
fifty feet high, an object appearing on the brow of those
in our rear could be seen projected against the sky.

McKay took no further notice of the strange visitor than to mention that he saw it and thought it was a wolf; but before we retired to our tents at 2 A.M. we noticed another figure, which he declared to be an Indian, appear near the same spot. Two of the party cautiously approached the foot of the hill, but before they reached it the figure crouched and slowly retired. The horses were gathered near the carts and a watch set, but daylight dawned without the re-appearance of the object of our suspicion. In the morning we endeavoured to discover tracks at the spot where it had appeared, but the hill being composed of gravel, the soil had received no impression which our most sharp-sighted half-breeds could detect. Having verified our observations on Polaris by a solar observation at noon, we started for a new camping-ground about twelve miles up Red Deer's Head River, where we proposed to take in a supply of wood for fuel, before crossing the great prairie to Fort Ellice. On our way thither the old hunter who had joined us at Prairie Portage said he *smelt fire;* we all strained our olfactories to the utmost, but without detecting any odour which might be supposed to proceed from a burning substance; nevertheless the old hunter persisted in the statement that he had " smelt it." We camped at sunset close to the river, and while at supper some of the party distinctly heard the distant neigh of a horse ; this of itself would have been considered sufficient warning, but when taken in connexion with the appearance of the object on the hill in the rear of our camp the night before, it was held to be conclusive evidence that we were watched by the Sioux, and that an attempt would be made in the night to steal our horses.

Our camp fires were put out immediately, the carts were placed close together, and a watch organized ; the half-breeds did not anticipate an attack until the approach of dawn, but the sudden galloping towards the carts soon after ten o'clock of several horses, who were feeding in the valley about one hundred yards from us, proved that Indians were near us. On hearing the horses approach, the men started up and ran to stop them, which they succeeded in doing before they passed the carts. Each horse was now tethered to a cart or stake, and the half-breeds crawling through the long grass arranged themselves in a half-circle about seventy yards from the carts, each with his gun loaded with buckshot. The night was dark, and perfect silence was maintained in the camp; towards morning one man came in to report: he stated that he had heard "something" cross the river and crawl through the grass within a few yards of him; he waited a few minutes for more to follow before he fired or gave the alarm, and then cautiously crawled through the long grass in track of the "something" which had passed near to him. The track led him to within thirty yards of our tents, and then turned towards the river, and evidently crossed it. Morning soon dawned, and the watchers came in; we examined the tracks described by the half-breed who had first heard the intruders, and they were pronounced to be those of an Indian. Further examination in full daylight showed that we had been surrounded by a band, who, however, perceiving we were on the alert, and that the horses were tethered, made no attempt to steal them. Had it not been for the old hunter's excellent nose, there is little doubt that we should have lost our horses during the night.

On the morning of the 4th, having loaded the carts with wood and taken a supply of water from Red Deer's Head River, which is here a rapid, clear stream, twelve feet broad, we started on a nearly due north course to cross the Great Prairie. The watermarks on the banks of Red Deer's Head River show that it rises fifteen feet during spring freshets, almost filling the low, narrow valley in which it flows. The banks are fringed with small balsam-poplar and aspen; patches of elm and oak occur on the points.

The prairie for many miles north of this river appears to be perfectly horizontal; in passing through it we always seemed to be in the centre of a very shallow depression, with a uniform and well-defined horizon in all directions. Early in the morning the distant outline meeting the clear sky was best defined; as the day wore on refraction magnified the tufts of grass and small willows into bushes and trees, destroying the continuity of the fine horizontal line where sky and earth seemed to meet. Occasionally the effects of mirage were very delusive, beautiful tranquil lakes suddenly appeared in the distance, and as quickly faded from our view. Fortunately the almost daily thunderstorms which had occurred replenished the marshes and small ponds, and gave us an abundant supply of water, but in some seasons the buffalo-hunters suffer much from the want of that necessary of life in crossing this vast treeless desert.

On the afternoon of the 5th we arrived at the northern limit of the burnt prairie, as far as we could judge; south of our point of view, the aspect of the vast level tract was of a dark green hue, with short grass of this year's growth; northwards the colour of the prairie was brown, from the old grass of last year which had not been consumed by the fires. Whenever we approached the old and shallow beds of brooks, boulders became numerous. Some of the little valleys contained ponds, occupying the shallow bed, all of which would probably be united, and form a river in the spring of the year.

Among the birds noticed during this monotonous journey were turkey buzzards, ravens, barking crows and black terns ; on the borders of several shallow ponds or marshes, which are often dry in the autumn, ducks were plentiful, and afforded us a grateful supply of fresh food.   We saw some herds of cabri, and McKay succeeded in killing a female after a long chase.   The grasshoppers were very numerous, and during four days filled the air like flakes of snow ; they rose simultaneously when about to take their flight, from areas of two to twenty acres in extent, first perpendicularly to the height of twelve or fourteen feet, then in a slanting direction, until they had attained an elevation of from two to three hundred feet, after which they pursued a horizontal course before the wind.   In a light breeze, the noise produced by their wings was like a gentle wind stirring the leaves of a forest.

Our half-breeds informed us that this great prairie west of the Souris continues treeless and arid for a distance of sixty miles, it is then crossed by a river, probably the Moose Mountain Creek, shown on Capt. Palliser's map ; beyond this river the prairie continues for eighty miles further without tree or shrub ; and as this was the utmost westerly limit to which any of them had journeyed in their buffalo-hunting expeditions, they could afford us no further information respecting its extent.   They were most of them familiar with the country south of the Great Prairie, the Grand Coteau de Missouri, where the buffalo range during the summer in vast herds.

On the 6th July we arrived at Pipestone Creek, and found the country swarming with a young brood of grasshoppers, with wings about a quarter of an inch long, showing that their progenitors had arrived in the preceding autumn in time to deposit their eggs in the soil. Innumerable hosts of these insects passed overhead during the day, and on looking up through an excellent marine glass, I could see them flying like scud at an immense height. Had it not been for the thunderstorms which daily refreshed and invigorated the herbage, it is probable that our cattle would have suffered seriously from the devastations of these insects.

Pipestone Creek is 20 feet broad at our crossing-place, with a swift current, and a depth of water varying from $1\frac{1}{2}$ to 3 feet. The valley is narrow, but rich and beautiful in comparison with the desolate prairie lying to the south. Among the trees fringing its banks the ash-leaved maple is most numerous, and the hop, together with the frost grape, is abundant on the edge of the stream. On the hills in its neighbourhood boulders are uniformly distributed, but on the highest a considerable number have been collected together by the Assinniboine Indians, and a rude monument erected in commemoration of a battle fought at a remote period.

The level character of the country disappears after passing Pipestone Creek ; the prairie is either undulating and sandy, or varied with low hills of drift, on which boulders are scattered. On the evening of the 6th we camped at Boss Hill Creek, which flows into the Assinniboine through a broad valley among low hills and gentle slopes. From a conical eminence near our camp, Boss Hill, Standing Stone Mountain, and the woods fringing Oak Lake are visible.

The " Standing Stone " is probably the same familiar object in these regions as mentioned by the Rev. John West, who traveled during the winter of 1821 to Brandon House and the mouth of the Qu'appelle on a missionary journey. Mr. West relates that he stopped to breakfast at the Standing Stone, where the Indians had deposited bits of tobacco, small pieces of cloth, and other trifles, in superstitious expectation that it would influence their Manitou to give them buffaloes and a good hunt. During Mr. West's journey, now forty years since, buffalo were very numerous in the winter months on this part of the Assinniboine, and many bands of the race of Indians bearing that name made this part of the country their winter quarters. Mr. West saw an Indian corpse staged about ten feet from the ground, at a short distance from Brandon House, a provision post now abandoned by the Hudson's Bay Company.

We arrived at the Assinniboine about ten miles southeast of the Two Creeks, after passing through a rolling prairie of light sandy soil, and in many places covered with boulders. Small "hummocks"* of aspens, and clumps of partially burnt willows, were the only remaining representatives of an extensive aspen forest which formerly covered the country between Boss Creek and the Assinniboine. So great had been the change during twenty years in the general aspect of this region that our old hunter, who had undertaken to guide us in a straight line across. the prairie from Red Deer's Head River, confessed that he did not " know the country" when within ten miles of the Assinniboine ; he nevertheless declared his conviction that we should strike the river at the point to which he had promised to lead us. He had not visited it for twenty years, and during that interval the timber, which formerly consisted of aspens and willows, had nearly all disappeared. The old man was correct; the face of the country had changed, the aspen forest had been burnt, and no vestige, beyond the scattered hummocks and burnt willow clumps, remained ; his " instinct," as he termed it, and that singular facility which practised prairie wanderers possess,

* A half-breed expression.

of journeying from point to point at great distances apart in a direct line, served him in lieu of memory or compass, for we struck the Assinniboine within two or three miles of the spot to which he had been directed to lead us.

The approach to this river is made by descending a steep slope, which forms the boundary of the prairie two or three miles from its present excavated valley. The plateau thus formed is covered with erratics, consisting of granite, gneiss, limestone, &c. The broad subordinate excavation in which the river flows is about one mile across, and from 200 to 250 feet deep. The narrow plateau covered with boulders points to a former condition of the Assinniboine valley, when a much larger river flowed in a wider and shallower valley 200 feet above its present level. The same remark applies to many other rivers in Rupert's Land, which, although now insignificant streams, yet flow through deep subordinate excavations in a broad but generally shallow and well defined trough with steep margins, erratics being dispersed over both margin and plateau. These records of former physical structure appear to indicate that the water once conveyed by these channels, must have been very largely in excess of the present supply.

On the morning of the 8th we passed through a good grazing country on the high prairie level, but being compelled to descend to the first plateau of the Assinniboine valley for water, we found our progress obstructed by a large number of erratics, which endangered the wheels of the carts. Here, however, we saw the first buffalo bull, and after a chase of half an hour's duration, succeeded in killing him. Although very tough and rather strong flavoured, he was an acceptable addition to our larder. On arriving at the second of the Two Creeks, cretaceous rocks were again recognised. They had the same litho-

logical aspect as those of the Souris ; organic remains
were scarce, but in sufficient numbers and variety to
establish their position. A band of soft yellowish-green
substance, resembling some varieties of soap-stone, was
observed forming a characteristic feature in the exposure
of the shales at this place. An analysis of the material
composing the green band is introduced in the Chapter on
the Cretaceous Series, Vol. II. In the low valley of the As-

Fort Ellice, Beaver Creek.

sinniboine, to which I descended, similar exposures arising
from land-slips were also seen. The section exposed was
capped by about ten feet of coarse gravelly drift on the
brink of the deep and broad excavation which now forms
the valley proper of the river. The shales resemble those
on the Souris, but contain fewer fossils, and are perhaps
more fissile and less impregnated with oxide of iron.

Three more bulls were seen on the following morning, but being anxious to reach Fort Ellice, and already

provided with meat, they were permitted to pass us
unmolested. The country in the neighbourhood of Beaver
Creek is undulating and attractive, but the soil is sandy,
capable only of supporting a short stunted herbage. We
arrived at the Fort on the morning of the 10th, and took
up our camping-ground on the banks of Beaver Creek,
close to the broad and deep valley of the Assinniboine.

Fort Ellice was at one period a post of considerable
importance, being the depôt of supplies for the Swan
River District, now removed to Fort Pelly. The buildings
are of wood, surrounded by a high picket enclosure. Mr.
McKay, one of the sub-officers, was in charge at the time
of our arrival. Some twenty years ago, before the small-
pox and constant wars had reduced the Plain Crees to
a sixth or eighth of their former numbers, this post was
often the scene of exciting Indian display. Formerly Fort
Ellice used to be visited by the Crees alone, now it numbers many Ojibways among the Indians trading with it.
The Ojibways have been driven from the woods by the
scarcity of game, the large animals, such as moose deer
and bear, having greatly diminished in numbers. Many
of the wood Indians now keep horses, and enjoy the
advantage of making the prairie and the forest tributary
to their wants.

On the 11th July, a number of hunters attached to Fort
Ellice came in with provisions, such as pemmican and dried
buffalo meat, which they had prepared in the prairies a
few days before, about thirty miles from the post, where
the buffalo were numerous. Fort Ellice, the Qu'appelle
post, and the establishment on the Touchwood Hills being
situated on the borders of the great Buffalo Plains, are
provision trading posts. The Hudson's Bay Company
obtain from the Plain Crees, the Assinniboines, and the
Ojibways, pemmican and dried meat to supply the brigades

of boats in their expeditions to York Factory on Hudson's Bay, and throughout the northern interior. Pemmican is made by pounding or chopping buffalo meat into small pieces and then mixing it with an equal quantity of fat. It is packed in bags made of the hide of the animal, in quantities of about ninety pounds each. Dried meat is the flesh of the buffalo cut into long, broad, and thin pieces about two feet by fifteen inches, which are smoked over a slow fire for a few minutes and then packed into a bale of about sixty pounds. We had many opportunities of seeing the Cree women on the Qu'appelle, cut, prepare, and pack dried meat.

At Fort Ellice (longitude 101° 48′, latitude 50° 24′ 32″, Captain Palliser) the thunder storms were as violent as on the Souris; not a day passed without lightning, thunder, and generally violent rain of half an hour's duration. The grasshoppers at this post had destroyed the crops last year, and, at the time of our visit, the young brood were well advanced, their wings being about one-third of an inch long. Full grown insects from the south were flying overhead or alighting in clouds around us, so that all hopes of obtaining a crop from the garden or potato fields were abandoned for this year. Provisions were very scarce at the post, and had it not been for the fortunate arrival of the hunters with some pemmican and dried meat, we should have been compelled to hunt or kill the ox.

From Mr. McKay I received a particular account of the "Great Bones" on Shell Creek, which had long been a source of wonder and awe to the Indians hunting on the left bank of the Assinniboine, and whose magnificent descriptions led me to suppose they might belong to a cetacean, and were worth a day's journey out of our track to visit and examine. They were seen many years ago

protruding from the bank of Shell Creek, 20 feet below
the prairie's level.   Mr. McKay instructed some of the
hunters attached to the post to bring them to him, but no
Indians would touch them, and the half-breeds only
brought a tooth and collar bone, which were stated by a
medical gentleman to whom they were shown to have
formerly belonged to a mammoth.   Mr. Christie, of Fort
Pelly, we were told, went to Shell Creek, with a view to
collect more specimens; he obtained some ribs, but in a
state of crumbling decay; they were sent to Red River
Settlement.   The Indians had long regarded these ancient
relics as the bones of a Manitou and worthy objects of
veneration.   An old Indian on Dauphin Lake, to whom
reference will be made hereafter, described similar bones
in the banks of Valley River leading to Dauphin Lake;
but the season was too late when exploring that part of
the country to permit of an examination.

On Monday, the 12th, preparations for continuing our
journey westward were completed, by engaging an Indian
to assist in paddling Mr. Dickinson down the Qu'appelle
or Calling River from the Mission to its junction with the
Assinniboine.   The half of his wages he stipulated to
have in advance.   Mr. McKay told me he was a bad
Indian and not to be trusted, but we could not succeed
in getting another.   When on the point of starting, a
young Ojibway, painted and adorned with feathers, galloped up to the post, entered the room, drew from beneath
his moose skin robe two moose tongues and a mouffle,
which he quietly handed to Mr. McKay, and, squatting
on the floor without speaking a word, lit his pipe.   After
a few minutes, he informed us that he and his father had
killed two moose, thirty miles off, and desired McKay to
send for them.   Two half-breed hunters also arrived at
this moment, in sad plight, hungry and tired, with worn

horses and torn clothes. They had come from Fort Union, on the Missouri, having been hunting on the Grand Coteau, where they met a war party of sixty Blackfeet. They fled to the fort, the Blackfeet pursuing them and insisting that the Fort Union people should give them up, a request which was promptly refused.

During the night, the Fort Union people furnished them with a small supply of provisions, and leading them out to the prairies, told them to run for it; they did so, and arrived in safety at Fort Ellice after a harassing journey.

At 4 P.M. on the 12th July, we left Fort Ellice and traveled due west through a pretty country near the banks of the Qu'appelle or Calling River. After breakfast on the following day we arrived at the Cross Woods, which, according to our half-breeds, extend as far as Pipestone Creek; they consist of aspen, with a splendid undergrowth. Here we observed during the morning the grasshoppers descending from a great height perpendicularly, like hail—a sign, our half-breeds stated, of approaching rain. Our route lay through a rolling country, the soil consisting of sandy loam with much vegetable matter in the valleys. Aspen groves are numerous, and many little lakes margined with reeds afford quiet breeding-places for duck. The road is good in summer, but wet and soft in the spring.

The grasshoppers were excellent prognosticators: a violent thunder storm in the afternoon commenced in the east (all preceding storms had come from the west), and was accompanied by exceedingly heavy rain and a very boisterous wind. The storm continued several hours. At 9 in the evening, the air was calm and the heavens clear and bright; at 10, the storm returned from the west, and a more terrific and sublime exhibition of elemental warfare none of us had ever before witnessed. Three times the

lightning struck the earth so close to us that there was no
perceptible interval between the flash and the shock.   It
was distinctly heard to *hiss* through the air, and, instead
of penetrating the ground at once, it seemed to leap from
bush to bush for a distance of 60 or 70 yards.   So close
did one flash approach that when we had recovered from
the shock, and our eyes had regained their power, several
of us met each other, groping from cart to cart, to see if
any of the party had been struck.   It is remarkable that
although the wind was blowing violently before and after
the two flashes just described occurred, yet, between them,
an interval of about three-quarters of a minute, there was
a dead calm, and a calm of short duration succeeded each
flash in our immediate vicinity.

The trail on the 14th continued through good land for
nine miles, with aspen groves on the crown of each undu-
lation, and willow bushes in the hollows, it was succeeded
by a prairie three miles across, but of much greater extent
longitudinally.   Ponds were numerous, abounding with
ducks and ducklings.   Another rain and thunderstorm
on the evening of this day lasted for about an hour.   On
the following morning we entered a treeless prairie marked
at its western extremity by a sandy ridge running N.W.
by S.E., known among the Indians as the Weed Ridge.
It was covered with the bear-berry from which kinni-
kinnik is made.   This was the first time we saw this
weed since leaving the Sandy Hills of the Assinniboine.
The Indians of the prairies generally use the inner bark
of the *Cornus sericea*, the red-barked willow as they term
it.   We also saw them smoke the inner bark of the dogwood, *Cornus alternifolia.*

The mode in which these barks are prepared is very
simple.   A few branches about three-quarters of an inch
thick and four or five feet long are procured, and the

outer bark scraped off after having been warmed over a
fire; a knife is then pressed against the inner bark and
drawn upwards, for a space of six or eight inches, until
the whole of the inner bark is gathered in curly clusters
round the stick, it is then thrust in the ground over the
embers and roasted until quite dry, when, mixed with
tobacco in equal proportions, it forms the favourite kinni
kinnik of the North-West Indians. I often saw them
smoke bark or the leaves of the bear-berry alone, when
their supply of tobacco was exhausted.

The Indian who accompanied us from Fort Ellice to
the Qu'appelle Mission, complained of weakness and pain
in the chest, he suffered much from cough, and was evidently consumptive; he was treacherous and indolent,
and, as will be shown hereafter, soon left us in the lurch.

Beyond the Weed Ridge the country is very undulating;
boulders of Silurian limestone and gneiss were strewed on
the flanks and summits of the hills. The white or whoop-
ing crane (*Grus Americana*) was first seen to-day. This
beautiful bird is common in the Qu'appelle Valley and in
the Touchwood Hill range. It is a dangerous antagonist
when wounded, striking with unerring aim and great
force with its powerful bill. When the bird is wounded,
the best way to avoid its attacks is to present the muzzle
of the gun as it approaches, it will fix its bill in the barrel
and may then be destroyed without danger. Instances
have been known of this bird driving his bill deep into
the bowels of a hunter when not successful in warding
off its blow. Magpies are numerous on the Weed Ridge,
and the cat bird is heard in every little wooded dell.

On the 15th we passed two streamlets flowing into the
Qu'appelle. Their banks were fringed with small timber,
and quite lively with birds, which are far more abundant
here than on the Souris. On the borders of all the wooded

brooks we saw magpies, cat birds, crows, and, occasionally, the solitary thrush; in the wet prairies, the rice bird, black tern, plover, the golden-legged and common, the yellow-headed blackbird, common meadow lark, chipping sparrow, and grackle; on ponds and in marshes, ducks of many species, bittern (*Ardea lentiginosa*), and cranes. In the morning, after a clear night, we always observed heavy dew; this phenomenon was not so frequently noticed on the Souris under similar circumstances. There can be little doubt that the sterility of the Great Prairie between the Qu'appelle and the 49th parallel is owing to the small quantity of dew and rain, and the occurrence of fires. North of the Qu'appelle, the country seemed to be far more humid, and the vegetation far richer and more abundant in many localities than south of that great valley.

Another prairie eight miles broad, bounded by ridges having a N. W. and S. E. direction, succeeded to the one last described, and introduced us on the 16th to a hilly country; the range is called the Indian Head Hills; it contains many beautiful lakes, is well wooded, and forms one of the northern spurs of the Moose Mountain, whose position is given on Captain Palliser's map. The northern slope of the Indian Head Hills is very abrupt, the southern, gentle and undulating. Here we met with Charles Pratt and party going to Red River. Charles Pratt is a half-breed catechist of the Church Missionary Society, well acquainted with the habits of Indians and of buffalo, but apparently scarcely sensible of the importance of his duties and the responsibility of his charge. He gave me a good deal of valuable information respecting the country, and, with characteristic generosity, if not Christian sympathy, told John McKay to take a young heifer belonging to him when we arrived at the Mission and kill it in honour of our arrival. Pratt showed me some specimens of lignite which he had taken from a bed two feet thick at the Wood Hills, about eighty miles south-west of the Hudson's Bay Company's post. He described the hill or range of hills as an island in the prairie, which, like the Stony Mountain near Red River, had escaped denudation, or, it may have been La Roche Percée, visited by Mr. F. W. Lander in 1853*, and by Dr. Hector in 1858.†

An old Indian accompanying Charles Pratt, born in this part of the country, told us that he remembered the time when the whole of the prairie through which we had passed since leaving Fort Ellice was one continuous forest, broken only by two or three narrow intervals of barren ground. They told us that the Plain Crees and the Blackfeet were at war with one another, and that the Crees were hurriedly "pitching east" in order to avoid the Blackfeet. This intelligence had considerable effect upon our old hunter; he made his will and gave it to Charles Pratt to give to Archdeacon Cochrane of Prairie Portage.

The view from the Indian Head range is exceedingly beautiful; it embraces an extensive area of level prairie to the north, bounded by the Aspen Woods on the borders of the Qu'appelle Valley. A portion of the old forest alluded to by the Indian still exists on this range. It consists of aspen of large growth and very thickly set. A few cabri (prong-horned antelope) were seen in the Indian Head range; they used to abound in the country drained by the Qu'appelle.

On Saturday the 17th we entered a very beautiful and fertile prairie at the foot of the Indian Head range, our course leading us in a northerly direction to the Qu'appelle Mission. The common yarrow was very abundant,

* Pacific Railway Explorations.
† Blue Book. Capt. Palliser's Explorations.

and with the hare-bell reminded us of other scenes far away. Six miles from the hills we arrived at a subordinate, shallow, and broad valley, parallel to that of the Qu'appelle. The aspect of its boundary suggested the shore of a lake or bank of a large river. The lower prairie consisted of a sandy loam, in which the Indian turnip was very abundant. We soon came up˙ with a group of squaws and children from the Qu'appelle Lakes who were gathering and drying this root, which the Crees call the *Mis-tas-coos-se-ne-na* or big grass root. The French half-breeds call it the pomme de prairie; the Sioux *Tip-si-na*. It is an important article of food in these regions. The botanical name is *Psoralea esculenta*. Many bushels had been collected by the squaws and children, and when we came to their tents they were employed in peeling the roots, cutting them into shreds and drying them in the sun. I saw many roots as large as the egg of a goose, and among those brought with me to Canada are some of even larger dimensions. The Crees consume this important vegetable in various ways ; they eat it uncooked, or they boil it, or roast it in the embers, or dry it, and crush it to powder and make soup of it. Large quantities are stored in buffalo skin bags for winter use. A sort of pudding made of the flour of the root and the mesaskatomina berry is very palatable, and a favourite dish among the Plain Crees.

We reached the Qu'appelle Lakes at 6 P. M., after passing through a magnificent prairie the whole day ; in fact the country north of the Indian Head and Chalk Hill ranges is truly beautiful, and will one day become a very important tract. The Chalk Hills are a continuation of the Indian Head range. In the language of the Indians they contain bands of " soft white earth or mud." The half-breeds call them " Chalk Hills." It is a matter of regret that the time at our disposal did not permit us to make an excursion to them, notwithstanding that no indications of rocks in position were seen on the Indian Head range; they were recorded as composed of drift which may or may not conceal rocks in position above the general level of the prairie north of them.

Great was our astonishment on arriving at the Qu'appelle Lakes to find that they were narrow bodies of water, occupying an excavated valley about one mile broad, 250 feet deep, and differing in no important particular from the same valley at its junction with the Assinniboine —120 miles distant by the river, or 134 by the trail. The importance of the Qu'appelle valley began to develope itself when the Crees at the Lakes informed us that it continued through to the Saskatchewan without losing its breadth, and maintained, except for a short distance, a great depth below the prairie level. I determined, therefore, to explore the whole valley from the South Branch of the Saskatchewan to the Assinniboine, and to ascertain the relation it bore to those rivers. With this view the canoes were put in order, the party and supplies divided, and the arrangements detailed in the following paragraph completed.

Mr. Dickinson, with a French Canadian and a Cree half-breed, was instructed to descend the Qu'appelle River from the first Fishing Lake to its mouth. Mr. Fleming and myself were to ascend it from the same starting place to its source, and follow up the valley to the South Branch of the Saskatchewan. Mr. Hime was to explore Long Lake and meet Mr. Dickinson at Fort Pelly. I intended, upon reaching the South Branch, to descend that magnificent river in canoe to the Grand Forks, and then by the main Saskatchewan to Lake Winnipeg and Red River, a distance of about 1000 miles canoe navigation.

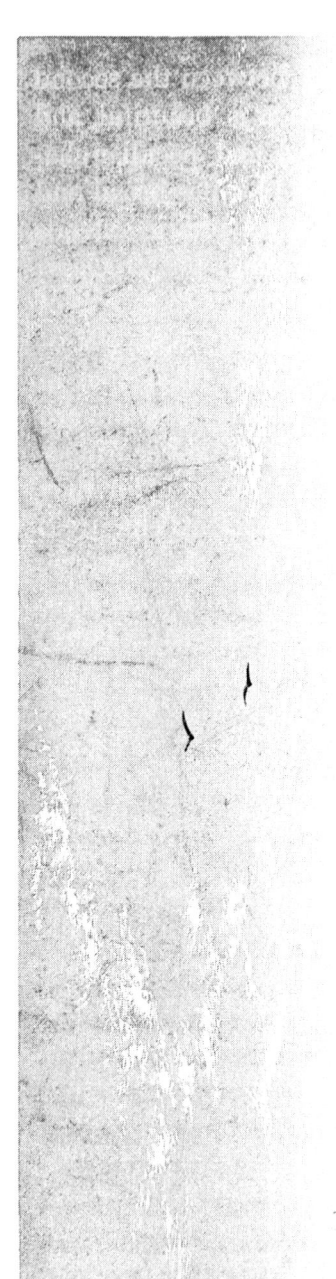

FISHING LAKES, QU'APPELLE RIVER.

Printed by Spottiswoode and Co.]

[New-street Square, London.

The Qu'appelle Mission is situated between the second and third Fishing Lakes. The situation is beautiful, and the country on all sides of a very novel and peculiar character. Here the Qu'appelle valley is $1\frac{1}{4}$ miles broad and 250 feet deep. On the south a vast level prairie extends to the Indian Head Hills, fertile, inviting, but treeless; towards the north the country is studded with groves of aspen over a light and sometimes gravelly soil. Most beautiful and attractive, however, are the lakes, four in number, which from the rich store of fish they contain, are well named the Fishing Lakes. A belt of timber fringes their sides at the foot of the steep hills they wash, for they fill the entire breadth of the valley. Ancient elm trees with long and drooping branches bend over their waters; the ash-leaved maple acquires dimensions not seen since leaving the Red River, and the Me-sas-ka-to-mi-na (la Poire) (*Amelanchier Canadensis*) is no longer a bush, but a tree eighteen to twenty feet high, and loaded with the most luscious fruit.

The Qu'appelle Mission was established last year (1858). For some time past, however, Charles Pratt, the catechist, has resided where the Mission is situated, and has constructed a comfortable log-house, fenced in a garden, and now possesses six or seven cows and calves. An old halfbreed, whose name is obliterated in my note-book, took up his residence with Pratt; he had been engaged for the better part of his life at different fishing stations belonging to the Hudson's Bay Company throughout Rupert's Land, and he declared that in all his experience he had never seen the white fish (*Coregonus albus*) so large, numerous, and well flavoured as in the Qu'appelle Fishing Lakes.

The Rev. James Settee, the missionary, a native of Swampy Cree origin, occupied Pratt's house; he arrived at the Mission last autumn. In the garden where we found him, Indian corn was growing, as well as potatoes, turnips, beans, and other culinary vegetables. The grasshoppers had not yet visited the Mission, but vast flights had passed over it. They were seen passing the Company's post, twenty miles south, on the 8th of the month; they were then flying to the east. They had passed the Mission in 1857, for they visited the Touchwood Hills forty to fifty miles north of it, depositing their eggs in the ground, and during the present summer the young brood, as I learned a few weeks afterwards, destroyed all garden crops at the Touchwood Hills, and on the 28th July took their flight to the south-east.

On Sunday we attended service in Pratt's house; the Rev. Mr. Settee read the prayers in English with great ease and correctness; he preached in Ojibway; and a hymn was sung in the Cree language. Before the sermon the missionary surprised us by waking up a drowsy Indian, who was enjoying a quiet nap in a corner of the room, and leading him to the temporary reading desk, commenced the ceremony of public baptism. My astonishment was not diminished when the reverend gentleman turning to me, without any preliminary notice, said abruptly, " Name this man!" After a moment's reflection I said " John," and without any unnecessary loss of time or words, "John" walked to his bench, and was soon apparently lost, in noisy slumber, to all consciousness of the privileges and blessings of which adult Christian baptism, duly received, had made him the inheritor.

When the Rev. James Settee arrived at the Mission last autumn, the Crees of the Sandy Hills having received intelligence that the bishop had sent a " praying man" to teach them the truths of Christianity, directed messengers to inquire whether " the great praying father had sent plenty of rum; if so, they would soon become followers of the white man's Manitou." The messengers returned with the intelligence that the great praying father had not only omitted to send rum, but he hoped that the Plain Crees would soon abandon the practice of demanding rum in exchange for their pemmican and robes. The messengers were directed to return to the missionary with the announcement, that " if the great praying father did not intend to send any rum, the sooner he took his praying man away from the Qu'appelle Lakes the better for him."

Encampment in the Qu'appelle Valley.

There are very few tents about the Mission at present, the Indians being in the plains engaged in hunting the buffalo. Mr. Settee speaks English very fluently, the field for his labour is extensive, but not at present promising. When conversing with the Crees of the Sandy Hills, many of them expressed a wish to have their children taught by white men, but they did not appear to like the idea of their being taught by a native of a different origin. This is an important point to be observed in the selection of native missionaries. The school, however, appears here, as elsewhere among Indian tribes, to be the only sure ground for establishing the true faith among them. "Teach my children for two or three years, but let me follow the ways of my fathers," said the son of the chief of the Sandy Hills to me. Many expressed a wish that their little ones should know the white man's cunning, and learn to cultivate the soil, but they would stipulate to remain themselves still the wild prairie Indians, hunting the buffalo, and occasionally tasting the savage excitement of war.

It is a wrong policy to send a Swampy Cree among the Plain Crees, or an Ojibway amongst the Crees, as a teacher and minister of religion. These highly sensitive and jealous people do not willingly accept gifts or favours which involve any recognition of mental superiority in the donor from one not of their own kindred, language, and blood; although he may be of their own race. An Ojibway remains always an Ojibway, and a Swampy Cree a Swampy Cree, in the eyes of the haughty and independent children of the prairies, and they will never acknowledge or respect them as teachers of the "white man's religion."

Several of my half-breeds appeared to think that Mr. Settee would have troublesome times, and that he would not be able to make much impression among the Plain Crees. What has to be done must be done soon, the time is short and the race is fast passing away; in another generation we shall probably lament the disappearance of a tribe which twenty years ago could muster one thousand mounted warriors, and who, in all the pride and barbarous pomp of Indian display, were accustomed to approach the Hudson's Bay Company's posts singing their Hi-he-ah, Hi-he-ah, or war. song, in savage unison.

On the 20th July the canoes were launched on the Third Fishing Lake, and having seen Mr. Hime *en route* for Long Lake, my carts and horses on the way to the Grand Forks of the Qu'appelle, and Mr. Dickinson started for the mouth of the river, I embarked with Mr. Fleming, an Ojibway, and a Cree half-breed, and paddled up stream with a view to trace out the valley to its junction with the South Branch of the Saskatchewan. The succeeding Chapter contains a narrative of this exploration, which is followed by Mr. Dickinson's description of his canoe voyage to the Assinniboine. We arranged to meet at Fort Ellice forty-three days after our simultaneous departure from the Third Fishing Lake.

CHAP. XV.

THE QU'APPELLE VALLEY. — FROM THE MISSION TO SAND HILL LAKE.

THREE quarters of a mile from the mouth of the little stream joining the second and third Fishing Lakes, the lead showed 44 feet of water. This great depth surprised us, as we had been paddling, since leaving the Mission, in shallows not exceeding four and five feet in depth. Cross sections subsequently made, showed that the lakes were generally deep on the north and shallow

on the south side. An abundant growth of green confervæ covered the surface, which, in its aggregations and
general distribution, reminded me of a similar profusion
on the Lake of Woods during August, in 1857. The hill
sides of the valley are deeply ravined and wooded, but
the hills they separate are bare ; we soon noticed too
that the north side began to show far less timber than
the south, and of more stunted growth. The snowberry
was seen in every hollow. Ash-leaved maple and elm
were numerous on the south side of the lake, together
with the mesaskatomina.

Two excellent photographs, taken near the Mission, of
the lakes and hills, display the chief characteristic of the
valley with the fidelity which can only be attained by
that wonderful art.

Soundings near the middle of the lake showed 56 feet,
which when added to 249 feet, the depth of the valley
below the prairie as ascertained by trigonometrical measurement, make the total excavation 305 feet. Another
sounding 200 yards from the N.W. point, gave 57 feet of
water. This was the greatest depth we obtained, but Mr.
Dickinson found the lower lakes to be 66 feet deep. The
shores of gravel are strewed with blocks of drift limestone
and the unfossiliferous rocks. Gulls are numerous about
these remote lakes, and a pair of eagles have had their
eyrie for many years in a fine elm tree, near the west
end of the Third Fishing Lake. The hop grows very
luxuriantly in the thin belt of woods on the south side,
and the frost grape hangs in beautiful festoons from the
drooping branches of the elm. The water mark shows
that this lake rises from six to seven feet above its present
level.

A low plateau, inundated every spring, separates the
Third from the Fourth Lake. It is the delta of two ravines

which in the spring and autumn, bring down a large quantity of water from the prairie above. Third Fishing Lake is connected with Fourth Fishing Lake by a rapid stream flowing through the plateau, about 100 feet broad. At its mouth we saw a large number of fish rising at the grasshoppers, which dropped from flights of these insects passing over at the time. In the same stream were many large fish, and among them several individuals of a species to which further reference will be made. Soundings in the Fourth Lake showed 54 feet; this depth was maintained for a long distance with great regularity. In fact, these lakes appear to be nearly uniformly deep, and point to an excavating force, or peculiarity of rock formation deserving of further inquiry. The deltas at the mouth of the ravines, coming in from the prairie at right angles to the general course of the valley, give a clue to the mode in which the lakes were separated one from the other. It is very probable that they were once all united.

Geese appeared in large numbers in the Fourth Lake, and at its western end we saw a splendid flock of pelicans containing thirty-five individuals; as we approached they sailed majestically round and round, but took flight before we arrived within gunshot. Magpies are very numerous in the thin woods fringing the lakes; so also are grackles, the cat bird, and many smaller birds. The Fourth Lake is very shallow at its western extremity, six feet being the greatest depth recorded. The hills on the north side are quite bare, and trees on the south side are found only in the ravines. The lake is full of weeds and its water emits a very disagreeable odour, but the watermarks show, that during spring freshets, its level is eight feet higher than in the summer season. This is an important fact when taken in connection with the alleged

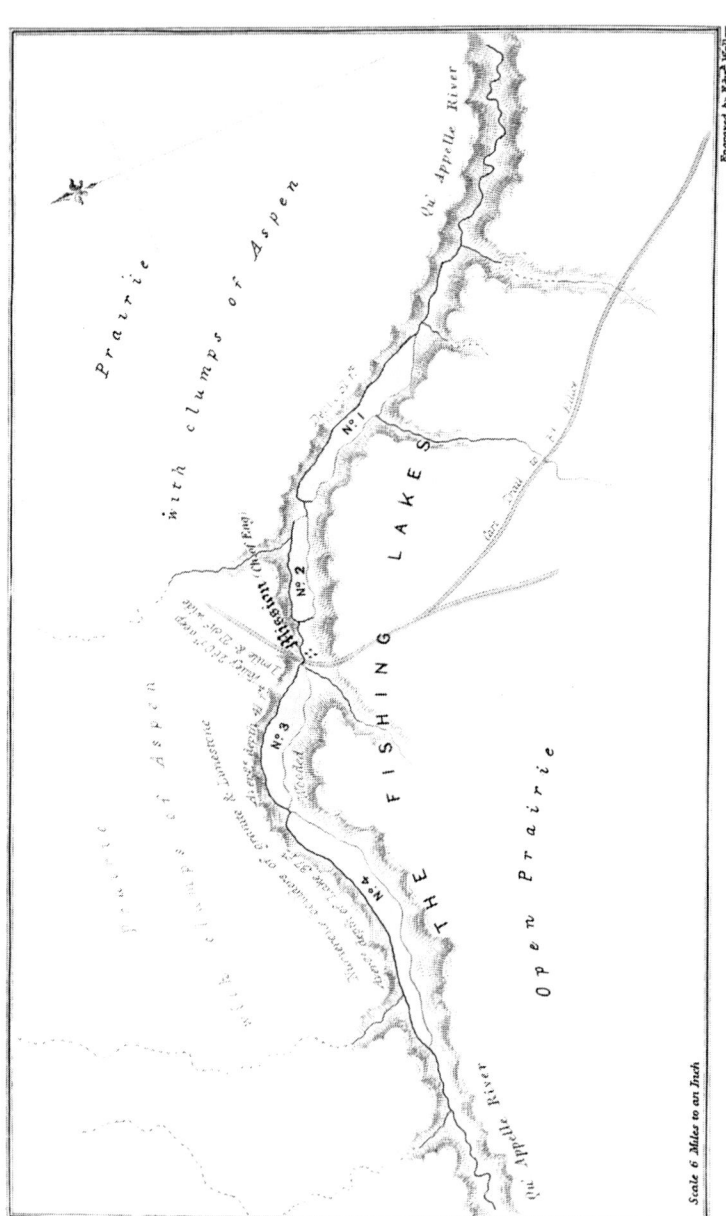

Prairie

with clumps of Aspen

Qu Appelle River

Nº 1

Nº 2

Plateau

1 mile R 2¼ mile

Clumps of Aspen

Nº 3

Nº 4

THE FISHING LAKES

Open Prairie

Qu Appelle River

Engraved by Edwᵈ Weller.

Scale 6 Miles to an Inch

Plan of the Fishing Lakes, Qu Appelle River.

appearance of the whole valley during wet springs; it is then said to resemble a broad river from a few miles east of the Saskatchewan to the Assinniboine. In 1852, a year memorable in Rupert's Land for the great floods which covered an immense tract of country, the Indians represent the Qu'appelle Valley as filled with a mighty river throughout its entire length, flowing with a swift current from the lakelets at the Height of Land, soon to be described, to the Assinniboine, and as a mountain torrent through the short distance of twelve miles, which separates them from the South Branch of the Saskatchewan.

After leaving the Fourth Lake and the marshes at its west extremity, we paddled, sailed, or tracked up a narrow swift stream, four or five feet deep, seventy feet broad, and winding through an alluvial flat in a valley of undiminished breadth and depth. The hill sides were now absolutely bare, not a tree or shrub was to be seen. We had reached the point where timber ceases to grow in the valleys of the rivers except in peculiar situations; the altitude of the banks could not be less than 280 or 300 feet. The prairie on either side is also treeless and arid. On the 21st, after spending a restless night owing to the attacks of multitudes of mosquitoes, we left the canoe in the hands of our half-breeds to track up the stream, and, ascending to the prairie, walked for some miles on the brink of this great excavation. We waited five hours for the canoe to reach us, the windings of the stream involving a course three times as long as a straight line up the valley. The hill sides here began to acquire a more imposing altitude, and probably exceeded 300 feet. White cranes appeared in flocks of four and seven together, but they were so wary that it was impossible to approach them.

The river was often seen to draw near to either side of

the Great Valley, having excavated a channel 10 to 12 feet deep in the alluvial flats through which it pursued its tortuous course. Its banks revealed the following section : —

    6 inches light vegetable mould with sand,
    4 inches yellow clay,
  10 inches light vegetable mould (former surface),
    9 feet yellow clay,
    2 to 3 inches ferruginous sand to the level of the river.

The last layer was hard, compact, and very coarse-grained. The river is here 60 feet broad, and flows at a rate of a mile and a half an hour. At the mouth of Long Creek, an insignificant affluent, the hills are covered with limestone and granitic boulders ; the north side is treeless like the vast prairie beyond it, the south side has aspens in the ravines and aspen groves on the prairie. The width of the valley remains uniform, never exceeding one mile and a quarter or being less than one mile. The pasturage in the flats is superb, the grass being long and very thickly set. Robins, magpies, and yellow birds enliven small aspen groves on the south side, or the thickets of cherry, mesaskatomina, dogwood, and snow-berry, which fill the hollows and ravines ; the cat bird is also common, and the tyrant flycatcher everywhere. In the river were vast numbers of ducks and geese ; the young birds frequently made us an excellent meal ; but no four-footed animals were seen, with the exception of one prong-horned antelope and one prairie hare.

In the afternoon of this day we made many miles by sailing before a strong east wind, notwithstanding a heavy rain and thunder-storm ; we were glad to be able to push on through this seemingly interminable and now mono-

tonous valley, as the air from the marshes on either side
of the river was fetid and oppressive. A scramble to the
summit of the steep hill bank, 300 feet high, though very
fatiguing, was amply repaid by the cool, pure and delightful breeze blowing over the desolate prairies around
us. Roses of three different varieties, red, white, and
variegated, were numerous on the upland; and in the
morning, when the dew was on them, or at night, when
it was falling, the fresh air from above came down in
puffs into our deep, hot valley with delicious and in-
vigorating fragrance. On the fourth day after our
departure from the lakes we sighted the Grand Forks;
leaving the canoe, I hastened on to a point where the men
with the carts and horses were to await our arrival, and
found them safely encamped on a beautiful meadow,
anxiously looking for us. An empty cart and a couple of
horses were despatched for the canoe, still some miles
below us, and in the evening we were joined by Mr.
Fleming and the two voyageurs.

During the day the temperature of the River was found
to be 74°. At the mouth of a dry bed of a stream which
we called Maple Creek, some very old trees of the ash-
leaved species were observed. Many of them showed
marks where they had been tapped. The willows which
fringed the banks of the Qu'appelle were barked by ice
eight feet above the surface of the water. Numerous
buffalo tracks began to appear before we reached the
Forks, and where these animals had crossed the river,
they had cut deep roads to the water's edge, and lanes
through the willow bushes. The bones of many a young
bull and cow were seen sticking out of the banks where
they had been mired.

The tortuous character of the stream before we took
the canoe out of the water, may be imagined from the

fact that eleven hours constant, steady tracking enabled us to progress only five miles in a straight line through the valley, and not less than 200 courses and distances were recorded in the canoe. Some little time was lost in crossing from one side to the other in order to avoid the willow bushes, which only grew on the inside of a bend, rarely or never on the outside or longest curve. The breadth of the river where we left it was forty feet, and the speed of its current one mile and a quarter an hour. The fetid air from the marshes made most of the party feel unwell, and I therefore determined to carry the canoe in a cart on the immediate edge of the prairie, keeping the valley in constant view, and occasionally descending into it and crossing it, to ascertain by levelling and measurement its leading dimensions.

No rock exposure was anywhere to be seen; drift appears to cover the country to a great depth. Where land slips have occurred and exposed an almost perpendicular section, yellow gravelly clay alone is visible. Some of the limestone erratics strewed over the sides of the ravines resemble those frequently seen on the south-east side of Lake Winnipeg.

Near our camp, on the 23rd, were six or seven log-houses, occasionally inhabited during the winter months by *freemen*, that is, men no longer in the service of the Company. The prairie above the freemen's houses slopes gently to the edge of the valley from the distant horizon on both sides. Clumps of aspen vary its monotonous aspect, and though clothed with green herbage, due to the late abundant rains, the soil is light and poor. Some distance back from the valley it is of better quality, the finer particles not having been washed out of it; the grass there is longer and more abundant, but the greatest drawback is the want of timber.

Since we have been on the Qu'appelle we have frequently noticed thunderstorms towards the north-west
and north, in the neighbourhood of the Touchwood Hill
range, which did not reach us ; the day before yesterday
(21st July) a very violent thunderstorm in the Qu'appelle
valley, which delayed us for several hours, did not wet
the carts ten miles to the south.   Rain clouds appear to
follow the Touchwood Hill range ; the frequency of
storms in that region is proverbial, and the richness of
the vegetation there proves that an abundant supply of
rain falls during the hot summer months.

Soon after sunset our camp received an unexpected
addition of six " Bungays," * who were on their way to
Fort Ellice with dried buffalo meat and pemmican.

They had been hunting between the two branches of
the Saskatchewan, and represented the season as very dry
and the buffalo scarce.   We passed a quiet and friendly
night with them, and on the following morning made
them a small present and pursued our way to the Grand
Forks.

I happened to be about 100 yards in advance of the
carts, after we had traveled for about a quarter of an
hour, when hearing a loud clatter of horses' feet behind
me, I looked round and saw the six Indians approaching
at a gallop.   One of them, who had represented himself as
a chief, seized my bridle, drew the horse's head round,
and motioned me to dismount.   I replied by jerking my
bridle out of the Indian's hand.   My people came up at
this moment and asked in Cree what this interference
meant.   " We wanted to have a little more talk," said the
chief ; " we are anxious to know the reason why you are
traveling through our country."   It turned out after a

* Crees and Ojibways of mixed origin.

little more " talk " that they wished to establish a sort of
toll of tobacco and tea for permission to pass through
their country, threatening that if it were not given they
would gather their friends in advance of us, and stop us
by force.   We knew that we should have to pass through
about 100 tents, so there was some little meaning in the
threat.   The old hunter, however, knowing Indian habits
and diplomacy well, at once remarked that we were
taking a large present to the chief of the Sandy Hills, and
we did not intend to distribute any tobacco or tea until
we had seen him, according to Indian custom.   They
tried a few more threats, but I closed the parley by
unslinging a double-barrelled gun from the cart, and
instructing the men to show quietly that they had theirs
in readiness.   Wishing the rascals good day, we rode on ;
they sat on the ground, silently watching us, but made
no sign.   In the evening one of them passed near us at
full gallop, towards some tents which we saw in the distance as we ascended the hill at the Grand Forks.

One rather significant statement they made proved to
be correct, namely, that the Plain Crees, in council assembled, had last year " determined that in consequence
of promises often made and broken by the white men and
half-breeds, and the rapid destruction by them of the buffalo they fed on, they would not permit either white men
or half-breeds to hunt in their country or travel through
it, except for the purpose of trading for their dried meat,
pemmican, skins, and robes."

We crossed to the north side of the Qu'appelle when
we arrived at the Grand Forks, and ascended the hill
bank to the prairie.   The Grand Forks consist of the
junction of two deep and broad valleys bearing a great
resemblance to each other ; the south valley is that in
which the Qu'appelle river flows, the other is occupied

by Long Lake, or Last Mountain Lake, forty miles in
length, and from one-half to two miles broad, being in
fact an exact counterpart of the Qu'appelle Valley and
Lakes. It is narrow, deep, filled throughout with water,
and is said to inosculate with the South Branch of the
Saskatchewan, some miles below the Elbow. In its general aspect Last Mountain Lake is similar to the Fishing
Lakes. A rapid, winding stream, 30 feet broad, runs
from it into the Qu'appelle. Both valleys are of uniform
breadth and depth, and very little narrower than, when
united, they form the main valley of the Qu'appelle river.
From the Grand Forks to the Souris Forks (Elbow Bone
Creek) the country is treeless, slightly undulating and
poor. The Indians say that the Souris River of the
Qu'appelle, coming from the Grand Coteau de Missouri,
inosculates with an arm of the Souris of the Assinniboine
before described, and that a canoe in high water might
pass from one river to the other without a portage. If
this be the case, the diversion of the waters of the South
Branch down the Qu'appelle valley would acquire additional importance, and give value to an immense extent
of territory, now comparatively inaccessible, and very in-
sufficiently watered.

A few miles west of the Souris Forks the Qu'appelle is
19 feet wide and 1½ feet deep, but the great valley is still
a mile broad and 200 feet deep. Here on the 25th we
caught a glimpse of the blue outline of the Grand Coteau,
with a treeless plain between us and the nearest part which
is called the "Dancing Point of the Grand Coteau;"
and has long been distinguished for the "medicine ceremonies" which are celebrated there. After passing these
forks, the country is more undulating, small hills begin to
show themselves ; the general character of the soil is light
and poor ; the herbage consists of short tufted buffalo

grass, and plants common in dry arid plains.   This afternoon we saw three fires spring up between us and the
Grand Coteau.   They were Indian signs, but whether they
referred to the presence of buffalo, or whether they were
designed to intimate to distant bands the arrival of sus-
picious strangers, we could not then tell, and not knowing
whether they originated from Crees, Assinniboines, or
Blackfeet, we became cautious.   In a few days we
ascertained that the fire had been put out* by Crees, to
inform their friends that they had found buffalo.

The grandeur of the prairie on fire belongs to itself.   It
is like a volcano in full activity, you cannot imitate it, because it is impossible to obtain those gigantic elements
from which it derives its awful splendour.   Fortunately,
in the present instance the wind was from the west, and
drove the fires in the opposite direction, and being south
of us we could contemplate the magnificent spectacle
without anxiety.   One object in burning the prairie at this
time, was to turn the buffalo ; they had crossed the
Saskatchewan in great numbers near the Elbow, and were
advancing towards us, and crossing the Qu'appelle not far
from the Height of Land ; by burning the prairie east of
their course, they would be diverted to the south, and
feed for a time on the Grand Coteau before they pursued
their way to the Little Souris, in the country of the Sioux,
south of the 49th parallel.

Putting out fire in the prairies is a telegraphic mode of
communication frequently resorted to by Indians.   Its
consequences are seen in the destruction of the forests
which once covered an immense area south of the
Qu'appelle and Assinniboine.   The aridity of those vast
prairies is partly due to this cause.   The soil, though
light, derives much of its apparent sterility from the

* A native expression ; "put out fire," signifies to set the prairie on fire.

annual fires. In low places and in shallow depressions where marshes are formed in spring, the soil is rich, much mixed with vegetable matter, and supports a very luxuriant growth of grass. If willows and aspens were permitted to grow over the prairies, they would soon be converted into humid tracts in which vegetable matter would accumulate and a soil adapted to forest trees be formed. If a portion of prairie escapes fire for two or three years the result is seen in the growth of willows and aspens, first in patches, then in large areas, which in a short time become united and cover the country, thus retarding evaporation and permitting the accumulation of vegetable matter in the soil. A fire comes, destroys the young forest growth and establishes a prairie once more. The reclamation of immense areas is not beyond human power; the extension of the prairies is evidently due to fires, and the fires are caused by Indians, chiefly for the purpose of telegraphic communication, or to divert the buffalo from the course they may be taking. These operations will cease as the Indians and buffalo diminish, events which are taking place with great rapidity.

The extension of the prairie country must have exercised a powerful influence upon the Indian population of Rupert's Land. By the progressive limitation of their hunting grounds during the winter season, hostile tribes would possess greater opportunities of destroying one another than when spread over the prairies. Migratory bands of Indians dependent upon wild animals for their support must diminish or increase with the area over which their sustenance extends, and it is apparent that the extension of absolutely treeless prairies and of sterile soil, the formation of "Plains," in a word*, is un-

* See the succeeding chapter for the distinction between Prairie and Plain.

favourable to the increase of the buffalo, the elk, the moose, the antelope and the bear,—animals which always seek the protection of "woods" during the terribly inclement winters of the north-western part of the American continent.

Wood began to be a great treasure in the prairie after passing the Moose Jaws Forks; we were compelled to go supperless to bed on the night of the 25th, because we had neglected to take a supply at the last aspen grove we passed, thinking that the bois de vache (dried buffalo dung) would be found in abundance, but the fires had burnt it also, and not even a fragment was to be procured. No tree or shrub, or even willow twig could be seen in any direction from our camp on the morning of the 25th. Our customary breakfast of tea and buffalo meat was impossible. We had to content ourselves with uncooked pemmican and water from a marsh.

Immediately on the banks of the Qu'appelle Valley near the "Round Hill" opposite Moose Jaws Forks, are the remains of ancient encampments, where the Plain Crees, in the day of their power and pride, had erected large skin tents, and strengthened them with rings of stones placed round the base. These circular remains were twenty-five feet in diameter, the stones or boulders being about one foot in circumference. They wore the aspect of great antiquity, being partially covered with soil and grass. When this camp ground was occupied by the Crees, timber no doubt grew in the valley below, or on the prairie and ravines in detached groves, for their permanent camping grounds are always placed near a supply of fuel.

Making an early start in search of wood, we came suddenly upon four Cree tents, whose inmates were still fast asleep; about three hundred yards west of them we found ten more tents, with over fifty or sixty Indians in all.

They were preparing to cross the valley in the direction of the Grand Coteau, following the buffalo. Their provisions for trade, such as dried meat and pemmican, were drawn by dogs, each bag of pemmican being supported upon two long poles, which are shaft, body and wheels in one. Buffalo Pound Hill Lake, sixteen miles long, begins near the Moose Jaws Forks, and on the opposite or south side of this long sheet of water, we saw eighteen tents and a large number of horses. The women in those we visited on our side of the valley and lake, had collected a great quantity of the mesaskatomina berry which they were drying. In gathering the mesaskatomina, which the Indians represented to be scarce in the valley of the Qu'appelle, they break off the branches of the trees loaded with fruit in order that they may collect the juicy berries with greater ease to themselves, never thinking that this practice continued from year to year must diminish and ultimately extirpate the shrub which they prize so highly, and which forms an important part of their summer food. They announced the cheering intelligence that the Chief Mis-tick-oos, with some thirty tents, was at the Sandy Hills impounding buffalo. Leaving the hospitable Crees after an excellent breakfast on pounded meat and marrow fat, we arrived at Buffalo Pound Hill at noon. The whole country here assumed a different appearance ; it now bore resemblance to a stormy sea suddenly become rigid ; the hills were of gravel and very abrupt, but none exceeded 100 feet in height. The Coteau de Missouri, particularly the " Dancing Point," is clearly seen from Buffalo Pound Hill towards the south, while north-easterly the last mountain of the Touchwood Hill Range looms grey or blue in the distance. Between these distant ranges a treeless plain intervenes.

Ponds and lakes are numerous on the Grand Coteau

side, and it is probably on this account that the buffalo cross the Qu'appelle valley near the Moose Jaws Fork and west of Buffalo Pound Hill Lake ; in the winter they keep towards the Touchwood Hills for the sake of shelter, and the excellent herbage which grows in the beautiful meadows between the aspen clumps. The prairies there too are not so often burned as south of the Qu'appelle, the valley of that river serving as a great barrier to prevent the onward progress of the devastating fires.

We now began to find the fresh bones of buffalo very numerous on the ground, and here and there startled a pack of wolves feeding on a carcass which had been deprived of its tongue and hump only by the careless, thriftless Crees. On the high banks of the valley the remains of ancient encampments in the form of rings of stones to hold down the skin tents are everywhere visible, and testify to the former numbers of the Plain Crees, affording a sad evidence of the ancient power of the people who once held undisputed sway from the Missouri to the Saskatchewan. The remains of a race fast passing away give more than a transient interest to Buffalo Pound Hill Lake. The largest ancient encampment we saw lies near a shallow lake in the prairie about a mile from the Qu'appelle valley. It is surrounded by a few low sandy and gravelly hills, and is quite screened from observation. It may have been a camping ground for centuries, as some circles of stones are partially covered with grass and embedded in the soil.

At noon on the 26th we rested for a few hours opposite to a large camp of Crees on the other side of the lake ; our sudden appearance at the edge of the prairie threw them into a state of the greatest excitement as evinced by their haste in collecting their horses and gathering in groups in the valley below. A few of them set out to ride round the lake but in the wrong direction, so that the chance of their overtaking us was highly doubtful. This magnificent sheet of water, never less than half a mile broad and sixteen miles long, shadowed forth what the Qu'appelle valley might become if a river like the Saskatchewan could be made to flow through it. As we neared the height of land, the physical structure of this great valley became a deeply interesting and almost exciting subject of inquiry. So far it had preserved its breadth and depth with astonishing uniformity all the way from the Mission, and we were within forty miles of the South Branch of the Saskatchewan. The hill banks of the river now became clothed with shrubs again, and the ash-leaved maple and elm appeared in the ravines, sustained, no doubt, by the presence of so large a body of water as Buffalo Pound Hill Lake.

Towards evening we arrived at another Cree encampment, where we were again hospitably treated to beaten buffalo meat and marrow fat. Birch bark dishes full of that nutritious but not very tempting food were placed on the ground before us and we were requested to partake of it. The Indians took a piece of the pounded meat in their fingers and dipped it into the soft marrow. A hunting knife which I employed for the same purpose excited their admiration, and after allowing them to examine it, I placed it as I thought securely in my coat pocket; on the following morning, however, the knife was not to be found, nor did I ever see it again. They were delighted to receive a small present of tea and tobacco, and while I was engaged in the tent with the men, the girls, children and old women assembled round the carts asking if we had any rum, and snuffed the boxes and bags containing provisions, in search of that odoriferous stimulant. We left our hospitable friends in the evening and camped about three miles from the last Cree tent. The chief of the band, an old man, accompanied us for some distance, expressing very amicable sentiments, and hinting that it would be as well to keep a watch over our horses during the night, for there were some young scamps among his band who would think it an honour to steal a white man's horse. Visitors came during the evening, and from their actions we thought it advisable to keep watch and tether the horses; observing these precautions they retired at an early hour after a friendly smoke. On the following morning when looking for my hunting knife I was very suspiciously reminded of the old chief's caution; it flashed upon me that the cunning fellow had himself secretly abstracted the knife while pressing his friendly advice.

At dawn we were *en route* again, and towards noon approached the Sandy Hills, the valley continuing about 140 feet deep and maintaining its width. Two days before our arrival the Indians had been running buffalo, and many carcases of these animals were scattered over the arid, treeless plain through which our route lay. Several herds of buffalo were visible wending their way in single file to the Grand Coteau de Missouri distinctly looming south of the Qu'appelle valley. After travelling through a dry, barren region, strewed with erratics, until 2 P.M., we arrived at the lake of the Sandy Hills, and on the opposite side of the valley saw a number of tents, with many horses feeding in the flats. When within a mile of the lake a buffalo bull suddenly appeared upon the brow of a little hill on our right. A finer sight of its kind could hardly be imagined. The animal was in his prime and a magnificent specimen of the buffalo. He gazed at us through the long hair which hung over his eyes in thick profusion, pawed the ground, tossed his head and snorted with proud disdain. He was not more than fifty yards from us, and while we were admiring his splendid proportions he set off at a gallop towards some low hills we had just passed over.

Our appearance on the brink of the valley opposite the tents surprised the Indians, they quickly caught their horses and about twenty galloped across the valley, here quite dry, and in a quarter of an hour were seated in friendly chat with the half-breeds. We kindled a fire with bois de vache, of which there was a vast quantity strewn over the plain, but no wood was near at hand. When the men were going to the lake for water to make some tea, the Indians told us it was salt, and that the only fresh water within a distance of some miles was close to their camp on the opposite side of the valley. We were therefore constrained to cross to the other side and erect our tents near to the spring. Advantage was taken of our passage across the valley to make an instrumental measurement of its leading dimensions. It was found to be 140 feet deep, estimating from the abrupt edge of the bank, and one mile five chains broad. The depth below the general level of the prairie is considerably greater, for there was a descent of fifty or sixty feet by a gentle slope not included in the foregoing measurement. A vast number of erratics strewed this slope, indeed it was with great difficulty that we steered the carts through the formidable accumulation of boulders which beset our path. The bed of the Qu'appelle is quite visible in the valley, but on account of the porous nature of the soil, the overflow from Sand Hill Lake penetrates it in dry weather, and reappears about half a mile below in the form of a little stream about ten feet broad, issuing from a marshy tract occupying the entire breadth of the valley. When crossing it the carts and horses sank deeply in the soft grassy bottom, already much cut up by the passage of a large number of buffalo during the week preceding our arrival.

Although still early in the afternoon, the difficulty of obtaining water and fuel, as well as a desire to procure a guide from the Indians, induced us to camp at the east end of Sand Hill Lake with the Crees by whom we were surrounded. Advantage was taken of the time at our disposal, and the opportunity offered at a large camp, to obtain some information respecting their habits and mode of life.

Scarcely had we made a distribution of tobacco and tea, when a buffalo bull appearing on the opposite side of the valley near where we had passed in the morning, afforded one of the young Indians an opportunity of showing his skill and bravery in attacking this formidable animal single handed and on foot, a conflict which is briefly described in the following chapter as seen through a good telescope from our camp on the south side of the valley.

CHAP. XVI.

Encounter with a Buffalo Bull.—Interior of Tents.—Barter.—Watchers.—
Dogs.—Eyebrow Hill.— Prairies.—Prairies and Plains.—Difference between Prairies and Plains.—Limits of the Prairie Country in the United
States.—Growth of Timber in River Bottoms.—Plains in Rupert's Land.
—Origin of Prairies.—Grand Coteau de Missouri.—Extent and Boundaries of.—Character of the Grand Coteau.—Elevation of.—Vegetation
of. — Eyebrow Hill. — Source of Qu'appelle. — Buffalo. — Character of
Qu'appelle Valley. — Water-marks. — Sandy Hills. — Distribution of
Boulders. — Section.— Rock Exposure.— Mis-tick-oos.— Sand-Dunes.—
South Branch.—The Qu'appelle Valley.—Cree Camp.—Height of Land.
—Section of Valley.—Levels.—Buffalo Pound.—Camp Moving.— "Dead
Men."—Old Buffalo Pound.—Horrible Spectacle.—New Pound.—Bringing
in Buffalo.—Slaughter in Pound.—"A Talk."—Objections to Half-breeds.
— To the H. B. Co.—Demeanour of the Indians.— The Wants of Mistick-oos.—His Tent.—His Wives.—Rock Exposure.—Boulders in Valley.
—Character of the South Branch.

ARMED with his bow from the bois d'arc, his arrows from
the mesaskatomina, neatly feathered with the plumes of
the wild duck, and headed with a barb fashioned from a
bit of iron hoop, the young Plain Cree threw off his leather
hunting shirt, jumped on a horse, and hurried across the
valley. Dismounting at the foot of the bank, he rapidly
ascended its steep sides, and just before reaching the top,
cautiously approached a large boulder which lay on the
brink, and crouched behind it.

The buffalo was within forty yards of the spot where
the Indian crouched, and slowly approaching the valley as
he leisurely cropped the tufts of parched herbage which

the sterile soil was capable of supporting. When within twenty yards of the Indian the bull raised his head, snuffed the air, and began to paw the ground. Lying at full length, the Indian sent an arrow into the side of his huge antagonist. The bull shook his head and mane, planted his fore feet firmly in front of him, and looked from side to side in search of his unseen foe, who after driving the arrow, had again crouched behind the boulder. Soon, however, observing the fixed attitude of the bull, a sure sign he was severely wounded, he stepped on one side and showed himself. The bull instantly charged, but when within five yards of his nimble enemy, the Indian sprang lightly behind the boulder, and the bull plunged headlong down the hill, receiving after he had passed the Indian, a second arrow in his flanks. As soon as he reached the bottom, he fell on his knees, and looked over his shoulder at his wary antagonist, who, however, speedily followed, and observing the bull's helpless condition, sat on the ground within a few yards of him and waited for the death gasp. After one or two efforts to rise, the huge animal drooped his head and gave up the strife. The Indian was at his side without a moment's pause, cut out his tongue, caught his horse, — an excited spectator of the conflict,—and galloping across the valley, handed me the trophy of his success.

We made ourselves acceptable to the Indians by offering them a present of powder, shot, tea, and tobacco, and in return they invited us to partake of pounded meat, marrowfat and berries. The chief of the band assured us that his young men were honest and trustworthy ; and in compliance with his instructions, property would be perfectly safe.

I visited the interior of most of their tents, and found the squaws almost exclusively engaged in drying buffalo meat. A couple of table-spoonfuls of tea, and a small plug of tobacco, always ensured a hearty welcome, and in return they generally presented me with a choice piece of buffalo meat from a fat cow, or a small skin of marrowfat. One of the young men took a fancy to a checked flannel shirt I was wearing at the time, and offered his saddle for it; declining the bargain, he added his bow and a quiver of arrows; I told him to bring the bow and arrows to my tent at night, and I would give him a new shirt for them. He said he should prefer a white one, and then the buffalo would not mind him; and when he came to complete his bargain, he selected a white jersey in preference to a showy coloured check. From time to time, scouts would come in and go out towards the Grand Coteau, on the look out for Blackfeet, and as nightfall approached, the wandering horses were gathered closer to the camp. The dogs, however, are their great protection; it is almost impossible for any stranger to approach a camp without arousing the whole canine population; and the passage of bands of buffalo during the night-time is signalised by a prolonged baying, which, however suggestive of sport and good cheer, is most wearisome to those who are anxious to rest. During the night a heavy rain filled the hollows with water, and gave us promise of an abundant supply until we arrived at the Sandy Hills, where the main body of Plain Crees were encamped. On the following day, the 28th, I rode to the Eye-brow Hill range, a prolongation of the Grand Coteau, and distant from the Qu'appelle Valley about four miles. It was there that the Indians told me I should find one of the sources of the Qu'appelle river. After an hour's ride I reached the hills, and quickly came upon a deep ravine at the bottom of which bubbled a little stream about three feet broad. I subsequently followed its course until it entered the prairie leading to the great valley, and traced it to its junction with the main excavation, through a deep narrow gully.

The Eye-brow Hill range is about 150 feet above the prairie, and forms the flank of a tableland stretching to the Grand Coteau, of which it is the northern extension. The recent tracks of buffalo were countless on the hill sides, and in the distance several herds could be seen feeding on the treeless plateau to the south. On the flanks of the Grand Coteau the true prairies may be said to terminate, and the plains to commence. It is doubtful whether the term "Plain" is not now applicable to a large portion of the country west and south of the Qu'appelle Mission. The destruction of " woods " by fires has converted into sterile areas an immense tract of country which does not appear necessarily sterile from aridity, or poverty of soil.

The Plains and Prairies of America occupy regions differing widely from one another in physical characteristics. The phraseology of the half-breeds tends to mislead a traveler not familiar with the precise application of the words they use. Such terms as " woods," " prairies," and " plains " are illustrations of this apparent want of precision, which if employed without explanation in a written narrative, would very probably cause considerable misapprehension, and lead to deductions wholly at variance with fact.* A tract of country may be described as a " wooded country," conveying the idea that timber covers the surface and is capable of affording a supply of that indispensable material for building purposes and fuel ; but in Rupert's Land, west of the Low Lake Region, and south of the 53rd parallel, the " woods " consist generally of

* Vide Col. Emory's Remarks quoted in the Chapter on the Climate of the Southern Part of Rupert's Land, Vol. II.

small aspens very rarely exceeding six inches in diameter
or twenty-five in altitude, and most frequently distributed
in detached groves, " bluffs," or belts. The same remark
applies to the use of the word " prairie," and to prairie
country ; prairies may be level, rich and dry, sustaining
luxuriant grasses and affording splendid pasturage ; they
may be marshy and wet, or undulating—" stony," " sandy,"
or " salt." Such indefinite terms as " open prairie," " rolling prairie," " alluvial prairie," not unfrequently employed
in describing without limit as to space, the vast unpeopled
wastes,—often beautiful and rich, often desolate and barren,—of the Prairies and *Plains* of America, are sometimes both physically and geologically wrong, and serve
to convey the impression that the large areas to which
they are applied possess, if not a fertile, at least not an
unkindly soil, or an arid climate, rendering husbandry
hopeless.

The difference between " Prairies " and " Plains " will
be best shown by describing their limits in the United
States and Rupert's Land. In the United States the true
Prairie region extends " over the eastern part of Ohio,
Indiana, the southern portion of Michigan, the southern
part of Wisconsin, nearly the whole of the States of Illinois and Iowa, and the northern portion of Missouri,
gradually passing, in the territories of Kansas and Nebraska, into the *Plains*, or the arid and desert region
which lies at the base of the Rocky Mountains. This
passage takes place in the region between the me-
ridians of 97° and 100°, west of which belt the country
becomes too barren to be inhabited and worthless for
cultivation. The passage from the heavily wooded region
of the north and east into the treeless plains of the west
is a gradual one, and the disappearance of the underwood and the predominance of " oak openings," or groves

of oak and other forest trees, not crowded together, but scattered over the surface at a considerable distance from one another, without any low shrubs or underbrush between them, is the characteristic of the border of the Prairie region." *

The growth of timber in the river bottoms in the United States disappears altogether at the borders of the Plains about the 98° of longitude.† In Rupert's Land, south of the 52nd parallel, trees cease in the river bottoms except near the Upper Qu'appelle Lakes, in longitude 104°; but they occur only on the *northern* aspect of the south side of the deep valley west of longitude 102°, or a few miles west of Fort Ellice.

The true limit of the Plains in Rupert's Land, east of the South Branch, is well shown by the Grand Coteau de Missouri. The country east of that natural boundary may be classed as Prairie country, over the greater portion of which forests of aspen would grow if annual fires did not arrest their progress. The plateau of the Grand Coteau forms the true Plains of Rupert's Land, where both soil and climate unite in establishing a sterile region. Mr. J. D. Whitney considers the absence of forest on the rich prairies of the United States to be mainly due to the physical nature of the soil; "the extreme firmness of the particles of which the prairie soil is composed, is probably the principal reason why it is better adapted to the growth of its peculiar vegetation than to the development of forests." ‡ The origin of the groves scattered over the Prairies is traced to ridges of coarse material, apparently deposits of drift, on which, from some local cause, there

---

* Report on the Geological Survey of the State of Iowa, by James Hall, State Geologist, and J. D. Whitney, Chemist and Mineralogist.
† Ibid.
‡ Geology of Iowa.

has never been an accumulation of fine sediment. The origin of the prairies and plains of Rupert's Land, as well as the present distribution of timber, will be discussed in another chapter. The Grand Coteau de Missouri, distinctly visible from the Eye-brow Hill, begins in latitude 45°, about sixty miles south-west of the head of the Coteau des Prairies in latitude 45° 55'; the intervening valley is occupied by James's River, an affluent of the Missouri. Its boundary pursues a course nearly due north, under the 99th meridian as far as the 47th parallel, when, turning north-westerly, it enters British territory near the 104th meridian, and still preserving a north-westerly direction comes on the South Branch of the Saskatchewan, a few miles from the Elbow, in longitude 108°.[*] The region east of the Grand Coteau belongs to the prairie region, the Grand Coteau itself and its prolongation towards Battle River, from its eastern boundary to the foot of the Rocky Mountains, constitutes the "Plains" properly so called of the north-western territories of the United States and of British America. From the character of its soil and the aridity of its climate, the Grand Coteau is permanently sterile and unfit for the abode of civilised man. The length of its abrupt flank, not including small bays or indentations, is about 650 miles from its commencement in the valley of James River to the Elbow of the South Branch of the Saskatchewan. For a distance of 380 miles the course of the Missouri is approximately parallel to the flanks of the Grand Coteau, and at an average distance of 50 miles from it. This vast tableland rises from 400 to 800 feet above the Missouri, and bears the same

---

[*] Vide Map accompanying Capt. Palliser's Reports; also Military Map of Nebraska and Dakotah, by Lieut. G. K. Warren, Top. Eng. U. S. Army.

altitude in relation to the high prairies through which the Shayenne meanders to the Red River of the north, on the one hand, and James River to the Missouri on the other.

The vegetation on the Grand Coteau is very scanty, the Indian turnip is common, so also is a species of cactus; no tree or shrub is seen, and it is only in the bottoms and marshes that rank herbage is found.*

In the afternoon I bade farewell to our Cree friends, and riding west joined the carts on the south side of Sand Hill Lake, on the brink of which we travelled until we arrived at the gully through which the stream from the Eye-brow Hill range enters the Qu'appelle valley. It was here nine feet broad, and three deep, having received accessions in a short course through the prairie from the hills where I had observed it scarcely three feet broad. We camped in the valley, and employed the evening in taking levels.

About four miles west of us we saw the Sandy Hills, and could discern the great valley passing through them, and containing, as the Indians had alleged, ponds which sent water both to the South Branch and the Assinniboine; an important physical fact which we afterwards verified instrumentally and by optical proof. We found the streamlet from the Eye-brow Hill range strike the Qu'appelle Valley eight and a half miles west of Sand Hill Lake, and four miles from the height of land where the ponds lie. The fall between the ponds and our camp was about five feet, and the valley 150 feet deep, and one mile seventy chains broad. The Eye-brow Hill stream had excavated a channel nine feet deep in the

---

* Vide Explorations and Surveys for a Railroad to the Pacific, by Governor Stevens.

bottom of the great valley, and was joined, by a sluggish brook coming from the ponds above, a few yards from our camp. Water marks on the hill banks showed that the entire breadth of the valley is flooded during spring freshets.

The Sandy Hills commence on the north side about two miles west of Sand Hill Lake as it appears in summer. They are drifting dunes, and many of them present a clear ripple-marked surface without any vegetation, not even a blade of grass. They have invaded the great valley, and materially lessened its depth. One feature in its banks is worthy of special notice. Many boulders or erratics are distributed over the western extremities of small hills or ridges into which the steep banks are broken, 70 to 120 feet above the level of the flats. These ridges have the form of long, narrow islands, their longitudinal axes being parallel to the sides of the valley, and the erratics are deposited and arranged on the top of each ridge, and at the west extremity. The form of these ridges is also peculiar, they are sharp at the west end where the erratics lie, and rounded at the east end. The slope is gentle at the west end, abrupt at the east end. This peculiarity is a constant feature of all the ridges seen on the sides of the banks of the valley. They vary in height from 10 to 30 feet, and in length from 60 to 140 feet, and in breadth from 20 to 80 feet. They have evidently some relation to the excavating force which has produced this great valley, and cannot be attributed to the long continued action of a small stream ; however competent running water may be to produce deep and long depressions in loose drift, or a soft friable rock.

A section of the bank of the Eye-brow Hill stream in its course through the flats, showed fine clay brought by recent rains from the hill banks, sand blown from the dunes, and loam produced by the blending of the two. Where it leaves the prairie the little river has exposed a section of a drift hill round the base of which it sweeps. Gravelly drift is seen to repose upon an ochreous stratified rock, seamed with veins of selenite. It exhibits a stratum of yellow and red ferruginous clay about six feet thick, and below it a hard greenish sandstone, in which gigantic concretionary masses are numerous. Veins of selenite penetrate the greenish coloured rock, but are most abundant in the ferruginous clay. This is the first rock seen in position above the Mission. Subsequent comparison with the rocks on the South Branch showed it to belong to the uppermost member of the cretaceous series.

On the morning of the 29th, we prepared to visit the main body of the Crees at the Sandy Hills, and with a view to secure a favourable reception sent a messenger to announce our arrival, and to express a wish to see Mistick-oos, their chief. Soon after breakfast we crossed the valley and threaded our way between sand dunes; one dune was found to be seventy feet high, quite steep on one side, beautifully ripple-marked by the wind, and crescent-shaped; from its summit we saw the woods and hills beyond the South Branch of the Saskatchewan, and what was more delightful to us, traced with the eye the Qu'appelle valley apparently with undiminished depth and breadth through the Sandy Hills, until it was lost as it dipped towards the South Branch.

At eight o'clock A.M. we came in sight of the Cree camp, and soon afterwards messengers arrived from Mistick-oos, in reply to the announcement we had transmitted to him of our approach, expressing a hope that we would delay our visit until they had moved their camp half-a-mile further west, where the odour of the putrid buffalo would be less annoying. We employed the time in ascertaining the exact position of the height of land, and soon found a pond from which we observed water flowing to the Saskatchewan and the Assinniboine. The pond was fed by a number of springs and small streams, a foot or two broad, issuing from the Sandy Hills at right angles to the valley.

Sand hills or dunes cover the country for a considerable distance on both sides. We selected this spot to level across it, and found the depth to be 110 feet below

Transverse Section of the Valley of the Qu'appelle at the height of land.

the first plateau, its breadth, although partially invaded by sand dunes, seventy-three chains, or nearly one mile. On the south side it rises much more gently than on the north, and shows two terraces, both of which are covered with drifting sand. Here we commenced taking the levels to the South Branch, twelve miles distant from us, an operation which we soon found necessary to close for the present, in consequence of the arrival of about sixty Cree horsemen, many of them naked with exception of the breech cloth, and belt. They were accompanied by the chief's son, who informed us that in an hour's time they would escort us to the camp.

They were about constructing a new pound, having literally filled the present one with buffalo, and being compelled to abandon it on account of the stench which arose from the putrifying bodies. We sat on the ground and smoked, until they thought it time for us to accompany them to their encampment. Mis-tick-oos had hurried away to make preparations for " bringing in the buffalo,"* the new pound being nearly ready. He expressed, through his son, a wish that we should see them entrap the buffalo in this pound, a rare opportunity few would be willing to lose.

We passed through the camp to a place which the chief's son pointed out, and there erected our tents. The women were still employed in moving the camp, being assisted in the operation by large numbers of dogs, each dog having two poles harnessed to him, on which his little load of meat, pemmican, or camp furniture was laid. After another smoke, the chief's son asked me, through the interpreter, if I would like to see the old buffalo pound, in which they had been entrapping buffalo during the past week. With a ready compliance I accompanied the guide to a little valley between sand hills, through a lane of branches of trees, which are called " dead men" to the gate or trap of the pound. A sight most horrible and disgusting broke upon us as we ascended a sand dune overhanging the little dell in which the pound was built. Within a circular fence 120 feet broad, constructed of the trunks of trees, laced with withes together, and braced by outside supports, lay tossed in every conceivable position over two hundred dead buffalo. From old bulls to calves of three months old, animals of every age were huddled together in all the forced attitudes of violent death. Some lay on their backs, with eyes starting from their heads, and tongue thrust out through clotted gore. Others were impaled on the horns of the

---

* A half-breed expression signifying "to drive the buffalo from the Prairie towards and into the pound," or Ponds, as they are termed in half-breed phraseology.

old and strong bulls. Others again, which had been tossed, were lying with broken backs two and three deep. One little calf hung suspended on the horns of a bull which had impaled it in the wild race round and round the pound.

The Indians looked upon the dreadful and sickening scene with evident delight, and told how such and such a bull or cow had exhibited feats of wonderful strength in the death-struggle. The flesh of many of the cows had been taken from them, and was drying in the sun on stages near the tents. It is needless to say that the odour was overpowering, and millions of large blue flesh flies, humming and buzzing over the putrefying bodies was not the least disgusting part of the spectacle. At my request the chief's son jumped into the pound, and with a small axe knocked off half a dozen pair of horns, which I wished to preserve in memory of this terrible slaughter. "To-morrow," said my companion, " you shall see us bring in the buffalo to the new pound."

After the first " run," ten days before our arrival, the Indians had driven about 200 buffalo into the enclosure, and were still urging on the remainder of the herd, when one wary old bull, espying a narrow crevice which had not been closed by the robes of those on the outside, whose duty it was to conceal every orifice, made a dash and broke the fence, the whole body then ran helter skelter through the gap, and dispersing among the sand dunes escaped, with the exception of eight who were speared or shot with arrows as they passed in their mad career. In all, 240 animals had been killed in the pound, and it was its offensive condition which led the reckless and wasteful savages to construct a new one. This was formed in a pretty dell between sand hills, about half a mile from the first, and leading from it in two diverging rows, the bushes they designate " dead men," and which serve to guide the buffalo when at full speed, were arranged.   The " dead men " extended a distance of four miles into the prairie, west of and beyond the Sand Hills.   They were placed about 50 feet apart, and between the extremity of the rows might be a distance of from one and a half to two miles.

When the skilled hunters are about to bring in a herd of buffalo from the prairie, they direct the course of the gallop of the alarmed animals by confederates stationed in hollows or small depressions, who, when the buffalo appear inclined to take a direction leading from the space marked out by the " dead men," show themselves for a moment and wave their robes, immediately hiding again. This serves to turn the buffalo slightly in another direction, and when the animals, having arrived between the rows of " dead men," endeavour to pass through them, Indians here and there stationed behind a " dead man," go through the same operation, and thus keep the animals within the narrowing limits of the converging lines.   At the entrance to the pound there is a strong trunk of a tree placed about one foot from the ground, and on the inner side an excavation is made sufficiently deep to prevent the buffalo from leaping back when once in the pound.   As soon as the animals have taken the fatal spring they begin to gallop round and round the ring fence looking for a chance of escape, but with the utmost silence women and children on the outside hold their robes before every orifice until the whole herd is brought in ; they then climb to the top of the fence, and, with the hunters who have followed closely in the rear of the buffalo, spear or shoot with bows and arrows or fire-arms at the bewildered animals, rapidly becoming frantic with rage and terror, within the narrow limits of the pound.

Printed by Spottiswoode and Co.

[New-street Square, London.

PLAIN CREES DRIVING BUFFALOES INTO A POUND.

A dreadful scene of confusion and slaughter then begins, the oldest and strongest animals crush and toss the weaker; the shouts and screams of the excited Indians rise above the roaring of the bulls, the bellowing of the cows, and the piteous moaning of the calves. The dying struggles of so many huge and powerful animals crowded together, create a revolting and terrible scene, dreadful from the excess of its cruelty and waste of life, but with occasional displays of wonderful brute strength and rage; while man in his savage, untutored, and heathen state shows both in deed and expression how little he is superior to the noble beasts he so wantonly and cruelly destroys.

Mis tick oos, or " Shortstick," is about fifty years old, of low stature, but very powerfully built. His arms and breast were deeply marked with scars and gashes, records of grief and mourning for departed friends. His son's body was painted with blue bars across the chest and arms. The only clothing they wore consisted of a robe of dressed elk or buffalo hide, and the breech cloth; the robe was often cast off the shoulders and drawn over the knees when in a sitting posture; they wore no covering on the head, their long hair was plaited or tied in knots, or hung loose over their shoulders and back. The forms of some of the young men were faultless, of the middle-aged men bony and wiry, and of the aged men, in one instance at least, a living skeleton. I inquired the age of an extremely old fellow who asked me for medicine to cure a pain in his chest; he replied he was a strong man when the two Companies (the Hudson's Bay and the North West) were trading with his tribe very many summers ago. He remembers the time " when his people were as numerous as the buffalo are now, and the buffalo thick as trees in the forest." The half-breeds thought he was more than 100 years old.

When Mis-tick-oos was ready to receive me, I proceeded to the spot where he was sitting surrounded by the elders of his tribe, and as a preliminary, rarely known to fail in its good effect upon Indians, I instructed one of my men to hand him a basin of tea and a dish of preserved vegetables, biscuit, and fresh buffalo steaks. He had not eaten since an early hour in the morning, and evidently enjoyed his dinner. Hunger, that great enemy to charity and comfort, being appeased, I presented him with a pipe and a canister of tobacco, begging him to help himself and hand the remainder to the Indians around us. The presents were then brought and laid at his feet. They consisted of tea, tobacco, bullets, powder, and blankets, all which he examined and accepted with marked satisfaction. After a while he expressed a wish to know the object of our visit; and having at my request adjourned the meeting to my tent in order to avoid sitting in the hot sun, we held a " talk," during which Mis-tick-oos expressed himself freely on various subjects, and listened with the utmost attention and apparent respect to the speeches of the Indians he had summoned to attend the " Council."

All speakers objected strongly to the half-breeds' hunting buffalo during the winter in the Plain Cree country. They had no objection to trade with them or with white people, but they insisted that all strangers should purchase dried meat or pemmican, and not hunt for themselves.

They urged strong objections against the Hudson's Bay Company encroaching upon the prairies and driving away the buffalo. They would be glad to see them establish as many posts as they chose on the edge of the prairie country, but they did not like to see the prairies and plains invaded. During the existence of the two companies, all went well with the Indians, they obtained excellent pay, and could always sell their meat, skins, robes, and pemmican. Since the union of the companies they had not fared half so well, had received bad pay for their provisions, and were growing poorer, weaker, and more miserable year by year. The buffalo were fast disappearing before the encroachments of white men, and although they acknowledged the value of fire-arms, they thought they were better off in olden times, when they had only bows and spears, and wild animals were numerous. They generally commenced with the creation, giving a short history of that event in most general terms, and after a few flourishes about equality of origin, descended suddenly to buffalo, half-breeds, the Hudson's Bay company, tobacco, and rum. I asked Mis-tick-oos to name the articles he would wish me to bring if I came into his country again. He asked for tea, a horse of English breed, a cart, a gun, a supply of powder and ball, knives, tobacco, a medal with a chain, a flag, a suit of fine clothes, and rum. The "talk" lasted between six and seven hours, the greater portion of the time being taken up in interpreting sentence by sentence, the speeches of each man in turn.

During the whole time we were engaged in "Council" the pipe was passed from mouth to mouth, each man taking a few whiffs and then handing it to his neighbour. It was a black stone pipe, which Mis-tick-oos had received as a present from a chief of the Blackfeet at the Eagle Hills a few weeks before.* When the pipe came round to me I usually replenished it, and taking a box of "vespers" from my pocket, lit it with a match. This operation was observed with a subdued curiosity, each Indian watching me with-

* See Chapter on Indian customs, superstitions, &c., Vol. II.

out moving his head, turning only his eyes in the direction
of the pipe. No outward sign of wonder or curiosity
escaped them during the " talk." On one occasion the
pipe was out when passed to the Indian sitting next to
me; without turning his head he gently touched my arm,
imitated the action of lighting the match by friction
against the bottom of the box, and pointed with one
finger to the pipe. They generally sat with their eyes
fixed on the ground when one of them was speaking,
giving every outward sign of respectful attention, and
occasionally expressing their approval by a low gurgling
sound. When the talk was over, I went witl Mis-tick-
oos to his tent; he then asked me to reproduce the
match-box, and show its wonders to his four wives. One
of them was evidently sceptical, and did not think it was
" real fire " until she had ignited some chips of wood
from the lighted match I presented to her. I gave a
bundle to Mis-tick-oos, who wrapped them carefully in a
piece of deerskin, and said he should keep them safely,—
they were " good medicine." I made his wives happy
and merry by distributing about a pound of tea amongst
them, and their hilarity knew no bounds when the incident narrated in the second volume* took place.

Mis-tick-oos apologised for the smallness of his tent,
remarking that he had collected twenty skins for a new
one, and when he had obtained two more he should get his
wives to make them up. I asked how it was that a chief
with four wives should have a smaller tent than many of
his young men around us. No reply was given to this
question, and the half-breed interpreter said that there
was a mystery attached to the tent formerly possessed by
Mis-tick-oos, which he would endeavour to solve, but

* Chapter on Indian customs and superstitions, &c., Vol. II.

none of the people present would talk about it, so he surmised that there might be some disappointment, or perhaps the death of a relative connected with it. From information afterwards obtained, and verified at the Hudson's Bay Company's post at the Touchwood Hills, it appears that a favourite son of Mis-tick-oos was killed while hunting for the post some months before, and that in accordance with the customs of the Crees, Mis-tick-oos had presented the officer in charge of the post with his tent of twenty-two skins as a " grief offering."

Early on the morning of the 30th I retraced my steps to examine an exposure of cretaceous rock forming part of the bank at the summit level of the Qu'appelle valley, while Mr. Fleming continued taking the levels to the South Branch. The rock is a sandstone, dipping very slightly to the south-west. The length of the exposure, east and west, is about fifty yards; it is covered with drifting sand. Near the summit the layers are highly fossiliferous, and almost wholly composed of *Avicula Linguæformis* (Evans and Shumard); above and below the fossiliferous portion there is a coarse greenish-coloured sand, interstratified with brown ferruginous layers. The thickness visible is about twelve feet. The rock occurs near the bend of the valley at its summit level; the exposure is perpendicular, and about sixty feet above the bottom of· the valley. Some of the beds, those which are unfossiliferous are very soft and friable, easily disintegrating, and may, farther west, be the origin of some of the sand dunes distributed over so wide an area in this part of the country. In descending the slope from the summit level to the Saskatchewan, the boulders on the ridges in the valley were found to be generally deposited upon the west side. The inclination of the boulders was towards the east, those forming the upper stratum were inclined against or superimposed upon the west side of those beneath, leading to the inference that the current which directed the course of the ice which bore them, came here, as on the other side of the summit level in the valley, *from the west.*

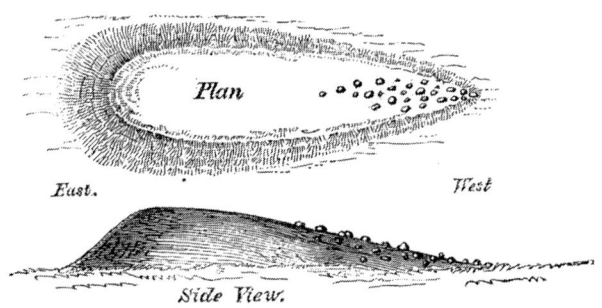

Ridges, with Boulders, on the East and West side of the height of land in the Qu'appelle Valley.

About fourteen miles from the South Branch there is a gigantic erratic of unfossiliferous rock on the south side of the valley. It is seventy-nine feet in horizontal circumference, three feet from the ground ; and a tape stretched across the exposed portion, from side to side over the highest point, measured forty-six feet. The Indians place on it offerings to Manitou, and at the time of our visit it contained beads, bits of tobacco, fragments of cloth, and other trifles.

At noon I bade farewell to Mis-tick-oos, and joining the carts we wended our way by the side of " the River that Turns," occupying the continuation of the Qu'appelle valley, to the South Branch of the Saskatchewan. The carts were accompanied by several Indians, who watched with much curiosity the progress of taking the levels, and were very anxious to know what " medicine " I was

searching for when sketching the position of the erratics in
the valley.

Now and then a fine buffalo bull would appear at the
brow of the hill forming the boundary of the prairie, gaze
at us for a few minutes and gallop off. The buffalo were
crossing the South Branch a few miles below us in great
numbers, and at night, by putting the ear to the ground,
we could hear them bellowing. Towards evening we all
arrived at the South Branch, built a fire, gummed the
canoe, which had been sadly damaged by a journey of
700 miles across the prairies, and hastened to make a distribution of the supplies for a canoe voyage down that
splendid river. We were not anxious to camp at the
mouth of "the River that Turns," in consequence of a war
party of Blackfeet who were said to be in the neighbourhood of the Cree camp, watching for an opportunity to
steal horses, and if possible to "lift a scalp."

The Indians who had accompanied us hastened to join
their friends as soon as they saw we were ready to em-
bark, and just as the sun set, the canoe containing Mr.
Fleming and myself, with two half-breeds, pushed off from
the shore ; the remainder of the party in charge of the
old hunter, retired from the river with the carts and horses
to camp in the open prairie, where they would be able to
guard against a surprise by the Blackfeet, or the thieving
propensities of treacherous Crees. Great precautions
were undoubtedly necessary, as sure signs had been observed within three miles of the Sandy Hills, proving that
a war party of Blackfeet were skulking about. The
Crees, always accustomed when on the South Branch to
their attacks, merely adopted the precaution of posting
watchers on the highest dunes, about a mile from their
camp, but in accordance with the friendly advice of Mis-
tick-oos, we embarked at this late hour in the evening

with a view to avoid surprise and mislead any watchful eyes that might have taken note of our movements. We drifted a mile or two down the river until we came to a precipitous cliff showing a fine exposure of rock, which proved a temptation too great to be resisted, so we drew the canoe on the bank and camped for the night on the east side of the river, making arrangements to watch in turns.

The first view of the South Branch of the Saskatchewan, fully 600 miles from the point where the main river disembogues into Lake Winnipeg, filled me with astonishment and admiration. We stood on the banks of a river of the first class, nearly half a mile broad, and flowing with a swift current, not more than 350 miles from the Rocky Mountains, where it takes its rise. We had reached this river by tracing for a distance of 270 miles, a narrow deep excavation continuous from the valley of one great river to that of another, and exhibiting in many features evidences of an excavating force far greater than the little Qu'appelle which meanders through it, was at the first blush, thought capable of creating. How were the deep lakes hollowed out? lakes filling the breadth of the valley, but during the lapse of ages not having increased its breadth, preserving too, for many miles, such remarkable depths, and although in some instances far removed from one another, yet maintaining those depths with striking uniformity. What could be the nature of the eroding force which dug out narrow basins 54 to 66 feet deep at the bottom of a valley already 300 feet below the slightly undulating prairies, and rarely exceeding one mile in breadth? It was easy to understand how a small river like the Qu'appelle could gradually excavate a valley a mile broad and 300 feet deep. The vast prairies of the north-

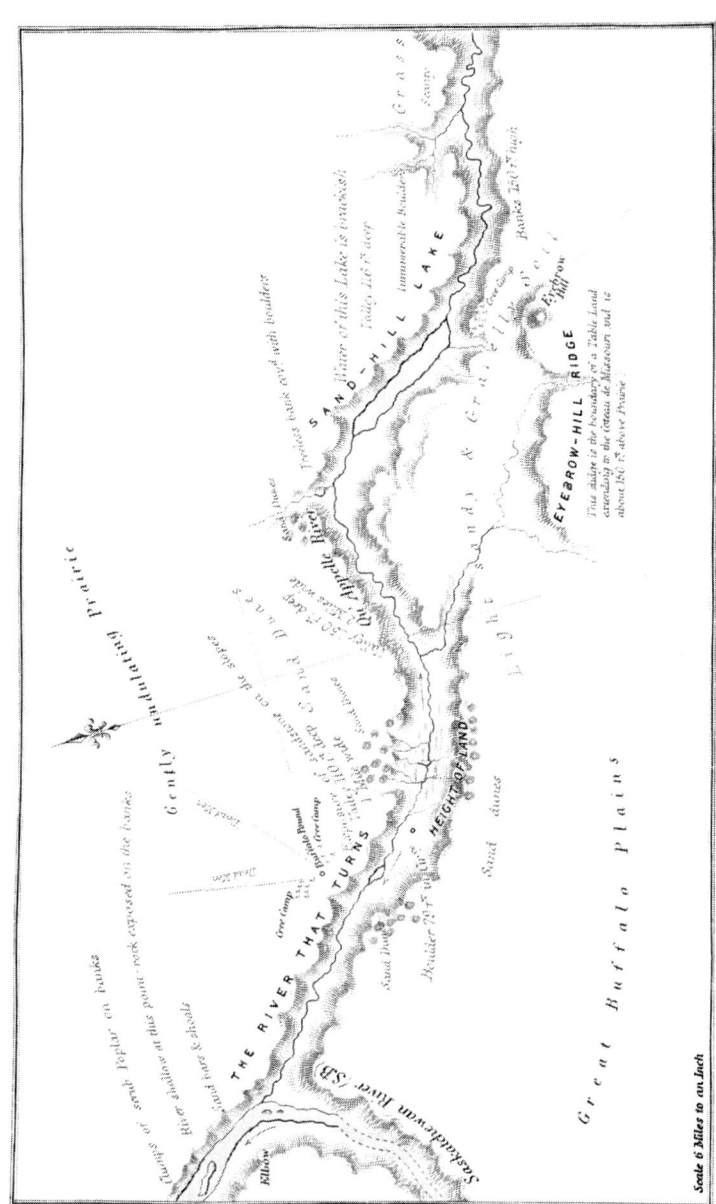

Scale 6 Miles to an Inch

Plan shewing the Junction of the Qu'Appelle Valley with the Saskatchewan.

Engraved by Edw? Weller

Great Buffalo Plains

EYEBROW-HILL RIDGE

This ridge is the boundary of a Table Land extending to the cotteau de Missouri and is about 150 ft above Prairie

SAND-HILL LAKE

Gently undulating Prairie

THE RIVER THAT TURNS

Saskatchewan River S.B.

Elbow

Light Sands & Grassy Hills

west offer many such instances; the Little Souris River, for example, in passing through the Blue Hills; the Assinniboine, for a 150 miles, flows through a broad deep valley, evidently excavated by its waters; the rivers in western Canada often flow in deep eroded valleys; but in no instance to my knowledge are deep and long lakes known to occupy a river valley, where nearly horizontal and very soft rocks preclude the assumption that they may have been occasioned by falls, without bearing some traces of the force which excavated their basins. They seem to point to the former existence of a much deeper valley now broken into detached lakes by the partial filling up of intervening distances. It was certainly with mingled feelings of anxiety and pleasurable anticipation that we embarked on the broad Saskatchewan, hoping during our long journey down its swift stream to find some clue to the origin of the curious inosculating valley of the Qu'appelle we had traced from one water-shed to another.

CHAP. XVII.

FROM THE QU'APPELLE MISSION TO FORT ELLICE, DOWN THE QU'APPELLE RIVER.

The Second Fishing Lake — Depth of.—Indian Map. — Origin of name Qu'appelle, or " Who calls River. ''—The First Lake, or Pakitawiwin.— Great Depth of First Lake. — Fish.— Confervæ. — Depth of Valley.— Width of River.— High-water Mark. — Valley flooded. — Affluents.— Depth of Valley.—Crooked Lake, or Ka-wa-wa-ki-ka-mac—Dimensions of.—Effects of Fires.— Trees in Valley.— Boulders.— Character of the Country.—Indian Surprise.—Indians.—Summer Berry Creek.— Dimensions of Valley.—Valley and Prairie Scene.—Camp Scene.—Character of Valley. — Ka-wah-wi-ya-ka-mac, or Round Lake — Dimensions of. — Stony Basin.—Granite Boulders.—Little Cut-Arm Creek.—The Scissors Creek.— Rock Exposure. -- Grasshoppers.— Big Cut-Arm Creek.— Dimensions of Qu'appelle. — Flooding of Valley.—Timber.—Undergrowth —Birds. — Minks. — Deer. — Uniformity of Qu'appelle Valley.— Table showing the dimensions of the Qu'appelle Valley and Qu'appelle Lakes.

MR. DICKINSON'S NARRATIVE.

DEAR SIR,—Soon after parting from you on the morning of July 20th, at the Church of England Mission in the Qu'appelle Valley, my instruments for surveying, with watch, a magnetic compass, a log line and sounding line, all arranged for ready use, and a cargo of kettles, pans, pemmican and blankets stowed away, our little canoe commenced its voyage down the river. In half an hour we reached the lake, which is generally called the second of the Fishing Lakes. Before venturing to go down it we were obliged to stop for the purpose of gumming the canoe, as it was leaking more than was desirable. To save time we took breakfast here. The distance between this lake and the one at the Mission is a mile and a half, while the actual length of the river is upwards of two miles.   Its width averages eighty feet, and its depth three feet; the rate of current, which is nearly uniform throughout its length, is one mile per hour.   The difference of level between these two lakes, obtained instrumentally on a previous day, is 1·50 feet.   These measurements, not valuable in themselves, are taken for the purposes explained in the "Rules for conducting the Exploring Survey," namely, as the means for calculating approximately the total fall in the river.   I may mention, that at every opportunity similar measurements and observations were made, with the assistance of Mr. John Fleming, from which we were able to deduce some general rules for guiding us in estimating the fall in rivers, and also, that the log line was found to be most invaluable in ascertaining the rate of the canoe on the rivers as well as on the lakes, being a much more accurate way than that of estimating it by the eye.

The canoe being now declared to be sea-worthy, we started on our way again.   The lake is three miles and a quarter long and three-quarters of a mile in breadth, extending between the slopes of the valley, and appearing to be merely an expansion of the river, but on trial found to be something more than that.   For some distance out from the mouth of the river it is only from three to four feet deep, but on trying it, when we were about half a mile distant, with a sounding line thirty feet long, to my great surprise, I could find no bottom; having added more line, the depth proved to be forty-two feet.   About the middle of the lake the depth is forty-eight feet.

A stream a quarter of a mile in length, flowing sluggishly through a marsh, connects this lake with the next, the first of the Fishing Lakes, or as it is in Cree, *Pakitawiwin*. All the Indian names of the lakes and tributaries of the

Qu'appelle I got afterwards on my arrival at Fort Ellice, from an old Indian seventy years of age, who had been once upon a time a great hunter and warrior, now in peace and comfort spending his remaining days at the hospitable Fort. With a piece of charred wood he drew on the floor a map of the Qu'appelle Valley from the Fishing Lakes to the Assinniboine, showing every little creek so accurately that I easily recognised them. Mr. McKay, who was then in charge of the Fort, kindly acted as interpreter on the occasion. The Cree name of the Qu'appelle river is *Katapaywie sepe*, and this is the origin of the name as told me by the Indian :—A solitary Indian was coming down the river in his canoe many summers ago, when one day he heard a loud voice calling to him ; he stopped and listened and again heard the same voice as before. He shouted in reply, but there was no answer. He searched everywhere around, but could not find the tracks of any one. So from that time forth it was named the " Who Calls River."

Pakitawiwin is six miles long and half a mile wide, and is most wonderfully deep. In one place, by means of putting together various pieces of cord, sashes, &c., the sounding line being too short, the depth was found to be about sixty-six feet. The mean of several depths is fifty-two feet. It is famous for the quantity and quality of its fish. For three miles we passed through a dense decaying mass of confervæ, which an east wind had driven to the upper parts of the lake. The smell of it was most unpleasant ; the men pushed through it as hard as they could, no easy matter, as it impeded the progress of the canoe considerably. The valley here is about the same depth as it is at the Mission, but the slopes are not so precipitous ; one of them, that on the south side, has been the whole way covered with a dense growth of young aspens, and the

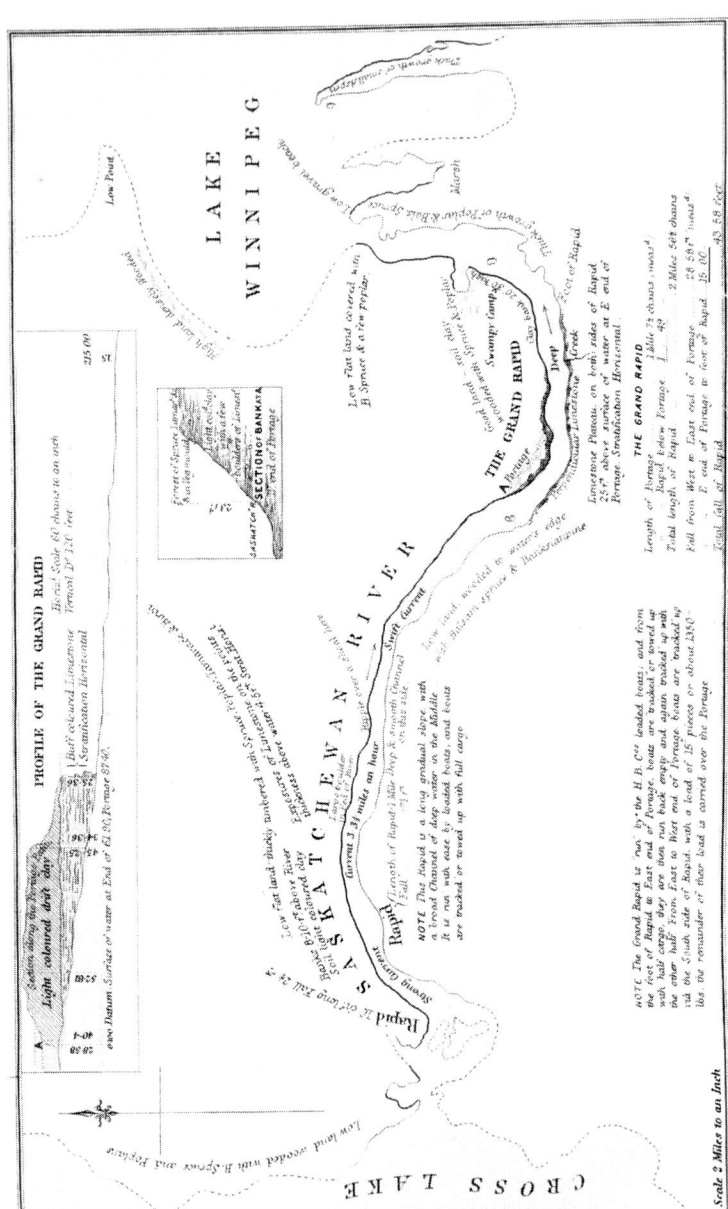

Plan shewing the Grand Rapid of the Saskatchewan

other has been bare of trees except in some of its many hollows and ravines.

Leaving the lake we now descend the river at an average speed of four miles an hour, the rate of current being generally about one mile and a quarter per hour. Paddling was easy work, but the steering by no means so, for the bends of the river are innumerable and very sharp, and the waters sweep round them with great velocity; oftentimes, but for the strong and dexterous arm of the steersman, the canoe would have been dashed against the bank; as it was he could not avoid sometimes getting entangled among the overhanging branches of the willows. The width varies from one chain to one and a half, and the depth from four and a half to two feet. The bed, for the most part, consists of soft mud, and is quite free from boulders, as is the case the whole way to the mouth, excepting in one place to be mentioned hereafter. The high water mark, very apparent on the willows growing along the banks, was eight feet over the present level of the water; the whole bottom of the valley, I was told, is often flooded to a depth of three feet.

Nineteen small creeks flow into this portion of the river, two only of them having names, the first and second Pheasant Creeks, called in Cree *Akiskoowi sepesis;* named after a hill which lies to the north some miles away, from near which they both take their rise. I took a cross section of the valley here, and found it to be 320 feet deep and seventy-eight chains wide; it is, I think, the deepest part of it. At noon, on July 23, we reached Crooked Lake, called in Cree *Kawawak-kamac*, the most picturesque of the Qu'appelle Lakes. Several streams draining the prairies on both sides have excavated deep and wide gorges opening into the main valley, which here sweeps in graceful curves, so that Crooked Lake seems to be embosomed amongst hills, and thus differs from the others which have very much the appearance of a gigantic canal. It is a little more than six miles in length, and its mean width is three-quarters of a mile. The greatest depth I found was thirty-six feet, and the mean of several soundings was thirty-one feet. The south slope, as before, is clothed with a dense foliage of young aspens, willows, and dogwood; a great contrast to the opposite side, on which only grows short and scanty grass, leaving the granite boulders which lie scattered over it, exposed to view; only in the ravines and the deep hollows are seen patches of young aspens and straggling oaks which have escaped the devastating fires.

For some time I could not understand why one side should be covered with trees and the other quite bare, the soil on both being exactly similar, until I discovered unmistakeable evidences of fire, which may be the cause of it. On inquiry afterwards I found that Indians often travel along the valley on the north of the river, which accounts for the fires being on that side.

Between the gravelly beach and the first of the slopes a fringe of willows runs all round the lake, and several points of low land jut out on both sides, on which grow oak, elm, and ash; not very large trees certainly, but healthy and thriving looking, and giving additional beauty to the landscape.

I ascended a bluff on the north side by a well-worn deer path, on which there were many foot-marks quite fresh, for the purpose of taking some observations connected with the survey and seeing the nature of the surrounding country. A gently undulating prairie, dotted with clumps of small poplars and willows stretched away on every side, and as far as I could see, the soil was a light sandy and gravelly loam, and in many parts strewed with boulders. I rather think that such is the character of a considerable extent of this section of the country.

As I stood upon the summit of the bluff, looking down upon the glittering lake 300 feet below, and across the boundless plains, no living thing in view, no sound of life anywhere, I thought of the time to come when will be seen passing swiftly along the distant horizon the white cloud of the locomotive on its way from the Atlantic to the Pacific, and when the valley will resound with the merry voices of those who have come from the busy city on the banks of Red River to see the beautiful lakes of the Qu'appelle. The view down the valley, where the river, after issuing from the lake, commences again its strange contortions, was doubtless very pretty, but it showed too the trouble that was before me, that there would be no rest for eye or finger, such as I had when taking long straight courses on the lake.

Again re-seated in the canoe we soon passed out of the lake into the river, the current of which for some distance is very strong and rapid, about two and a half miles per hour according to the log-line, and the width averages 70 feet, and the depth 3 ft. 6 in. A little way down it, as we swiftly and noiselessly glided round a sudden bend, we were borne by the current very close indeed to a group of Indian women who were enjoying the pleasures of a bath, quite as much to our astonishment as to theirs. First a loud chorus of screams arose, and then there was a rushing about for blankets and other apparel, which they adjusted with most wonderful rapidity, and then away they scampered to their wigwams laughing heartily as they went. Presently men and boys came trooping down to us simply arrayed in blankets, some worn in rather a *negligé* fashion, for the day was very hot. The chief man of the party, which consisted of six families, invited me in the most polite and hospitable manner to go to his lodge and have something to eat; but I had to decline as he had told me previously, in answer to a question as to how many days' journey it was to Fort Ellice, that we would have to sleep four or five times before we reached it, and this was now our fourth day from the Mission; and, moreover, I thought that the interior of a wigwam would not be a very agreeable place on such a hot day.

While we were speaking, the young ladies, whom we had so unintentionally disturbed, came down one by one to see us. Although their toilets were quite completed, so very modest were they that they remained behind the bushes and peeped at us through the branches. Having given the men some tobacco, and received in return a large supply of Pembina berries (High-bush cranberries), we wished them good-bye and resumed our journey. We went at the average rate of four miles an hour for two hours and a half, and camped before sunset at the foot of a bluff on the south side of the valley, of which I had taken a bearing from the end of the lake, and close to a creek about ten feet wide called *Nipimenan sepesis*, or Summer Berry Creek.

The valley is here of the same breadth as heretofore, that is, about one mile, and its depth is from 250 to 300 feet. The bottom is covered with willows interspersed with young sugar maples, with here and there an open patch of long luxuriant grass. With some difficulty I made my way to the level of the prairie through a dense and tangled mass of aspens and underwood of willows, dog-wood, and rose trees; but the beauty of the glorious sunset, and the cool refreshing breeze that came across the plains, more than repaid the trouble. I need not try to describe the exceeding beauty of the scene, for I could not; I will merely state what the components of the picture were. The sun just merged from behind a bank of crimson clouds reflected in the waters of Crooked Lake; part of the valley in deep shade and part brightly illuminated. The vivid green of the young poplars on one side, and on the other large granite boulders lying on the bare and rugged surface of the slope. The blue smoke of the wigwams rising up high and straight from the bottom of the valley. The river, with its complicated coils, gliding among the willow bushes. To the south the great prairie, ocean-like, with its many islands of aspens and single trees, looking in the distance, and by twilight, like becalmed ships. As this view just dissolves away, another rises very pleasant to see,—our camp fire is now burning brightly below, and over it swings a kettle, and passing round and about it are my two men, one busily engaged in preparing supper, the other in spreading out the blankets on the ground between the fire and the canoe.

Next morning (24th) we started as soon as it was daylight, glad to escape from our insatiated tormentors the mosquitoes and black flies, that would not let us rest or sleep all night. While at breakfast at eight o'clock a great thunder-storm from the south-west came upon us. Having thrown an end of the tarpaulin over the canoe, and resting the other end on the paddles stuck into the ground, we got beneath it and very soon fell fast asleep, and slept till one o'clock, when I was awoke by the sudden calm, for the storm had apparently only just then ceased.

The valley and river still retain their old character and dimensions till we come to the lowest of the lakes, called *Ka-wa-wi-ga-kamac*, or Round Lake, which varies from one mile to half a mile in width, and is nearly five miles long.

The name is by no means an appropriate one, as it is far from being round. The mean of some soundings I took was twenty-eight feet, the greatest being thirty feet. On the sand banks which are at the head of the lake were my-riads of duck, and large numbers of geese were swimming about in every direction, and a few great northern divers or loons. We camped at a place about two and a half miles down the river, called the Stony Barrier, the Cree of which is *Asinni-pichigakan*. For about 100 yards in length the river is full of large and small granite boulders, rendering it quite impassable for the smallest canoe when the water is low ; at this time the water was just high enough to admit of us passing over it.

Two miles down the river from this spot a little stream brings in its gatherings from the prairies on the south, re-joicing in the name *Isquawistequannak Kaastaki*, which means, " where the heads of the women lie." A long time ago two women, one a Cree and the other an Ojib-way, were killed by the Mandans on the banks of this stream ; their bodies were left unburied, and their skulls are still lying there, from which circumstance the stream derives its name. This was all my informant at Fort El-lice knew of the story. The next creek which is dignified with a name is the " Little Cut-arm," or *Kiskipittonawe sepesis*, the origin of which I could not find out ; it flows in from the north.

A few miles further down, another creek, ten feet wide and very rapid, joins the Qu'appelle on the other side ; its name is *Pesquanamawe sepesis*, which may be rendered into English " the Scissors Creek ; " it is not a very literal translation, but is the best that can be given. The incident to which it owes its name exhibits a peculiar habit of the Indian, but is one that cannot be told. Near this spot there is an exposure of rock on the north slope of the valley, which on examination proved to be a shale similar to that on the Little Souris, but so decomposed that the amount or direction of its dip could not be ascertained. There are several extensive patches where the surface of the rock has been reconverted into soft mud, very much cracked, and on which no grass grows. On digging into it I found the mud to be three inches thick, then fragments very small and soft, and gradually increasing in size and hardness to a depth of about two feet, where the rock is perfectly hard but very much shattered. About fifteen miles to the east of this the rock is again to be seen on the south slope of the valley, also much broken.

On the 26th vast clouds of grasshoppers, flying towards the east, passed high over our heads, without intermission, for nearly two hours. It was the last large flight I saw.

Big Cut-arm Creek, or *Kichekiskapettonano sepesis*, the last to be noted, joins the Qu'appelle about twenty miles from its mouth, and is the largest of its affluents. It is twenty-five feet wide and three feet deep where it issues from a wide ravine on the north side. The Qu'appelle thence to its mouth is from eight to twelve feet deep, and varies in width from seventy to ninety feet, and the rate of current is a mile and a half per hour.

There is much good land in the valley from the Fishing lakes to the Assinniboine, but as it is flooded every spring, it is questionable whether it will ever be of much importance. For ten miles up it there is an abundance of timber, consisting of aspens, balsam-poplars, elm, black ash, oak, birch, and sugar maple; none, however, exceeding one foot six inches in diameter, and few so large. The underwood is chiefly composed of dog-wood, roses, cherries, and pembinas, intertwined with convolvuli and vetches. In this wooded part the birds are innumerable. Kingfishers, blue jays, and Canada jays, cat-birds, and American magpies, flitted from tree to tree uttering their discordant notes. Cherry-birds and pigeons were calmly and listlessly perched on the dense trees, having eaten plentifully of their favourite fruits, while the tyrant flycatcher, when alone or with some companions, chased and worried the crows, ravens, hawks, and eagles, who tried in vain to escape from them. The beautiful white-bellied swallow swiftly skimming the surface of the river, helped in addition to enliven the valley. Ducks and geese crowded the river for several miles; there were enough of them, I should think, to supply all the markets in Canada. Minks were perpetually crossing and re-crossing the river in front of the canoe. I was told that deer are sometimes very numerous in the valley, but I was only fortunate enough to see two jumping deer who were coming down to the river to drink, but the moment they got a glimpse of us away they bounded up the slope. The only other animal we saw was a little prairie wolf, Togany, as he is called by the Indians, that was standing by the edge of the river, and who was so much astonished at our sudden appearance that he never thought of run-ning away, but stood staring at us incapable of motion.

The wonderful uniformity of the valley, or that part of it which I have described, necessarily causes a great deal of repetition in the description of it; so similar is its character throughout, that my two men, half-breeds, well accustomed to mark any peculiarities in the features of a country, said, that though they might pass up and down it several times, they thought they would often be at a loss to know in what part of it they were. The length of the valley from the second Fishing Lakes to its junction with the valley of the Assinniboine is 110 miles, while the river itself is about 270 miles long, which will give an idea of its tortuous course. We arrived at its termination on the evening of July 27th, and having hauled up the canoe on the bank, walked across to Fort Ellice, distant about three miles, where I was kindly received by Mr. McKay.

Very truly yours,

J. A. DICKINSON.

H. Y. Hind, Esq., &c. &c. &c.

CHAP. XVIII.

FROM THE ELBOW OF THE SOUTH BRANCH OF THE SASKAT-CHEWAN TO THE NEPOWEWIN MISSION ON THE MAIN SASKATCHEWAN.

Rocks on the South Branch.—Cretaceous.—Altitude of Exposure.—Character of.—Selenite.—Fossils.—Concretions.—Mesaskatomia Berry.—Character of River.—Drift.—Rock Exposures.—Fibrous Lignite.—Treeless Prairie.—Cree Camp.—Mud Flats.—Rock Exposure.—Concretions.—Treeless Banks and Prairie.—Low Country.—Driftwood.—Ripple Marks.—Dimensions of the South Branch.—The Moose Woods.—Water and Ice Marks.—Forest Timber.—Character of River.—Treeless Prairie.—Boulders.—Soundings.—Buffalo.—"The Woods."—Rate of Current.—Boulders, Arrangement of.—Artificial Pavement.—Tiers of Boulders.—Temperature.—Balsam Spruce.—Former Aspen Forest.—Good Country.—Water Marks.—Soundings.—Stratified Mud.—Fall of River.—Character of River.—Colour and Temperature of North and South Branch.—The North Branch.—Absence of Indians.—Absence of Animal Life.—Grizzly Bear.—Aridity of Country through which the South Branch flows.—Current of North Branch.—Coals Falls.—Boulders.—Trees.—The Grand Forks.—The Main Saskatchewan.—Fort à la Corne.—The Rev. Henry Budd.—The Nepowewin Mission.—Cubic Feet of Water in North and South Branch and Main Saskatchewan, or Ki-sis-kah-che-wun.—Opening and Closing of the River.

THE first rock in position on the South Branch below the Qu'appelle valley is a cretaceous sandstone, exposed on the river bank, for some miles. The altitude of the highest part of the exposure is sixty feet above the level of the water. It is capped by about seven feet of drift (*a*), which rests on twenty feet of soft and easily disintegrated sandstone of a pale yellowish-grey colour (*b*), containing a large number of small, pale yellow, spheroidal bodies, varying from one-tenth of an inch to one inch and a half in diameter, and composed of sand. Below this soft stratum there occurs a layer of sandstone (c) about three feet six inches thick, which is broken into an irregular projecting outline by the protrusion of a series of immense concretions of a flat spheroidal form, like that of a lemon slightly compressed parallel to its longest diameter. The concretions vary from three feet to six feet in horizontal dimensions. They are very hard in the centre, and show concentric rings for at least six inches from their outer casing, which is a shell of gypsum, often passing into selenite. Selenite is found in this and lower strata in

Section on the South Branch of the Saskatchewan, showing layers holding Avicula Linguæformis and Avicula Nebrascana.

veins and fragments. Some of the concretions thrust out their rounded forms from the face of the cliff, others have been broken off and show their internal structure. A grey sandstone (d) with a slight tinge of green, soft and friable, then occurs for a space of four feet; it is succeeded by five feet of hard sandstone (e) containing a vast number of cylindrical forms, slightly conical, and showing traces of organisation. Below this stratum a layer of sandstone six feet thick occurs (*f*), holding spheroidal forms, which vary in size from six inches to two feet in diameter; they are formed of yellow sand containing a hard central calcareous nucleus often four inches to one foot in thickness, and composed almost altogether of an aggregation of *Avicula Nebrascana* (Evans and Shumard). The stratum in which they are embedded holds *Avicula Linguæformis* (Evans and Shumard).

A second layer (*g*) of large concretions follows, similar in external aspect to those already described. Below them there is a persistent layer of hard calcareous sandstone about four feet thick (*h*), containing *Avicula Linguæformis*.

The lowest stratum exposed is a soft sandstone (*i*) about six feet above the river, and passing beneath its level. This rock is worn into caves by the action of water. The formation is nearly horizontal, with a slight north-westerly dip. For several miles this upper cretaceous* rock continues to form the river bank. The concretionary masses are persistent, bold, and prominent, and about three miles in a north-westerly direction from the point where they were first observed, those of the lower stratum are nearly on the same level as the water, thus showing a north-westerly dip of about three feet in the mile.

The banks of the river slope gently from the prairie on the south-west side to an altitude of about 250 feet above it, they then assume the form of steep declivities. On the north-west side the sandstone cliff rises abruptly from the river to a height varying from thirty to sixty feet, when it meets the foot of an undulating slope which extends to the prairie level. Trees, consisting chiefly of aspen and the mesaskatomina, are found in patches on both sides. The river continues for many miles about 700 yards

* See Chapter on the Cretaceous Series, Vol. II.

broad, with numerous sand-bars and low alluvial islands. The drift above the sandstone is gravelly, and many small sand dunes occur on the hill bank sloping to the prairie, into which they have progressed to a considerable distance. A treeless prairie, boundless and green, except where the patches of drifting sand occur, is visible on either hand from the top of the bank; below, the river glides with a strong current, two, and two and a half miles an hour, filling the broad trench it has eroded. The mesaskatomina is very abundant; shrubs or trees eighteen to twenty feet high, loaded with fruit perfectly ripe and of excellent flavour are numerous in every grove, the berries being of the size of large black currants, and very juicy, sweet, and wholesome.

During the morning of the 31st three Crees from a camp on the east bank came to the river; they shouted to us, asking us to land, an invitation we declined. They were "pitching eastward" to avoid the Blackfeet. About twelve miles below the Qu'appelle the river becomes narrower, being not more than a quarter of a mile broad, but full of mud flats and shoals. The banks are more sloping, and frequently broken into two terraces, the upper one being the prairie. The lower terrace is studded with small groves, the intervals consisting of pretty grassy areas, smooth as a lawn.

About fifteen miles from the Qu'appelle Valley the drift is occasionally exposed in cliffs, which disclose its structure twenty to thirty feet above the river. It consists of coarse sand stratified in curves, and often containing beds of gravel; it is also frequently capped by the same material enclosing small boulders. The dip of the rocks to the north-west, and the aspect of the drift, appear to indicate a depression, which may have been the seat of a large lake during earlier periods.

Some exposures of sandstone are visible on the river at intervals lower down, and the drift above them is stratified, containing layers of boulders of the same character as the sandstone below, and so regularly placed as to lead, when viewed from a small distance, to the belief that they are part of rock in position. Thirty miles from the Qu'appelle the rock appears on the south-west side, and consists of a white sandstone, with impressions of fragments of leaves, and some brown, fibrous lignite.

A treeless prairie, with a few sand dunes forms the country on either side for a distance of thirty-eight miles, which comprised the extent of our voyage during the day. As evening began to close upon us we came to a camp of Crees just after they had crossed the river. They numbered nineteen tents, and in order to avoid them we drifted several miles further down, and built our fire close to the river at the mouth of a small gully leading from the prairie, 100 feet above us. Mud flats and sandbars continue as before, but the river is not more than a third of a mile broad.

A narrative of a canoe voyage down a river flowing through a prairie country must necessarily involve numerous descriptive repetitions, which will appear perhaps less tedious and more readable in the form in which they were registered at the time in my note book, than if I were to attempt a connected narrative. I shall therefore strictly follow the daily record of what we observed, at the risk of its being nothing more than a dry enumeration of not very interesting facts.

August 1st.—Morning revealed a fine exposure of rock on the river bank where we camped last night. There is a change in the aspect of some of the strata, they occur massive, in rusty red and greenish-gray sandstone layers, with the concretionary bands as before described. A belt of sandstone twelve feet from the river level is capped by brown and red argillaceous layers, forty feet thick in the aggregate. Drift sand, ten feet thick to the prairie level, succeeds. The upper portion of the drift is hard and reddish coloured; as it approaches the clays below, it partakes of an argillaceous character. The upper stratum of the sandstone weathers reddish brown, with bands of deep red and purple. Below this a greenish-grey stratum occurs enveloping concretions of a reddish-brown colour. The concretions are hard and argillaceous. The greenish-grey matrix is soft when weathered, otherwise hard, and may be split without difficulty into thin layers. The concretions occur in the sandstone in forms easily detached, and often contain abundance of *Avicula Linguæformis.* If the clays above the sandstone are rock in position, the exposure has an altitude of about sixty feet. Fragments of fibrous lignite, dark-brown and sometimes approaching to black in colour, occur in the sandstone. The attitude of the rocks is nearly horizontal, and the greenish-grey sandstone is identical with the formation seen on the south bend of the Qu'appelle above Sand Hill Lake; the red layers are similar lithologically to those observed at the height of land in the same valley, holding the same species of shells. Sometimes layers of grey sandstone occur which are easily split; they contain the impressions and remains of plants. The position of these rocks is about forty-five miles from the Qu'appelle valley.

The river banks and the whole country is now much lower; this subsidence began about four miles south of our camp. The banks at our second camp are not more than 100 feet in altitude, and are becoming lower as we proceed north. They are treeless areas, and so is the prairie on either side, with few detached exceptions.

The river is about half a mile broad, with a current in the lead fully two miles and a half an hour. Large drifted trees are sometimes seen on the beach, and one pine was noticed this morning. They have probably traveled from the flanks of the Rocky Mountains.

About 60 miles from the Elbow, small forests of aspen begin to show themselves on the banks, after passing through a low country, which is an expansion of the river valley. Ripple marks are numerous on the fresh mud, the furrows lying parallel to the course of the stream, they are quite recent and similar to those observed on Red River in the spring. The ash-leaved maple begins to show itself, the aspen being still the prevailing tree, but the " woods are not continuous, and the prairie on either side of the river remains bare; it is fast regaining its former altitude. Sand hills are visible in the distance from the top of the bank, and on its side groves of the mesaskatomina are very abundant and the fruit fine flavoured. The exposed cliffs consist of reddish loam, but rock in position is no longer seen below them. At a point fifty-three miles from the Elbow we made a careful section of the river, and found its breadth to be nearly one-third of a mile (28 chains); its greatest depth was ten feet on the east side, but on the west side there is another channel with nine feet of water.

Approaching the Moose Woods we passed for several hours between a series of low alluvial islands from ten to twelve feet above the water. They sustain some fine elm, balsam-poplar, ash, ash-leaved maple, and a vast profusion of the mesaskatomina. The river valley is bounded by low hills leading to the prairie plateau four to eight miles back. The country here furnishes an excellent district for the establishment of a settlement. The spot where we encamped for the night is an extensive, open, undulating meadow, with long rich grass, and on the low elevations rose-bushes in bloom grow in the greatest profusion. It is only ten feet from the water, yet it does not appear to be flooded in the spring; water-marks and ice-marks are nowhere seen above four feet from the present level of the broad river.

August 2*nd.*—The region called the Moose Woods, which we entered last evening, is a dilatation of the Saskatchewan flowing through an extensive alluvial flat six miles in breadth, and cut into numerous islands by the changing course of the stream. This flat is bounded by sand hills, some of which are nothing more than shifting dunes. The woods are in patches, and in the low land consist of balsam-poplar, white wood, and aspen. Small aspen clumps cover the hills, but no living timber of importance has been seen as yet, although many fine dead trunks are visible, probably destroyed by fire. The river continues to flow through a broad alluvial flat for about twenty-five miles. Its water is very turbid, like that of the Mississippi, holding much solid matter in mechanical suspension.

Beyond the Moose Woods the banks close upon the river, and have an altitude not exceeding sixty feet. The breadth of the stream contracts to 250 yards, with a current fully three miles an hour. On the east bank the prairie is occasionally wooded with clumps of aspen, on the west side it is treeless, and shows many sand hills. During the afternoon we landed frequently to survey the surrounding country. Nothing but a treeless, slightly undulating prairie was visible; many large fragments of limestone not much water-worn lie on the hill banks of the river, which are about 100 feet in altitude. The river continues very swift, and maintains a breadth of 250 yards. Frequent soundings during the day showed a depth of ten to twelve feet. A little timber displays itself occasionally on the east bank below the level of the prairie. The dead bodies of buffalo are seen floating down the stream, or lodged on sand-bars in shallow water. The banks expose occasionally yellow drift clay with numerous boulders ; the soil of the prairie appears to improve as we progress northwards, and the grass is no longer stunted and withered. Little rapids occur at the bends of the river, but there is always deep water on the other side. A heavy thunder-storm compelled us to camp two hours before sunset.

August 3rd.—The river at our camp is not more than 200 yards broad, but deep and swift; the volume of water it carries here, about eighty miles from the Grand Forks, is much less than at the Elbow, where it is half a mile broad. No doubt evaporation during its course through arid plains is competent to occasion a large diminution. Recent water-marks show a rise of five and eight feet, but near the top of the lowest bank stranded timber occurs twenty-five feet above the present level of the river. On both sides a treeless prairie is alone visible. There is a remarkable absence of animal life, no deer or bear have been seen, the tracks of buffalo are everywhere, but they have already passed to the east. The nights are cold but fine, and dew is very abundant. The prairie level is not more than eighty feet above the river.

At 8 A.M. we arrived at a part of the river where it showed an increase in breadth, it is now about a quarter of a mile broad, still flowing through a treeless prairie, in which only one low hill is visible. This character continues for many miles, the hill banks then begin to increase in altitude, and are about 100 feet high, but the river still flows through a dreary prairie for thirty miles from our camp, after which " the Woods," as they are termed, begin ; they consist of a few clumps of aspen on the hill flanks of the deep valley. The face of the country is

changing fast, and is becoming more undulating, patches of
aspens showing themselves on the prairie ; here and there,
however, the remains of a heavier growth are visible in
clusters of blackened trunks ten to fourteen inches in
diameter.   During the afternoon we anchored to measure
the rate of the current.   The river is 200 yards broad,
and it flows three miles and a half an hour ; its average
depth is seven and a half feet.

Some remarkable exposures of drift, consisting of clay,
enclosing horizontal tiers of boulders, (c, f,) often occur
after entering the wooded parts of the South Branch of the
Saskatchewan.   The drift is exposed in cliffs at the bends

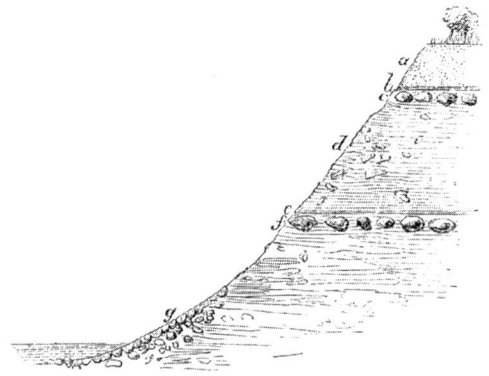

Horizontal tiers of Boulders in drift on the South Branch, with polished Boulder
Pavement at the edge of the River.

of the river, from fifty or eighty feet in altitude.   The
fragments of shale, slabs of limestone, and small boulders
imbedded in the clay (d) are not arranged according to
the position they would assume if dropped by floating ice ;
some of them stand in the drift with their longest axis
vertical, others slanting, and some are placed as it were
upon their edges.   They have the same forced arrangement
and position as the shale, &c., in the blue clay at Toronto.

In many places close to the water's edge and rising from
it in a slope for a space of twenty-five to thirty feet, the
fallen boulders are packed like stones in an artificial pavement (*g*), and are often ground down to a uniform level
by the action of ice, exhibiting ice grooves and scratches
in the direction of the current.   This pavement is visible
for many miles in aggregate length at the bends of the
river.   Sometimes it resembles fine mosaic work, at other
times it is rugged, as where granite boulders have long
resisted the wear of the ice and protected those of softer
materials lying less exposed.

Polished and Grooved Pavement of Boulders on the South Branch.

Two tiers of boulders (*c*, *f*) separated by an interval of
twenty feet, are visible in the clay cliffs lower down the
river.   When first noticed they (*f*) were about fifteen feet
above the water; as we descended the stream they were
seen to rise above its level, preserving evidently a nearly
horizontal position. The lower tier (*f*) consists of very large
fragments of water-worn limestone, granite, and gneissoid
boulders; above them is an indurated sand containing
pebbles; this is superimposed by an extremely fine stratified
clay (*e*), breaking up into excessively thin layers, which
envelope detached particles of sand, small pebbles, and
aggregations of particles of sand.   Above the fine stratified
clay, yellow clay and unstratified sand occur.   The fine
clay must have been deposited in very quiet water, a
microscopic examination subsequently made, failed to re-

veal any diatomaceæ. The polished pavement at the foot of the cliff was observed this afternoon inclined at a high angle, so much so indeed, that it was difficult to walk upon it.

Towards evening the country began to improve, and the timber to include a few elm and birch. In the prairie are clumps of aspen; on the flats, which occur regularly at the inside of each bend of the river opposite steep clay cliffs on the outside of the curve, fine aspens are common and the herbage is very luxuriant. Seventy-five miles from the Grand Forks we leveled the most rapid part of the river seen during the voyage thus far, and found the fall to be 2·85 feet per mile.

August 4th. — Temperature of air at 8 A.M. 61°, of the South Branch 67°. The balsam-spruce begins to appear in groves. The river winds between high wooded banks, with low points and wooded bottoms on one side, high cliffs also wooded with aspen and spruce groves on the opposite bank. The flats are covered with a rich profusion of vetches, grasses, and rose bushes. There are traces everywhere of a former fine aspen forest, with clumps of elm and ash; the dead trunks of these trees, eighteen inches in diameter, being frequently concealed by the undergrowth, offer a rude and stubborn obstacle to progress on foot through the tangled mass of vegetation which covers the rich flats. A view obtained from a low hill coming down to the banks of the river, continues to show a deep valley about three-quarters of a mile broad, through which the river winds from side to side in magnificent curves. The polished pavement on the banks was frequently seen during the day scarred with ice furrows and scratches. During the whole afternoon we passed swiftly through a good country, well fitted for settlement, as far as we could judge from soil and vegetation. Low islands are numerous in the river, and extensive alluvial flats spread out in an expansion of the valley, but water-marks are well preserved seven and nine feet above the present level. The banks of loose clay, when not protected by the pavement before described, being slowly undermined, fall bit by bit into the river. A violent thunder-storm at 5 P.M. compelled us to camp. Soundings during the day showed ten to fourteen feet water in the channel; the current maintaining its speed of three to three miles and a half an hour.

August 5th.—The early part of the morning was employed in examining the surrounding country, which gave evidence of an excellent soil, and timber sufficient for the first purposes of settlers. Much of the timber, however, has been burnt, and the country is fast becoming an open prairie land.

The stratified layers of fine mud (*b*, *e*) before described, were found this morning forty feet from the water's edge, above the horizontal layer of boulders (*c*, *b*) which has again made its appearance. The small aggregations of sand are still distributed between the thin layers of fine clay. A great change is coming over the character of the stream; its fall, as ascertained by leveling, exceeds at some of the bends two feet in the mile, with a very rapid current, sometimes six miles an hour. Large boulders are numerous in the bed of the river, but there is always a passage from fifty to sixty yards broad, often, however, very tumultuous, and for a small heavily freighted canoe very rough, and at times hazardous. The hill banks are getting higher as we approach the North Branch, and the balsam-spruce appears in patches and strips. The river sweeps in grand curves at the foot of high bluffs, in whic fine exposures of the drift may be seen, while on the opposite side are low alluvial points covered with aspens, thick and impenetrable. Six miles from the Grand Forks yellow clay cliffs, 120 feet high, appear at the outside curve of the bends, and where the adjoining flat begins, balsam-spruce two feet in diameter is not uncommon.

At half-past two P.M. we arrived at the North Branch, coming upon it suddenly and finding ourselves in its waters almost before we were aware of its proximity. The temperature of the South Branch was 67°, of the North Branch 62°; an important difference at this season of the year. It is perhaps a fair standard by which to estimate the climatic character of the regions of country through which these rivers flow. The water of the South Branch is yellowish-brown in colour, and turbid; of the North Branch a shade lighter and clearer; the one more resembled the waters of the Mississippi, the other those of the St. Lawrence. The South Branch is the larger river of the two at the Grand Forks.

Throughout the entire length of our voyage we have been surprised at the extraordinary absence of animal life. Of quadrupeds, we have seen half a dozen wolves, two or three badgers, several beaver, skunks, minks, foxes, and a number of dead buffalo; of birds, eagles, geese, a few ducks, kingfishers, cliff martins, pigeons, crows, cranes, plover, hawks, and a few of the smaller birds; but no deer, or bear, or live buffalo; and if we had been compelled to depend altogether upon our guns for a supply of provisions, it is probable that our voyage of 250 miles down the South Branch would have been attended with some inconvenience and delay. Early in spring and late in the autumn game is more abundant, but during the summer season the smaller rivers in the prairies, the ponds and lakes which abound throughout the country north of the Touchwood Hills, to be afterwards described, are the haunts of vast numbers of aquatic birds and of the larger four-footed animals which now form the small remnant of the earlier representatives of animal life in these wilds, before the fur trade led to their destruction, either for the sake of their flesh or skins.

With the exception of the Cree encampments passed during the first and second days of our voyage, we did not meet with a single Indian or half-breed. Once or twice "smokes," which, from their being soon answered in another quarter, we presumed to be signals, and might be raised by Blackfeet in the distant prairies, appeared on the west side of the river.

Once only were we disturbed in camp, and this may have been a false alarm. Both of our half-breeds came into the tent some time after we had retired to rest, and in a low tone whispered "a grizzly bear," at the same time seizing a rifle and a double-barrelled gun which were purposely placed at the foot of the tent ready for any unwelcome intruder upon our repose. The night was dark and the fire nearly out. Our men declared they had seen a large animal within ten yards of us, and pronounced it to be a grizzly bear; the alarm they testified was the only proof of the presence of that terrible animal, for the patient watching of the whole party during the greater part of the night, and a careful search for tracks next morning failed to satisfy me that we had been disturbed by this deservedly dreaded monster of the western plains.

That the grizzly bear is sometimes found far down the South Branch is a well known fact, and he is such a daring and formidable antagonist, that proper precautions are always advisable. A large camp fire often fails to deter this animal from making an attack, and when a

large fire might attract the attention of wandering parties
of Blackfeet which were known to be following the
Crees, who had crossed the river some distance above us,
it would not have been wise to have availed ourselves of
this doubtful security. Our camp was at the edge of a
cliff, we therefore were sure of not being attacked in our
rear, and the greater part of the night was passed in
quietly watching the open space in front of us. It was
the steady determination of the half-breeds to watch
after a fatiguing day, that led me to suppose they had
really seen a grizzly bear, for under ordinary circumstances no people are so unwilling to deprive themselves of sleep during the night in the prairies as those
who have lived the greater part of their lives in them,
without they have the best reasons for keeping themselves
awake.

The very small number of tributaries received by the
South Branch between the Elbow and the Grand Forks is
a remarkable proof of the aridity of the region through
which it flows. For nearly 200 miles the South Branch
receives but one small affluent from the east, and on the
west side, where the water-shed is of much greater
breadth, but where we should expect to find a more arid
climate, it receives eight insignificant brooks. From the
Lumpy Hill to the Grand Forks, a distance of about sixty
miles by the course of the river valley, four streamlets cut
its eastern bank. The water-shed on the east side has not
an average breadth exceeding twelve miles, and two of the
tributaries proceed from ponds in valleys cutting the low
dividing ridge, which, like those of the Qu'appelle, are
tributary to Long Lake or the main Saskatchewan, as
described in the succeeding chapter.

After resting for some time at the junction of these
mighty rivers, the South Branch being about 180 yards,

the North Branch 140 yards broad, their currents meeting one another at the rate of three and a half miles an hour, we turned our canoe up stream, and attempted to stem the tide of the North Branch of the Saskatchewan in search of the " Coal Falls."

During the afternoon of the 6th and the morning of the 7th of August we occupied ourselves in dragging the canoe up the North Branch. Paddling was quite out of the question, the current being from six to seven miles an hour a few hundred yards above the Forks, and continuing rapid for a distance of seven miles, that being the furthest limit of our exploration up the north Saskatchewan. This rapid current is maintained for eighteen miles above the Grand Forks ; the valley of the river, as far as we saw it, resembles in almost all particulars the last ten miles of the South Branch ; but the river channel is much more obstructed by boulders, and the depth and volume of water considerably less. It is doubtful whether in its present condition a steamer drawing more than two feet of water could ascend it, and in dry seasons the boulders and rapids would probably present an insuperable obstacle. The river was high at the time of our visit, nevertheless in descending we had a few narrow escapes from striking against huge boulders just concealed by the water. If some of these were removed, the chief difficulties during low summer levels to steamers of shallow draft and great power would vanish.

The character of the Coal Falls, above the point we reached, is described by the people at Fort à la Corne to be similar to the part we saw. The hill banks expose drift in which large masses of cretaceous rock, containing fish scales, are imbedded. Fragments of lignite are numerous, but no rock was seen in position. The breadth of the valley is about half a mile, and 150 feet deep, through which the river winds from side to side like the South

Branch. The low points are covered with aspen, the hill banks with white spruce, aspen, Banksian pine, and poplar. Just below the junction of the two branches, after they unite to form the main Saskatchewan at the Grand Forks, there is an extensive flat, on which the remains of an old post of the Hudson's Bay Company is situated.

The main Saskatchewan, or Ki-sis-kah-che-wun, as the natives call it, is a noble river, sweeping in magnificent curves through a valley about one mile broad, and from 150 to 200 feet deep. We paddled rapidly round eight points, making a distance of sixteen miles in three hours, and towards evening sighted Fort à la Corne, with the Nepowewin Mission on the opposite or north side of the river. As the description of the Saskatchewan and the valley in which it flows at Fort à la Corne applies equally to the river between it and the Grand Forks, it is unnecessary to incur the risk of needless repetition by enumerating the features of each of the eight points or bends we passed, and of the valley through which the river flows. At Fort à la Corne we made measurements of its leading dimensions, a section of the bed of the river, ascertained its rate of current, examined the cliffs, points, and flats, which are so curiously reproduced at every bend, which will be amply sufficient to illustrate the most interesting and important features of this noble stream between the Grand Forks and a short distance below Fort à la Corne; from that point the country begins to assume a different aspect, and will require an independent notice.

The Saskatchewan opposite Fort à la Corne is 320 yards broad, twenty feet deep in the channel, and flows at the rate of three miles an hour. The mean depth of the river is fourteen feet, but it is in the memory of those living at the fort, when it was crossed on horseback during a very dry season.

An approximate estimate of the number of cubic feet of water passing down the South Branch, North Branch, and main Saskatchewan, gives the following numbers :—

	Cubic feet per hour.
South Branch	123,425,616
North Branch	91,011,360
Main Saskatchewan, at Fort à la Corne . .	214,441,200
Main Saskatchewan near Tearing River . .	206,975,000

The following table will show the comparative magnitude of the Saskatchewan * :—

NAMES.	Area of drainage in square miles.	Length in miles.	Discharge in cubic feet per second.		
			Low water	Mean.	High water.
Amazon	2,400,000	4,000	. . . .	. . . . . .	1,700,000
Mississippi . . .	1,226,000	4,400	447,200	. . . . . .	1,270,000
St. Lawrence . .	565,000	2,600	. . . .	900,000	
Niagara	237,300	. . .	370,589	389,000	406,000
Ganges.	432,000	1,680	36,300	207,000	494,200
Main Saskatchewan	240,000	. . .	. . . .	59,289	
South Branch . .	. . . . . .			34,284	
Nile	520,200	2,240	23,100	220,000	
Ohio, at Wheeling .	25,000	. . .	1,400	. . . . . .	260,277
Thames	5,000	215	1,330	. . . . . .	7,900
Rhone	38,000	560	7,000	21,000	204,000
Rhine	88,000	700	13,400	33,700	164,000
Ottawa (Grenville)	80,000	700	35,000	85,000	150,000
French River . .	4,700	. . .	9,500		

Subjoined is the period of opening and closing of the Saskatchewan for several years at Fort à la Corne.

River Closed.				*River Opened.*				
1851	. .	November	6th.	1852	. . .	April	12th.
1852	. .	„	11th.	1853	. . .	„	14th.
1853	. .	„	13th.	1854	. . .	„	14th.
				1856	. . .	„	9th.
				1857	. . .	„	16th.
				1858	. . .	„	20th.

---

* This table, with the exception of the Main Saskatchewan and the South Branch, is from the Report of the Ottawa Survey, by T. C. Clarke, Esq., C.E.

The Nepowewin Mission is situated on the north bank
of the Saskatchewan, opposite to Fort à la Corne.

Nepowewin Mission and Fort à la Corne.

The Rev. Henry Budd, the native resident missionary,
was ordained priest in 1853 at Cumberland Station, near
the mouth of the Pasquia river. He had long laboured
at this missionary outpost as a catechist. So early as
1840 he set out from Red River Settlement to make
preparations for erecting a church and establishing a
station at the mouth of the Pasquia. In 1852 Mr. Budd
started from Christchurch, Cumberland station, for the
Nepowewin, where he arrived on the 8th of September,
and on the 13th of the same month commenced to clear
a small piece of ground on the river bank, opposite to
Fort à la Corne, for the erection of missionary buildings.
The name Nepowewin is derived from an Indian expression signifying " the standing place," where the natives
are accustomed to await the arrival of the Hudson's Bay
Company's boats, as they are tracked up the north side of
the river. Mr. Budd's house, garden, and little farm is a

pattern of neatness, order, and comfort, yet it is difficult
to say whether the prospects of this mission are favourable or otherwise. I cannot resist the impression that
the selection of a permanent missionary station should be
determined more by the agricultural capabilities of the
locality, than for any special advantages which it may
possess as a fishing station, or as lying on a main line
of communication. The area of fertile land at the
Nepowewin is limited to the points of the river, and
perhaps does not exceed 400 or 500 acres at each point,
these areas being separated from one another by the
river, or by a bend which sweeps the foot of the banks
of the deep valley, and which involve the ascent and
descent of the bank, perhaps 250 feet high, in order
to effect a communication between those which lie on the
same side.

The valley of Long Creek, five miles south of the
Nepowewin, appears to furnish a very large area of land
of the best quality, and will probably yet become the
seat of a thriving community, while the Nepowewin will
remain a mere fishing station or landing-place. But when
these events take place, the wild Indians will have passed
away, and the white race occupy the soil, yet it is to be
hoped that the descendants of some of those heathen
wanderers who have here the opportunity of hearing of
Christ and His kingdom, may find a permanent home
near the Nepowewin, so long distinguished for the medicine feasts which are celebrated in the pine woods
crowning the banks of the Saskatchewan, and whose
remains in the form of painted idol posts, I saw almost
within sight of the mission station, on the opposite side of
the swift flowing river.

CHAP. XIX.

FROM THE NEPOWEWIN MISSION ACROSS THE COUNTRY TO FORT ELLICE.

Sandy Strip on the Saskatchewan.—Banksian Pine.—Indian Idols.—Medicine Feasts.—Rev. Henry Budd.—His Journal.—Fine Country.—Long Creek.—Old Forest.—Fires, Extent of.— Extension of the Prairies. — Former Extent of wooded Country.—Effect of Fires.—Long Creek.—Hay Ground.—Moles.—Humidity of Climate.—A Bear.—Source of Long Creek.—The Birch Hills.—Flowers.—Aspect of Country.—Carrot River. —The Lumpy Hill of the Woods.— Lakes. — The wooded Country. Former Extent of.—Limits of good Land.—Raspberries.—Mosquitoes.— The Height of Land.—Continuation of the Eye-brow Hill Range.—Valley inosculating with the South Branch and Main Saskatchewan. — Grasshoppers.—Character of the Country.—Birds.—Destruction of Forests.— The Big Hill.—Boulders.— Limit of wooded Country.—Belts of Wood. —Great Prairie.—Character of the Country.—Salt Lakes.—The Touchwood Hills.—Beautiful Country.—Excellent Soil.—The Quill Lakes.— Flowers.—White Cranes.—The Heart Hill.—The Last Mountain.—The Little Touchwood Hills.—Lakes.—Touchwood Hill Fort.—Ka-ou-ta-at-tin-ak.—Touchwood Hills Range.—Long Lake.—Devil's Lake.—Garden at the Fort.—White-Fish in Long Lake.—Burnt Forest.—Grasshoppers. —Winter Forage for Horses.—White-Fish.—Buffalo.—Climate of Touchwood Hills— Humidity of.— Medicine Man. — "Wampum."—Trail to Fort Ellice.—Marshes.—Little Touchwood Hills.—Character of Country Changes.—Depressions.—Pheasant Mountain.—File Hill.—Character of the Country.—Heavy Dews.—Cut-Arm Creek.—Willow Prairie.—Little Cut-Arm Creek.—Rolling Prairie.— Attractive Country.— Spy Hill.— Boulders.—Aspen Groves increasing.—Sand Hills.—The Assinniboine.

THE trail from Fort à la Corne to the old track leading from Fort Ellice to Carlton House ascends the hills forming the banks of the deep eroded valley of the Saskatchewan in the rear of the fort. It passes through a thick forest of small aspens until near the summit, when a sandy soil begins, covered with Banksian pine and a few small oak.

This sandy area occupies a narrow strip on the banks of the river, varying from half a mile to four miles broad. South of the sandy strip the soil changes to a rich black mould distributed over a gently undulating country where the pine gives place to aspen and willows in groves, the aspens occupying the crest of the undulations, the willows the lowest portion of the intervening valleys. On the slopes the grass is long and luxuriant, affording fine pasturage. The general aspect of the country is highly favourable for agriculture, the soil deep and uniformly rich, rivaling the low prairies of Red River and the Assinniboine. Our course lay along the banks of Long Creek, which flows in a small depression parallel to the South Branch of the Saskatchewan, and enters the main river near Fort à la Corne.

The large poles of a great medicine tent, erected in the spring to celebrate the annual goose dance, were standing on the top of the hill sides of the valley which the Saskatchewan has excavated. Four painted posts, about five feet high, remained, two on the outside and two on the inside of the ring of the medicine tent. These were the images of Manitou the Indians invoked during the celebration of important ceremonies. The features of a man were roughly carved on each post, and smeared with patches of vermilion and green-coloured paint over the cheeks, nose, and eyebrows. When decorated with fresh paint, feathers, strips of leather, and a painted robe of elk, moose, or buffalo skin, these idols inspire the most superstitious awe among the untutored savages who carve and ornament them. But the awe of many becomes terror, and the superstition absolute idolatry, when illumined by fires at night, and invoked as the representatives of all-powerful Manitou, the whole assemblage jumping in time to the wild song and monotonous drum of the conjurors, circle round their idols, and join in chants to the praises of the spirits they represent.

The Rev. Henry Budd thus records in his journal the progress and continuances of these annual idolatrous ceremonies :—

"*March* 31*st.*—The Thickwood Crees—so named to distinguish them from the Plain Crees — having formed a considerable party, are preparing to commence their spring feasts, &c. ; they have brought some of the best of their last winter's hunt for the purpose. They will be feasting and dancing for several days and nights together.

"*April* 1*st.*—The Indians are busy this morning putting up a large tent, where they intend to keep their feast and dance. The first feast to be kept up is in honour of the God Pahkuk, for having preserved, as they believe, the Indians the whole of last winter, and given them plenty of animals of all kinds to live upon. I hear there is to be no Mitawin kept up here this spring, for what reason I have not yet learnt. Whether it is because we are here, I do not know, but it is the first spring for a long time that that ceremony is not to be kept up here. Old Mahnsuk has arrived to-day from his spring hunt : he has been away nearly one month ; but the river is so dangerous to cross now that we shall not be able to go and see him. The drum is going the whole of the evening in preparation for to-morrow.

"*April* 2*nd.*—The feast has commenced betimes this morning, and the drum has had no rest the whole of last night. The dance does not commence until there have been some long speeches put forth, and the feast over.

"*April* 3*rd : Lord's-day.*—The Indians have been dancing and drumming the whole of yesterday and last night, and this very likely will continue for some time yet.

" *April 4th.*—The Indians are still carrying on their dance and feasting : they are preparing some more places for dancing in. Their great dance, the goose dance, is not yet commenced. This dance is repeated every spring and fall, in honour to the gods for preserving the Indians.

" *April 6th.*—The Indians are now at the height of the goose dance : that over, there will be several ceremonies of less importance to be performed, before the Indians are considered to be in a proper state for enjoying their summer.

" *April 12th.*—We could see some of the Indian tents stripped of their covering, nothing but the bare poles standing, which intimates that those are going away."

During the whole of the afternoon of the 10th we passed through a good farming country. The remains of aspen forests, in which trees of large growth are numerous, are still to be seen standing in groves, or with blackened trunks lie hidden in the long luxuriant herbage until rudely encountered by the carts and horses as they push their way through the rank and tangled grass. Raspberries were abundant in patches but not yet ripe ; they were fully ripe a fortnight since on the Qu'appelle, 200 miles south.

Some of the small aspens had been nipped at the extremities of the branches by frost when in full leaf; the tops of many were black and drooping.

About four miles from Long Creek, and perhaps ten from the South Branch, a low range of hills running north-east and south-west, are still covered with an aspen forest of the same age as the blackened poles which stand in clumps on all sides. These poles are from nine to twelve inches thick ; the young aspens are from four to six inches in diameter. " The fire" was here last year,

and we have now traced the extent of that vast conflagration from Red River to the South Branch, and over
four degrees of latitude. The Rev. Henry Budd states
that in the autumn of 1857, north, south-east, and west
of the Nepowewin Mission the country appeared to be in
a blaze. The immediate banks of Long Creek, with the
exception of a narrow strip in the prairie south of the
Qu'appelle, is the only part of the country in which we
have not yet recognised traces of last year's fire. The
annual extension of the prairie from this cause is very
remarkable. The limits of the wooded country are
becoming less year by year, and from the almost universal
prevalence of small aspen woods it appears that in former
times the wooded country extended beyond the Qu'-
appelle, or three or four degrees of latitude south of its
present limit. It must however be borne in mind that
the term wooded country south of the Saskatchewan is
applied to a region in which prairie or grassy areas pre-
dominate over the parts occupied by young aspen woods.
The southern limit of the wooded country is some distance north of the Touchwood Hills range, but there are
areas north and south of the Qu'appelle where the remains of aspen forests of large dimensions exist, and
young forests are in rapid process of formation perhaps
soon to be destroyed by fire.

This lamentable destruction of forests is a great drawback to the country, and a serious obstacle to its future
progress. It appears to be beyond human power to
arrest the annual conflagrations as long as the Indians
hold the prairies and plains as their hunting grounds. Their
pretexts for "putting out fire" are so numerous, and their
characteristic indifference to the results which may follow
a conflagration in driving away or destroying the wild
animals, so thoroughly a part of their nature, that the

annual burning of the prairie may be looked for as
a matter of course as long as wild Indians live in the
country.   A fire lit on the South Branch of the Saskatchewan may extend in a few weeks, or even days, to Red
River, according to the season and the direction and force
of the wind.

Long Creek maintains a breadth of six feet, flowing
clear but sluggishly through a broad shallow depression,
where wild hay is as abundant as if the whole valley were
one continuous beaver meadow.   The burrows of moles
are very numerous, indeed wherever the soil is very rich
these little animals are to be found in large numbers;
they form excellent indicators of the fertility of a soil,
being never seen where the land is poor and sterile.
Ponds and lakes are abundant, this extensive distribution
of water pointing to a much more humid climate than that
of the country south of the Qu'appelle.   In the morning
I killed a black bear which was leisurely cropping the
willows on the banks of Long Creek.

Our trail on the 11th lay through an equally fertile
country.   The burrows of foxes and badgers have only
twice shown a light gravelly substratum on low ridges,
otherwise the black mould is everywhere distributed.
A chain of lakes, lying westerly from our course, give
rise to Long Creek.   The Lakes are from 200 yards to
a third of a mile broad, and form a continuous series
connected by a small rivulet for a distance of ten miles.
A hill range, called the Birch Hills, whose western flanks
we turned, is said by Indians to extend to the rear of
Fort Pelly.   The Birch Hills form the dividing ridge
between the water which flows into the main Saskatchewan and the Assinniboine, or Red Deer and Swan Rivers.
The remarkable profusion of flowers gives extraordinary
beauty to large open areas; they generally occur in

parterres of several acres in extent occupied by one species, here the yarrow, there the fire weed, then a field of a species of helianthus, followed by *Liatris scariosa*. When viewed from an eminence, the country appeared to be clothed with pink, white, yellow, and blue, in singular contrast to the uniform tint which prevails on the great prairies of the Little Souris. The valley of Long Creek offers by far the most attractive features for settlement of any part of the country through which we have passed since leaving Prairie Portage.

Our course now followed the windings of a shallow brook which runs into the South Branch. It meanders through a fine broad and rich valley with hills on its south-eastern side gently sloping towards it, and covered with the dead yet standing trunks of burnt aspens. The soil of this valley differs in no particular from that of Long Creek; the flowers are equally numerous and showy, consisting of the same varieties, and distributed in large patches occupied by a single species.

We passed near to the source of a river which flows into the main Saskatchewan at the Pas, about 140 miles distant from us. It is called Carrot River or Root River, and rising within twelve miles of the South Branch, drains an extensive area of wooded country, passing also in its course through numerous lakes. The head waters of Root River being within ten or twelve miles of the South Branch show that the height of land between the two water-sheds maintains the same distance as on the Qu'appelle, and at the north fork of that valley near the Moose Woods.

The valley through which the small tributary of the South Branch flows, separates the Lumpy Hill of the Woods from the west flank of the Birch Hills, it is rich in alluvial meadows, ponds and lakes. A view from the

Lumpy Hill, which I ascended on the evening of the 11th is very extensive. The altitude of this eminence is about 400 feet above the general level, and from its summit an undulating open country, dotted with lakes and flanked by the Birch Hills is visible towards the east; south and south-west is a lake region, also north and north-east. These lakes are numerous and large, often three miles long and two broad. Seventeen large lakes can be counted from the Lumpy Hill; low ranges of hills can also be discerned in several directions. The most important of these are the Bloody Hills, the Woody Hills, far in the prairie west of the South Branch, and the chain of Birch Hills running from the Lumpy Hill easterly. The view extends to the borders of the wooded land; beyond is a treeless prairie. The so-called wooded land now consists of widely separated groves of small aspens, with willows in the low places. The Cree Indian guide we took from the Lake of the Sand Hills states that formerly the woods extended in one unbroken range to the borders of the prairie, which may be twenty-five miles south-east of the Lumpy Hill.

Much of the soil on the south and east of the Lumpy Hill is sandy and poor, in fact we have reached the limit of the good land, and are about to enter a comparatively sterile country. Low hills and long ridges running north-east by east, and south-west by south, diversify the general level character of the prairies, as seen from the Lumpy Hill. This eminence consists of drift sand and clay as far as my opportunities of observation enabled me to judge, with boulders on its summit; the western side is very steep, and partially covered with a burnt forest of birch. Raspberries of large size abound on the west side, but the mosquitoes start from the bushes in such countless myriads that it is next to impossible to linger five

minutes to pick the delicious fruit. I offered the Cree
guide a piece of tobacco for a tin cup full of raspberries,
he tried to win it, but after a short struggle with these
terrible insects he rushed from the hill side and buried
his face in the smoke of the fire we had lit in the hope
of expelling them from the neighbourhood of our camp ;
the horses became quite frantic under the attacks of
their tormentors, holding their heads over the smoke, and
crowding together in a vain endeavour to avoid the
clouds of insatiable insects which surrounded us. Both
man and beast passed a miserable, restless, and sleepless
night.

The early part of the morning of the 12th was spent
on the summit of the Lumpy Hill. A strong breeze
drove the mosquitoes away, and permitted me to enjoy
a quiet view of the country, which lay mapped about 400
feet below. After breakfast the trail, taking a direction
nearly due east, passed over a series of hills and through
intervening valleys, constituting a height of land. This
range may be from thirteen to fifteen miles from the South
Branch ; it appears to be a continuation of the Eye-brow
Hill range on the Qu'appelle, before described, receiving
in its easterly prolongation the name of the Birch Hills,
which limit the valley of the north Saskatchewan, as far
as the rear of Fort Pelly. As soon as we had passed the
crest of this range and entered a small prairie east of the
hills, a valley through the range became apparent to our
right. From lakes in this shallow depression water passes
during spring freshets, to the South Branch and also to
the North Branch by a tributary of Carrot River.

Grasshoppers were seen during the day flying to the
north-east. They are the first that have been observed since
leaving the Fishing Lakes on the Qu'appelle. The vegetation still continues luxuriant, lakes are numerous, aspen

groves scattered here and there, and flowers abundant. Wild-fowl are found on all the lakes; and brown and white cranes, together with smaller waders of many species in the marshes. As we approach the great prairie the country becomes more undulating, and the soil light-coloured and poor; the aspens, which cap some of the low hills, are still large, although many are nothing more than dead trunks. The "wooded" country through which we passed during the day is only so called in remembrance of former forest growth. If the devastating fires continue for a few more years it will become a treeless prairie to the Lumpy Hill, and the aspen and birch woods will then be limited to the country between that eminence and the North and South Branches of the Saskatchewan. A young brood of grasshoppers has been seen to-day, showing that these destroyers reached this part of the country last autumn.

After traversing a very undulating country, in which are low ranges of hills and conical mounds with limestone boulders on their summits, we arrived about noon on the 13th at the Big Hill, a point of some interest, for south and south-east of it lies a boundless, undulating prairie. The summit of the Big Hill is covered with huge granite or gneissoid and limestone boulders, indeed on all the hills which surround the Big Hill boulders are very generally distributed. The limit of the so-called "wooded country," is about seventy miles from the North Branch in an air line, and thirty miles from the South Branch.

From the summit of the Big Hill the "Buffalo Cart Plain," and "Lake where the moose died," are visible; both noted localities in the wild history of these regions. South-east of the Big Hill the trail winds through a dreary labyrinth of dome-shaped hills, many of them covered with boulders. One or two small streams flowing

into Ashes' Lake, a large sheet of water to the north of
the trail, meander through this dreary prairie; timber is
only found in the form of small aspens, on low ridges or
near ponds.

The 14th brought us to a better country, still undulating, yet containing many beautiful lakelets fringed
with aspens. The soil is light and the herbage scanty, a
fit introduction to the " Carry Wood Plain," which lies at
the foot of the Touchwood Hills. From a low gravelly
ridge forming the north-western boundary of the Carry
Wood Plain, the Indian guide pointed out a hill towards
the south-west, which he said was close to Long or Last
Mountain Lake, already stated in a former chapter to
join with the Qu'appelle at the Grand Forks.

August 15*th.*—In journeying from the Lumpy Hill we
crossed three " belts of woods," as the Indian guide termed
them, before arriving at the great prairie west of the
Touchwood Hills. These belts, which consist of groves
of small aspen following a low gravelly ridge about a
mile broad, and having a north-east and south-west direction, are separated by prairie valleys which sustain in
their lowest parts a good soil and fine pasturage. Each
belt diminished to a point some ten or fifteen miles south-
west of our track. The points or termination of these
belts are visible from the summit of mounds on our
trail, not more than fifty feet high; beyond them is a
treeless prairie, stretching away to the South Branch.
The " belts of wood " become broader in a north-easterly
direction until they merge into the wooded country between the Birch Hills and the Saskatchewan. There are
many delightful spots in the belts, the herbage is clean
as a well shaven lawn, the clumps of aspen are neatly
rounded as if by art, and, where little lakes alive with
water-fowl abound, the scenery is very charming, and

appears to be the result of taste and skill, rather than the natural features of a wild and almost uninhabited country.

In the prairie valleys, and often when surrounded by conical hills, the ponds are fringed with boulders, while water-marks show that during the spring a large area is flooded. This is particularly the case at the foot of the Touchwood Hills. This great extent of pond and marsh affords food and shelter to vast numbers of aquatic birds. Grey geese were seen here for the first time; the Canada goose is very abundant; and duck, teal, cranes, and bittern are numerous. The lakes and marshes all contain salt or brackish water, which we found to our discomfort was not suitable for culinary purposes, or for slaking thirst. Tea made from it had a nauseous taste, and possessed the medicinal effect which might be supposed to result from preparing that beverage with a weak solution of Epsom Salts. The Touchwood Hills, as seen from this open "salt prairie," present a bold outline gently rising from the vast level, and maintaining a course nearly due east and west for ten or twelve miles.

In the afternoon we began the ascent of a gradual slope at the foot of the Touchwood Hills, following for some distance against the stream the course of a small brook which comes from the summit of the range, bright, cool, rapid, and sweet. At 6 P.M. we reached the summit plateau, and then passed through a very beautiful undulating country, diversified with many picturesque lakes and aspen groves, possessing land of the best quality, and covered with the most luxuriant herbage. From the west side of the summit plateau the Quill Lakes are seen to the north-west; these bodies of water have long been celebrated for the large numbers of goose quills which were occasionally collected there by Indians and brought to the fort for exportation. There is no timber visible on the west side of the range, with the exception of small aspen and burnt willow bushes. All the wild flowers, so numerous and beautiful in the valley of Long Creek, are met with on the summit plateau of the Touchwood Hills, of even larger growth and in greater profusion. Little prairie openings fringed with aspen occur here and there, through which the trail passes; we then come suddenly on to the banks of a romantic lakelet, in which ducks with their young broods are swimming, and flocks of white cranes start from their secluded haunts at the unexpected intrusion. The breadth of this beautiful plateau is about four miles, its level above the "salt prairie" to the west may be about 500 feet. Our course lay diagonally across it, so that we had to pass through seven miles of this delightful country. The Heart Hill, with others not seen before, come into view as we approach the eastern limit and begin a descent to Touchwood Hills Fort. The Last Mountain is visible in the west, but blue in the distance; the Little Touchwood Hills lie before us, the trail to Fort Ellice stretching towards their eastern flank. The country between the two ranges is dotted with lakes and groves of aspen. From a small hill near the fort I counted forty-seven lakes.

Touchwood Hills Fort, August 16th.—Arrived at the fort after sunset last evening. It is situated on the southeast flank of the range near the foot of a hill from which an extensive view of the country is obtained. Heart Hill, or *Ka-ou-ta-at-tin-ak*, is about 700 feet above the general level of the plain, and seven miles in an air line nearly due north (true) of the post. The general direction of the range is N. 45° E. (true). It appears to consist of a series of drift hills, many of which rise in rounded

dome-shaped forms from the summit plateau. The Last
Mountain bears S. 44° W., about twenty-five miles distant
from the post, and the end of Long Lake, as it was pointed
out to me by the guide, bears S. 70° W., distant from the
fort a good day's journey, or about thirty miles. The
Little Touchwood Hills bear S. 27° E., and have a general
direction parallel to the main range. At the foot of the
Heart Hill and on its northern flank is a lake about five
miles long, running east and west close to its foot, and is
said to contain white fish. Devil's Lake, which is connected with Last Mountain Lake, lies about forty miles due
west of the post.

The guide who accompanied Mr. Hime to Long Lake
from the Qu'appelle Mission * describes Long Lake to be
broader at its northern extremity than elsewhere, but
preserving throughout its length of forty to fifty miles a
breadth of one mile on an average.

The garden, or rather the remains of a garden, in the
rear of the fort, produces every variety of vegetable grown
in Canada, but the efforts to cultivate it are almost abandoned in consequence of the depredations committed by
Indians from the prairies, when they arrive in autumn with
their provisions for trade, such as buffalo meat and pemmican. A few of the lakes near the fort are known to contain fish, and it is probable that all of the large fresh water
lakes in this beautiful region also abound in them. The
officer in temporary charge of the post stated that the
people here had only known of the existence of white-fish
in the Last Mountain Lake for three years; they are now
taken there in the fall, and it is probable that the fishery
recently established will become of great importance to
this part of the country. The Plain Crees are not fisher-

* Page 325.

men like the Ojibways, they did not know how to catch fish when the attention of the people at the Touchwood Hills Fort was first directed to the treasures of Last Mountain Lake. Mr. Hoover, the officer in charge at the time of my visit, told me that he had first observed white-fish under the ice in November of 1854, and since that period they have established a fishery which provides the fort with an ample supply for winter consumption. The white-fish weigh on an average 7 lbs., but 10 lbs. each is not uncommon.

The timber on the Touchwood Hills is nearly all small and of recent growth, fires years ago having destroyed the valuable forest of aspen which once covered it. The remains of the forest are still seen in the forms of blackened poles, either standing erect or lying hidden in the rich covering of herbage which is found everywhere on the southwest flank of the range. So luxuriant and abundant is the vegetation here, that horses remain in the open glades all the winter, and always find plenty of forage to keep them in good condition. The cows are supplied with hay, the horses are worked during the winter, either journeying to Fort Pelly or to the Last Mountain Lake to fetch fish. Buffalo sometimes congregate during the winter in the beautiful prairie south of the fort in vast numbers.

On page 319 mention is made of a descent into a lower prairie before reaching the Qu'appelle Valley, whose boundary bore the aspect of a lake shore. On the north side of the Qu'appelle, and distant from it about fifteen miles, Mr. Hime saw a low ridge during the whole day's journey west from the Fishing Lakes towards Last Mountain Lake. This ridge was traced parallel to the Qu'appelle Valley for a distance of twenty miles in a westerly direction from the Fishing Lakes, how far east it extends is unknown.

These opposite ridges seem to show that a former lake or arm of the sea, from thirty to forty miles broad, has left a record behind.

Last year the grasshoppers visited the Touchwood Hills and deposited their eggs; this year the new brood consumed every green leaf in the garden, and made local ravages in the surrounding country. They took their flight on the 28th July for the south-east, and during the period of my visit but few were to be seen.

Snow falls on the Touchwood Hills to the depth of two feet and a half in the woods, and in the prairie where aspen groves are numerous it is not unfrequently found one foot and a half deep. In the great treeless prairie to the south where the herbage is short, the snow is drifted off by winds. The climate of the Touchwood Hills is evidently very humid; thunder storms appear to travel in the direction of this range and occasion a copious precipitation as they pass over it. Not only are lakes very numerous and well supplied with water, but there are several living streams flowing from the range. Indeed the whole country from the Touchwood Hills to the Riding Mountain, including the country about the head waters of the Assinniboine is dotted with innumerable lakes, annually replenished by summer rains.

The humidity of the climate of this part of British America contrasts in a remarkable manner with the variable character of that of the region under the same meridian lying six or seven degrees further south, in the territory of the United States.

When Lieutenant Warren reached Fort Pierre in latitude 44° 24', longitude 100° 30', in 1855, he was informed that there had been no rain or snow there for *more than a year*. The appearance of the vegetation confirmed the statement, as there was scarcely a green spot anywhere to be seen. In the summer of 1855 there was abundance of rain. At Fort Union in July, 1856, no rain had fallen during that year, and in many places there was a great scarcity of grass; after the 15th July rain was abundant. Again, in 1856, very little rain fell at Fort Pierre, so that on the 5th of October the grass had all dried up, although at the same period of the previous year it was everywhere green.*

During the two nights we remained at this post we were disturbed by a noted conjuror who was performing his ceremonies over the suffering form of a complaining woman who lay in his medicine tent near to the fort. His drum and song were heard nearly the whole of the night, and his incantations are described in another chapter as well as the remedy for the sickness of the poor squaw, which the conjuror suggested as infallible.

While at the Touchwood Hills, and indeed whenever I thought there was a chance of obtaining a "specimen," or information respecting it, I made inquiries respecting the celebrated "wampum." I was often told that a real Indian wampum belt is now extremely rare in Rupert's Land, and no one with whom I conversed on the subject had been fortunate enough even to see one.

The name "*wampum*" is applied to little cylinders made from sea-shells perforated through their longest diameter, and generally strung upon leather or sinews. The shells, which the Indians inhabiting the interior formerly obtained by traffic from the tribes on the Atlantic coast, were broken into fragments and ground by friction into the necessary cylindrical form. The little cylinders were strung into bracelets, belts, earrings, and

* Explorations in Nebraska and Dakotah.

ornaments of different kinds, and they were also used as money, as tokens of peace and friendship, and as records of important transactions.*

* "Les *Colliers* et les *Branches de porcelaine* étaient un agent universel en usage de tout tems chez les sauvages dans presque toute l'Amérique. Ils les employaient comme monnaie dans les transactions commerciales, comme ornement et parure dans les fêtes, comme annales pour l'histoire, comme gage et sanction dans les traités, comme satisfaction dans la réparation d'une injure ou d'un crime.

Les grains qui les composaient et qui portaient bien improprement le nom de *porcelaine*, provenaient de certains coquillages marins connus sous différens noms. On les a appelés *vignols, escargot de mer, concha venerea*, et chez les Italiens *porcella*. C'est de ce nom, dit le P. Lafiteau (t. ii., p. 200), qu'on a fait *porcelaine*.

Les sauvages les brisaient en morceaux, et en les frottant sur des pierres, ils leur donnaient la forme de petits cylindres applatis ou allongés. On en trouvait de blancs et de violets ; ceux-ci étaient les plus estimés. Les sauvages les perçaient par l'axe du cylindre, et les enfilaient sur des lanières de cuir. Dans cet état on les appelait *Branches de porcelaine*. Les *Colliers*, sous la forme d'une ceinture de deux pieds de long environ, étaient composés de plusieurs branches, dont les grains étaient liés entre eux comme dans un tissu, et disposés avec art, de telle sorte que le mélange des couleurs produisait des dessins variés. Les colliers ordinaires avaient douze rangs de 180 grains chacun. Les sauvages en fesaient des ceintures, des bracelets, des pendants d'oreille, et quelquefois des plaques qu'ils suspendaient sur la poitrine et sur le dos. (Sagard.)

Les grains ainsi travaillés recevaient généralement le nom de *Wampum*. Les coquillages se tiraient surtout des côtes de la Noüvelle-Angleterre, et de la Virginie.

Les Andastoes étaient célèbres pour ce genre de commerce. Champlain mentionne cette spécialité dans sa carte.

On en recueillait aussi beaucoup sur la côte de Long-Island. Les Hollandais, habitans de ces parages, se livraient à cette spéculation. Van Tienhoven la présentait à ses compatriotes en 1650, comme un motif pour eux de venir coloniser ces rivages, *dont elle était*, dit-il, *la richesse*. On voit en effet que jusqu'en 1673, il y avait peu de monnaie en circulation, même parmi les colons hollandais. Six grains de Wampum blanc, et trois noirs valaient deux sols. En 1683, le maître d'école de Flutsburg recevait encore son traitement dans cette monnaie.

_ Jacques Quartier, qui précéda Champlain de plus d'un demi-siècle sur les rives du St. Laurent, fut curieux de connaître comment les peuples de ces contrées, qui paraissaient si peu développés sous le rapport des arts, pouvaient se procurer cet ornement, à qui il donne le nom d'*Esurgny* et

In reply to a speech delivered by Lieut.-Colonel Mackey to the Indian tribes assembled at the island of St. Joseph, Lake Huron, in 1829, reference was frequently made by the speakers to "wampum." Minutes of the speeches were made at the time*, and the following extracts will show the different application of this significant token. *A Minominie chief:* " I beg of our father to view this Wampum as a pledge of our being faithful children." " With this Wampum we also make a road to his newly kindled fire." " This Wampum I expect to see next year" (delivering the Wampum and pipe). *A Chippewa chief:* " The great Master of Life gave us pipes and Wampum for the purpose of conveying our ideas from man to man." " This Wampum reaches to Penetan guishene" (delivering the Wampum). *A Chippewa chief*

quelquefois d'*Esurguy.* " C'est leur richesse," dit-il, " et la chose qu'ils estiment être la plus précieuse, comme nous faisons de l'or." (Troisième Voyage, c. i.)

Il recueillit la tradition, et voici ce qu'il raconte de cette pêche merveilleuse: "Quand un homme a deservi la mort, ou qu'ils ont pris aucun ennemi à la guerre, ils le tuent, puis l'incisent sur les cuisses et par les jambes, bras et épaules à grandes taillades; puis ès-lieux où est le dit Esurgui, avalent le dit corps au fond de l'eau, et le laissent 10 ou 12 heures, puis le retirent à mont et trouvent dedans les dites taillades et incisions les dits cornibots, desquels ils font des pâtesnostres et de ce usent comme nous faisons d'or et d'argent, et le tiennent la plus précieuse chose du monde." (III*. Voy. c. vii.)

On est tenté de regarder cette histoire comme fabuleuse. Peut-être les sauvages auront-ils voulu abuser de la crédulité d'un étranger, ou cacher leur secret, en lui racontant cette pêche étrange. C'est sur son autorité que beaucoup d'autres historiens ont adopté le même récit sans nouvelle recherche. Cependant quand Lescarbot publia son *Histoire de la Nouvelle-France,* en 1609, cette pêche ne se faisait plus dans le St. Laurent, et les peuples de la contrée n'avaient plus ces coquillages. " Peut-être," ajoute Lescarbot, " ils en avaient perdu le métier; car ils se servent fort de Matachaiz (grains de rassade) qu'on leur porte de France."—*Relation abrégée de quelques Missions des Pères de la Compagnie de Jésus dans la Nouvelle-France, par le R. P. F. J. Bressany, de la même Compagnie. Traduite de l'Italien, par le R. P. F. Martin, de la même Compagnie. Montreal,* 1852.

* Sessional Papers. Legislative Council. Canada, 1847.

from Lake Superior : " With Wampum like this my ancestors made a road to Montreal many years since ; one
end of this string is tied to my village at Sha-qua-me-cong
and the other I wish you to tie at Penetanguishene to
your new fire" (delivering the Wampum). *An Ottawa
chief :* " As a proof of our determination we make a road
with this Wampum, the end of which we expect to see
to-morrow* (meaning next year) at Penetanguishene
(where the presents were to be delivered), and trust it
will continue clear for generations to come."

On the 17th August we left the Touchwood Hills,
and followed the trail leading to the lesser range bearing
the same name.

A chain of hills joins the Greater and Lesser Touchwood Hills, having a course nearly north-west and south-
east, or at right angles to those of the main ranges. In
this subordinate range there are many conical elevations,
some of them well wooded up to their summits, but the
forest trees are small. The trail to Fort Ellice winds
round the base of dome-shaped hills, past small lakes and
aspen bluffs, through luxuriant herbage, and over an
excellent soil. About nine miles from the Fort it begins
to ascend the eastern flank of the Little Touchwood
range, and gently winding up it for several miles, it finally
reaches an extensive marsh which occupies a portion of
the summit plateau. The marsh is but the introduction
to numerous lakes, which continue to diversify the
country in all directions.

On the following day we entered a region differing in
many points from the rich tract we had left. Gravelly
hills and areas of coarse drift sand form the surface of
the country for a few miles, and apparently continue in

---

* "To-day" signifies, this year; "to-morrow," next year; "night,"
winter; "morning," spring; "play," war; "milk," rum. In the minutes
of the speeches the metaphorical allusions are curious and interesting.

a south-west direction to the Qu'appelle. They are succeeded by a number of curious depressions or hollows, circular or oval in form, and varying from one quarter to one mile in diameter, often with a lake in the centre, but without visible outlet. The land is hilly in which they occur, and the elevations form a ridge running nearly north-west and south-east, towards File Hill, like the general direction of the hill ranges before described, but the country is so undulating that it is difficult to ascertain the true character of the surface until we arrive at the summit plateau. Here boulders are seen, the sand is coarse and mixed with a little clay, so as to resemble a gravelly loam on the ridges and hills, as well as on their flanks, but in the hollows and valleys the soil is excellent, and the herbage very luxuriant.

In returning from Last Mountain Lake, in an easterly direction, Mr. Hime crossed a ridge supporting clumps of poplar, and then struck into an open prairie country, which soon became a series of high gravelly knolls, with numerous boulders on them. About fifteen miles east of Last Mountain Lake, he ascended a high range of gravelly knolls, running from north to south, and then came to a valley 150 feet deep ; in the bottom of this valley was a small creek, expanding into a chain of ponds, each about one quarter of a mile long, and three chains broad. Ascending the opposite bank, another ridge of gravelly knolls was passed, and a descent made into the prairie, which continues rolling, and interspersed with willow and aspen clumps and gravelly ridges, until File Hill is approached, when a more humid tract begins, dotted with marshes and ponds, in which innumerable hosts of duck find shelter and breeding places. On nearing File Hill, the soil improves in character, and the country becomes very picturesque and attractive.

August 19*th.*—The view this morning from the summit of a mound revealed a rolling, treeless prairie, stretching on all sides, and bounded only by the horizon. The wooded range of Pheasant Mountain appears low in the south-west, serving only to destroy the uniformity of the general outline. Numerous lakes, ponds, ·and marshes, covered with wild fowl, are visible in every direction. The soil in low situations is good, supporting long grass which afforded fine pasturage for our cattle. The ridges and mounds are gravelly, and a few boulders of the un-fossiliferous rocks are seen here and there. It is remark-able that east of the Touchwood Hills no limestone boulders have been yet noticed, but limestone gravel is common.

The Pheasant Mountain runs north-east and south-west, and may be from fifteen to twenty miles long. Like its western companion, File Hill, it is wooded with aspen, and full of ponds and lakelets. At its foot the half-breeds report a lake ten to fifteen miles in length, on the south-east side, which we thought we could see from our point of view. The Greater and Lesser Touchwood Hills, the Pheasant Hill, and the File Hill, all appear to be rich, humid tracts, which will become important centres when civilisation in conjunction with population reaches these solitudes. North of the Carlton trail, and in the direction of Fort Pelly, the country is marshy, and abounds in ponds and wet places, which emit a very disagreeable odour when disturbed in passing through them. Beaver Mountain, a continuation of the Touchwood Hills, is seen from this wet prairie. The wet grass reminded me that the dews in the Touchwood Hills are very heavy and abundant at this season of the year. Last night dew was deposited a few minutes after the setting of the sun, although the sky was cloudy, and prevented direct radiation. This phenomenon has been noticed several times; the setting of the sun appears to admit of the cooling of the air sufficiently to allow the dew point to be quickly attained on the surface of vegetables, notwithstanding the screen of clouds which must necessarily obstruct radiation into space, but it would also appear to show that the temperature of the clouds must be very low. With the thermometer at 65° in the air, ten minutes after sunset, and under a cloudy sky, I have three times observed dew form since leaving Fort à la Corne. On clear nights dew has always been deposited during the summer, often so copiously as to wet the tents.

On the 20th we crossed a rapid stream with a swift current, ten feet broad and one and a half deep flowing towards the Qu'appelle. It was thought to be Cut Arm Creek; it meanders through a prairie covered with low willows, and named the Willow Prairie, which embraces an extensive area of excellent land, sustaining fine pasturage. Limestone boulders were seen again to-day, but the country preserves a uniform and level character, with a few gravelly ridges and mounds; neither lakes nor marshes are numerous, and wood for fuel is very scarce. Little Cut Arm Creek, which we crossed during the morning, flows in a ravine about 80 feet deep and 400 broad. The next day lakes began to appear again, the prairies to become more rolling and intersected by ridges, which preserve a certain amount of parallelism, generally from north-east to south-west. The aspen replaces the willow in small clumps, and after passing Big Cut Arm Creek, the country is decidedly undulating, attractive, and very well watered. Large hills appear near the Big Cut Arm, which flows in a valley 1200 feet broad and 180 feet deep, resembling that of the Qu'appelle, from which we are not now far distant. We camped in the

evening near to Spy Hill, called also *Ka-pa-kam-a-ou*, or
" Some one knocked;" from a tragic incident which
occurred during the early history of the fur trade. A
war party of Assinniboines were following a body of
Crees, and had despatched a spy to ascertain the nature
of their camping ground. The spy placed himself on the
summit of " Spy Hill " during the night time, in order to
examine the camp of the Crees at the early dawn. A
Cree warrior started at nightfall to Spy Hill for the
purpose of watching the Assinniboines. When the
morning dawned the Cree observed that he was lying
within a few yards of the Assinniboine spy; he approached him stealthily, and succeeded in dealing a fatal
blow before his enemy was aware of his presence, hence
the name *Ka-pa-kam-a-ou*, or " Some one knocked."

Before reaching Spy Hill we saw a tent in the
distance, prettily pitched on the banks of a small stream.
Riding thither with an interpreter, I found an Ojibway
family, consisting of one man, two wives, and several
children. The Indian was a celebrated hunter, and
showed me the produce of his summer hunt, which he
was taking for barter to Fort Ellice. It consisted of
twenty-two dressed moose skins, some bundles of sinews,
and one or two red deer skins. He had killed the moose
and deer on the west flanks of the Riding and Duck Mountains, and appeared to be proud of his success, which was
certainly extraordinary at the present day. The money
value of the twenty-two moose skins would be about
eleven pounds sterling at Fort Ellice, and the Indian would
receive in trade articles not exceeding in their aggregate
value one-third of that sum. When one Indian during the
summer kills twenty-two large animals like the moose,
which would be at least three times as much as he would
require to feed his family, there can be no room for as-

tonishment that the influence of the fur trade has been
mainly instrumental in reducing many parts of the country,
once very thickly stocked with wild animals of the deer
tribe, to a comparative desert, scarcely able to support the
few wandering savages who depend upon the chase for
their subsistence.

August 22nd.—The Blue Hills across the Assinniboine
are visible from Spy Hill, so also are those on the
Qu'appelle.   Spy Hill is a gravelly eminence about
120 feet above the prairie; near to it boulders of the unfossiliferous rocks are very numerous, and of large dimensions.   One of gneiss measured thirteen feet in diameter.
Our old hunter remarked that the aspen groves were
much more numerous west of Spy Hill at the present
time, than when he first remembered the country, forty-
three years ago.   After crossing a sandy prairie flanked
on our left by numerous bare sand hills, we reached the
Assinniboine at the mouth of the Qu'appelle early in the
afternoon, and having forded that river in preference to
the Qu'appelle, we had the pleasure on the following day
of meeting Mr. Dickinson within a mile of the ferry, on
his way to Fort Ellice, our place of rendezvous.   The
distance from Fort à la Corne to Fort Ellice by the route
we followed is about 330 miles.

CHAP. XX.

THE QU'APPELLE VALLEY.—FORT PELLY TO THE SETTLE-MENTS ON RED RIVER.

THE QU'APPELLE VALLEY.—Leading Dimensions.—Character of the Great Plain it intersects.— Elevation above the South Branch. — Lakes in the Qu'appelle Valley—Depths of.—Timber on.—Valley flooded.— Effects of a Dam across the South Branch.—Diversion of the Waters of the Saskatchewan.—Table showing leading Dimensions of the Qu'appelle River, Lakes, and Valley.—THE ASSINNIBOINE.—The West Bank.—Fort Pelly.—White Sand River.—Manitou Lake.—Little White Mud River.—Leech Lake.— Character of Country. — Crops at Fort Pelly.— Swan River. — Snake Creek.—Fertility of Swan River.—Mr. Dawson's Description.—Thunder Mountain.—Porcupine and Duck Mountains.—Dividing Ridge between the Swan and Assinniboine Rivers.—Miry Creek.—Riding Mountain.—Shell River.—River Terraces.—Indian Graves.—Little Saskatchewan.— Cretaceous Shales on Birds-tail Creek.—On Rapid River. —Termination of Riding Mountain.— White Mud River. — Ancient Beach. — Beauty of White Mud River.—Rat River.—Prairie Portage.—The Settlements.

THE QU'APPELLE VALLEY.

THE valley of the Qu'appelle River joins the Assinniboine about five miles above Fort Ellice. It is 269 miles long, and appears to be a former continuation of the South Branch, in a direction nearly due east, to the low regions now occupied by Lakes Manitobah and Winnipeg. Its western extremity issues from the South Branch at the Elbow, or the point where that river, from a south-easterly course, suddenly takes and preserves for 250 miles a north-easterly course, until it joins with the North Branch.

The narrowest breadth of the bottom of the Qu'appelle valley is half a mile ; its greatest breadth about one mile and a half. Its shallowest part is about 120 feet below the level of the prairie, and its greatest depth is between 350 and 400 feet. It cuts a gently sloping plain extending from the South Branch to the Assinniboine. The surface of this plain is slightly undulating, and in a few localities broken by elevations which have a general direction from the north-east to the south-west; the north-western sides being abrupt and steep, the south-eastern descents gradual and undulating. The Touchwood Hills, Lumpy Hill, the Pheasant Mountain, the File Hill, &c., are among the most prominent of these elevations. So gradual is the general slope of this great plain, and so extensive is its surface that Elbow Bone Creek or the Souris Forks* inosculates with the Little Souris River, which after a course leading it sixty miles south of the boundary line, returns north and joins the Assinniboine about 115 miles in a south-easterly direction from Fort Ellice.

The highest part of the bottom of the Qu'appelle valley is only 85 feet above the South Branch at its summer level, and from 75 to 78 feet above it during the spring elevation of its waters. This occurs at a point distant $11\frac{1}{2}$ miles from the junction, where a lake is found, which discharges itself both into the Saskatchewan and Assinniboine. Before connecting with the Assinniboine, it falls about 280 feet in 256 miles, or 1 ft. 1 in. per mile. The difference of level between the South Branch at one end of the Qu'appelle valley and the Assinniboine at the other, does not exceed, according to our estimate, 200 feet.

In its long, deep, and narrow course there are eight

---

* On the map accompanying Captain Palliser's Reports, Moose Jaws Forks is shown by a dotted line to inosculate with the Little Souris, and Elbow Bone Creek is named "Many Bone Creek." This difference in names does not affect the remarkable fact of an inosculation of the Little Souris with the Qu'appelle.

lakes, having an aggregate length of fifty-three miles. Most of these lakes abound in white-fish of great size and the finest quality. They are connected with Last Mountain Lake, occupying another valley running northwesterly, a counterpart of that of the Qu'appelle, inosculating with it at the Grand Forks, and, as reported by Indians, with the South Branch some thirty miles north of the Elbow. Numerous soundings of the Qu'appelle Lakes, showed them to hold from forty to sixty-six feet of water, which depths are maintained with great regularity. Timber ceases in the valley about 168 miles from the Assinniboine. It appears again at the Grand Forks and Moose Jaws Forks, 194 miles from the Assinniboine, and occurs also in small quantities at the Sandy Hills, near the height of land. Moose Jaws Forks is well wooded for a considerable distance : it comes from the flanks of the Grand Coteau de Missouri, whose blue outlines are distinctly visible from this point of the Qu'appelle valley.

It is stated elsewhere, that we frequently found watermarks eight feet above the level of the Qu'appelle River in August, 1858. When the snow melts in the spring there is a continuous water communication from Fort Garry to the Sandy Hills of the Qu'appelle, down which large bateaux might drift without necessarily touching land. According to the testimony of the Crees who hunt on this river, the whole valley from the Sandy Hills to the Assinniboine was converted into a lake in 1852, a year memorable in Rupert's Land for its extraordinary humidity.

The construction of a dam 85 feet high and 800 yards long would send the waters of the South Branch down the Qu'appelle valley and the Assinniboine into Red River, thence past Fort Garry into Lake Winnipeg.

The same result would be produced if a cutting were made through the height of land in the Qu'appelle valley

to the depth of forty or fifty feet, and a dam some thirty or
forty feet high thrown across the South Branch. A second
low dam below the mouth of the Souris Forks would
send these waters through the valley of that river into the
Little Souris, thence into Lake Winnipeg by the Assinniboine and Red River. The time may yet arrive when the
future population of Rupert's Land and Dakotah territory
will find it advantageous to construct these or similar
works, even if they should be for the purposes of irriga-
tion or inland navigation.*

LEADING DIMENSIONS OF THE QU'APPELLE, OR CALLING, RIVER VALLEY,
AND OF THE LAKES WHICH OCCUPY IT.

*Table showing the length, breadth and depth, of the Qu'appelle Valley
at different points.*

					Miles.	Chains.
Length of Valley from the South Branch of the Saskatchewan to the Assinniboine	.	.	.	.	269	0
Breadth of valley 70 miles from the Assinniboine	.	.	0	78†
Do.	120	do.	.	.	1	21
Do.	177	do.	.	.	1	30
Do.	239	do.	.	.	1	5
Do.	253	do.	.	.	1	70
Do.	258	do.	.	.	0	73‡

At its junction with the Assinniboine its breadth exceeds one mile.
At its junction with the Saskatchewan its breadth exceeds one mile and a
quarter.

						Feet.
Depth of the valley 70 miles from the Assinniboine	.	.	320
Do.	120	do.	.	.	.	265
Do.	177	do.	.	.	.	250
Do.	239	do.	.	.	.	220
Do.	253	do.	.	.	.	140
Do.	258	do.	.	.	.	110
At its junction with the Assinniboine the prairie slopes to the valley of that river, and its depth here is	.	.	.	.	240
At its junction with the Saskatchewan the prairie also slopes to the valley of the Saskatchewan, and its depth was estimated to be	.	140

---

* Vide "On the Qu'appelle or Calling River and the diversion of the
waters of the South Branch of the Saskatchewan down its valley, with a
view to the construction of a steamboat communication from Fort Garry,
Red River, to near the foot of the Rocky Mountains;" Report on the Assinniboine and Saskatchewan Exploring Expedition by the Author.

† One mile less 44 yards.   ‡ One mile less 154 yards.

Table showing the length, breadth, mean depth, greatest depth, and distance from the Assinniboine, of the Lakes in the Qu'appelle Valley.

Name of Lake.	Length.	Breadth.	Depth. Mean.	Gr.	Distance from Mouth.
	m. ch.	Chains.	Feet.	Feet	m. ch.
Round Lake, or Ka-wa-wi-ya-ka-mac	4 56	60	28	30	41 20
Crooked Lake, or Ka-wa-wa-ki-ka-mac	6 10	60	31	42	56 0
Fishing Lake, No. 1, or Pa-ki-ta-wi-win	6 0	40	52	66	108 0
„ „ No. 2	3 25	40	32	48	114 20
„ „ No. 3	4 30	60	41	57	119 20
„ „ No. 4	8 50	60	37	54	124 12
Long Lake	. . . .	60 as far as seen.	. . .	. .	168 0
Buffalo Pound-hill Lake	16 0	40	. . .	. .	194 20
Sandhill Lake	4 50	45	. . .	. .	239 50
Total length of the Lakes . .	53 61				

NOTE.—The breadths and depths are the *means* of several measurements. The distances are taken along the centre of the valley.

FORT PELLY TO RED RIVER.

We spent two days in the valley of the Assinniboine near Fort Ellice, being occupied in making a section of the valley. We found its breadth to be one mile and thirty chains, and its depth 240 feet below the level of the prairie on either hand. The river is 135 feet broad, with a greatest depth of 11·9 feet, a mean depth of eight feet, and a current flowing at the rate of one mile and three-quarters per hour.

After drifting down the Qu'appelle from the Mission in canoe, Mr. Dickinson crossed the country to Fort Pelly, passing for the first fifteen miles through a very sterile region, the soil being a light sandy clay, and in many places consisting of pure sand, covered principally with a low growing creeper, bearing berries like the juniper; the grass is very short and scanty, and the aspens, which are the only trees, are very small.   North of Wolverine

Creek the country improves very much as to its soil and
vegetation, but it abounds in marshes, swamps, and ponds
of various sizes, around which grow willows and young
aspens, and this character continues for about sixty miles.

Thence to Fort Pelly the country is densely covered
with aspens from five to fifteen feet high, and with
willows of different kinds, the trail winding through the
Beaver Hills as far as White Sand River. There are
open spaces to be seen now and then, where the luxuriance of the vegetation is remarkable. Lakes and ponds
are very numerous throughout, encircled with large aspens
and balsam-poplars.

Several rivers and creeks flow into the Assinniboine
from the west, into which many of the marshes and
swamps might be easily drained. White Sand River,
which is the largest of them, is seventy feet wide, four
feet deep, and very rapid.

The Indians say that White Sand River rises in a small
lake in the Touchwood Hills, named "Manitou Lake;"
so called, it is alleged, in consequence of a whirlpool it
contains which carries the water of the lake round four
times in twenty-four hours. During the winter season
this whirling motion is attended with noise and com-
motion beneath the ice, which forms first round the edges
of the lake, and then slowly narrowing the area of open
water, finally closes it, the whirling motion still continuing below the surface.

Little White Mud River reveals a curious feature in
the topography of this region. It rises in Leech Lake,—a
marshy sheet of water,—half way between Round Lake,
in the Qu'appelle valley, and Fort Pelly. From the east
end of Leech Lake, a tributary of Big Cut Arm Creek
runs into the Qu'appelle, and from its west end Little
White Mud River rises, which joins the Assinniboine
through White Sand River, near Fort Pelly. Near

Leech Lake, the Indians who hunt in this part of the country (Ojibways and Swampy Crees) cultivate small patches of potatoes.

The country drained by the lower portion of White Sand River is very low and wet; swamps and bogs are numerous, together with innumerable small marshy lakes. At the crossing place, on the trail from the Qu'appelle Mission to Fort Pelly, the bank is high and very steep, being about fifty feet above the water. The river runs at a speed of five miles an hour, and was about four and a half feet deep at the end of July, 1858. An exposure of shale ten feet thick occurs at the crossing; it is capped by forty feet of yellow clay. The shale is probably cretaceous, and of the same age as that on the Qu'appelle and the Riding Mountain, to be described in the proper place. On the north side of the river, which is low and alluvial, an abundance of white sand has given a name to this tributary of the Assinniboine. Grasshoppers were numerous on the north side of the White Sand River.

The crops at Fort Pelly had been beautiful at the beginning of the season, but were all, excepting the potato, completely devoured by the grasshoppers in July. After a short stay at Fort Pelly, Mr. Dickinson visited Swan River, by the valley of Snake Creek, with Mr. Macdonald, the gentleman in charge at Fort Pelly, and Mr. Hime. This beautiful valley contains all the requirements necessary for a settlement. The timber is very plentiful and of a good size; the balsam-spruce is abundant, and averages two feet in diameter five feet from the ground. There is also some fine tamarack, varying from 1 ft. 6 in. to 2 feet in diameter. The balsam and aspen-poplar grow to a large size, and are everywhere to be seen. The land, for the most part, is good sandy loam, and is watered by numerous streams.

Snake Creek is about 13 feet wide, and 1 ft. 6 in. deep; it yields plenty of fish, so also do one or two small brooks running into it. Swan River is from 90 to 100 feet wide, and 14 feet deep, its current is very rapid, being about three miles an hour. The valley, which is from 80 to 100 feet below the general level of the country, is most rich and fertile, but almost altogether filled up with trees, such as poplar, balsam-spruce, and willows.

Mr. Dawson, who in the spring of 1858 ascended Swan River in canoe, thus describes the country through which he passed:—

"From Winnipegoos Lake to Swan Lake the distance is about six miles. The stream which connects them, here appropriately enough called Shoal River, varies in width from 150 to 300 feet. It is shallow, and has a very swift course.

"About Swan Lake the country is highly interesting. Numerous islands appear in the lake: to the north an apparently level and well wooded country extends to the base of the Porcupine Range, while to the south the blue outline of the Duck Mountain is seen on the verge of the horizon.

"Ascending from Swan Lake for two miles or so, the banks of Swan River are rather low. In the succeeding ten miles they gradually become higher, until they attain a height of nearly 100 feet above the river. The current is here remarkably swift, and the channel much embarrassed by round boulders of granite mixed with fragments of limestone, which latter is the rock proper to the country, although it does not crop out, so far as we could see, on any part of Swan River. Landslips occur in many places where the banks are high, exposing an alluvial soil of great depth resting on drift clay or shale, of a slightly bituminous appearance.

"About thirty miles above Swan Lake the prairie region fairly commences. There the river winds about in a fine valley, the banks of which rise to the height of eighty or one hundred feet. Beyond these an apparently unbroken level extends on one side for a distance of fifteen or twenty miles to the Porcupine Hills, and for an equal distance on the other, to the high table-land called the Duck Mountain. From this south-westward to Thunder Mountain the country is the finest I have ever seen in a state of nature. The prospect is bounded by the blue outline of the hills just named, while, in the plain, alternate wood and prairie present an appearance more pleasing than if either entirely prevailed.

"On the 10th of June, the time at which we passed, the trees were in full foliage, and the prairie openings presented a vast expanse of green sward.

"On approaching Thunder Mountain, which seems to be a connecting link between the Porcupine range and the Duck Mountain, the country becomes more uneven. Some of the ridges on the shoulder of the Thunder Mountain even show sand, but there are wide valleys between them.

"On leaving Swan River to cross to Fort Pelly, the land rises rapidly to a plateau elevated about 250 feet above the level of the stream. The road then follows for some distance a tributary of Swan River, which runs in a beautiful valley, with alternate slopes of wood-land and prairie. Numbers of horses were quietly feeding on the rich pasture of this valley when we passed, and what with the clumps of trees on the rising grounds, and the stream winding among green meadows, it seemed as if it wanted but the presence of human habitations to give it

the appearance of a highly cultivated country. The
Hudson's Bay Company keep a guard here to take care
of the numerous horses attached to their establishment of
Fort Pelly."

A very low and narrow dividing ridge separates the
waters of the Assinniboine from those of Swan River,
affording another instance of the remarkable character of
the water-partings and of the uniform inclination of the
great prairie-plains of Rupert's Land. Miry Creek which
flows into Snake Creek, an affluent of Swan River, is not
more than three miles from the Assinniboine. The one
river running far south along the edge of the escarpment
which forms the abrupt north-eastern boundary of these
table lands, the other breaking through it at right angles
and reaching the Winnipeg by a course which curiously
enough, has the same direction and turnings on the east
side as its counterpart has on the west. A bateau or
canoe may descend the Assinniboine from Fort Pelly and
reach Lake Winnipeg by Red River; a bateau or canoe
may also descend Miry Creek in the spring and passing
down Swan River reach Lake Winnipeg by the Little
Saskatchewan or Dauphin River.

On the 26th August we set out on our return to the
settlements, taking the trail on the east bank of the Assinniboine. Our route lay on the flanks of the Duck and
Riding Mountains, and through a country admirably
adapted for farming purposes. On the morning of the
27th the herbage was covered with hoar frost, but without any injury to vegetation. Ponds and lakes are very
numerous on the flanks of the Riding Mountain, but as
far as our opportunities enabled us to judge, the whole
country, with the exception of narrow ridges, possesses
a rich black fertile mould, supporting very luxuriant

herbage, and on the mountain an ample supply of timber, consisting chiefly of aspen of large dimensions. The Riding and Duck Mountains consist of a succession of slopes and terraces on their south-western sides, the ascent being almost imperceptible to the thick impenetrable forest which covers the highest plateau. On Birdstail Creek cretaceous shales identical with those on the Assinniboine crop out in different places, but organic remains are scarce and indistinct.

Shell River separates the Duck from the Riding Mountain, and its valley affords a very interesting illustration of river terraces. Between the present bed of the river and the prairie plateau it cuts near the crossing place, it discloses three distinct terraces, visible on both sides of the river, but more distinctly marked on the left bank than on the right. On the prairie banks of Shell River are several gravelly ridges resembling in most particulars the Big Ridge ; the elevation is thirty feet above the level on which they are formed. The length of the slope to the summit varies from 100 to 180 feet ; the course is nearly north and south, or nearly at right angles to the point where they touch the bank of the stream. The ridges die away in the general rise of the prairie as they recede from Shell River. They are thickly covered with boulders.

Although several landslips have revealed the structure of the banks of Shell River, yet no rock in position was observed. The banks consist, according to Mr. Dickinson, of clay, sand, and gravel mixed with a few boulders. Near the confluence of Shell River, or A-se-sepee as it is termed by the natives, with the Assinniboine, some Indian graves are placed on the edge of the bank, sad memorials not unfrequently met with in travelling through these beautiful wastes.

On Saturday, 28th August, we arrived at the Little Saskatchewan or Rapid River, which Mr. Dickinson had explored for a distance of one hundred miles from its source. The valley of this river is extremely beautiful and fertile until within a few miles of its junction with the Assinniboine; it offers probably the most attractive and desirable place for settlement in any part of the country west of Red River. The stream abounds in fish, the flats in the valley are covered with the richest herbage; timber, consisting of aspen, poplar, and oak, is abundant; the prairies on either side are clothed with the greatest luxuriance of vegetation, the scenery is very attractive, and the river navigable down stream for canoes and bateaux to the Assinniboine. Where the Rapid River enters the Riding Mountain balsam and white spruce appear, and our explorations on the east flank of the range showed that large birch, spruce, poplar, and aspen flourished on the summit plateau.

Fires here as elsewhere have damaged the forest which once covered the country. Vast numbers of young oak and aspen are springing up in all directions on the prairie fringing the river near the trail. Birds are very numerous throughout this region, and every lake contained duck with their young. The aspen groves and willow clumps were alive with grackle and yellow birds congregating in flocks. Humming-birds were also observed as well as the American cuckoo and the solitary thrush. While in the marshes, herons, cranes, and bitterns were disturbed in groups as we cautiously approached in search of duck.

In a brook emptying into Rapid River, I found an exposure of the cretaceous shales before described as occurring on the Assinniboine and the Little Souris. The rock was very fragile, and contained a few fossils in an imperfect state of preservation.

On the 29th we reached the south-eastern termination of the Riding Mountain, and obtained a fine view of the successive steps of which it is composed. These were three in number, each step being separated by a gently sloping terrace; subsequent observation showed that we were encamped on a fourth terrace which is continuous with Pembina Mountain. The entire mountain appeared, from our point of view, to be densely covered with forest trees, and effectually resisted all attempts to reach the summit on the west side with horses on account of the fallen timber and thick growth of aspens. The country through which we passed during the day was very wet and swampy in many places, but on the ridges the soil is dry and gravelly; we were in fact, descending the Pembina Mountain, which being here extended over a great breadth, is not easily recognised. In the afternoon we arrived at a beautiful ridge, running N. 12° W and S. 12° E. One side of this ridge is partly excavated by the White Mud River, and exhibits finely stratified gravel, consisting almost altogether of small limestone pebbles, with a few belonging to the unfossiliferous rocks. The ridge is gently sloping towards the east, and precipitous towards the west, having on either hand a level country, higher on the west side than on the east. I have no doubt this ridge is a continuation of a former lake boundary at a higher level than the Big Ridge of the Assinniboine. Some fine oak grows on the banks of White Mud River near the ridge, and ash-leaved maple begins to show itself again.

Our course on the 30th lay through the prairies drained by White Mud River. This tract of country is second only in beauty and fertility to the valley of Rapid River. Not only is the herbage of surprising luxuriance, but the trees in the river bottoms are of very large dimensions, and consist of oak, elm, ash, maple, aspen, and poplar. Near the crossing place there is a fish weir, where large quantities of pike, suckers, gold-eyes, and other species, are taken by the people of Prairie Portage, who have established a fishing station here, as well as one at Lake Manitobah, some miles further east.

The woods fringing the river at the crossing place are very important. The oak and elm are of the largest size, being often found 2 ft. to 2 ft. 6 in. in diameter, with tall, clean trunks. The hop and vine twine around the underbrush, and give a very attractive appearance to the belt of woods which fringe White Mud River

Wishing to ascertain the character of this stream to its outlet, we gummed the canoe, and once more launched it for a short voyage down the White Mud River, to the fishing station on Lake Manitobah. Mr. Dickinson proceeded down the river, the carts, with Mr. Hime, journeyed on towards Prairie Portage, while I rode to the fishing station, in company with a half-breed who was familiar with the history and progress of the station since its commencement.

We soon arrived at Rat River, a stream of much interest in connection with the floods of the Assinniboine. Down its valley the water of that river flows into Lake Manitobah during freshets, and by making a very shallow cut, a permanent communication in time of high water, could always be maintained. The fishing station at the mouth of White Mud River consists of about half a dozen houses, which are only tenanted during the fishing season. Very large quantities of white-fish are caught here, and no doubt when the demand requires it the station at the mouth of White Mud River will become an important source of supply. The Assinniboine prairies extend to the banks of Manitobah Lake, and their elevation as seen here and at

Oak Point is not twelve feet above the level of that extensive but shallow sheet of water.

We camped on the banks of Rat River, and the following day made a nearly due south course through a rich but treeless prairie to Prairie Portage on the Assinniboine. In making this traverse we passed the shallow, winding, but dry bed of a brook several times, a tributary of Portage River. In wet seasons this bed is occupied with drainage water from the Bad Woods, while Rat River rises within three miles of the Assinniboine in the same locality. The valley of Rat River and of the dry watercourse may yet become of vast importance if it should ever happen that the commercial inducements for effecting a communication with the South Branch by way of the Qu'appelle valley, should lead to the construction of works for that purpose.

On the 31st of August we arrived at Prairie Portage, and reached the settlements at Red River on the 4th of September, after an absence of nearly three months. Our course from Prairie Portage lay through the prairies already described.*

* Chapter XIII.

CHAP. XXI.

FROM FORT À LA CORNE, DOWN THE SASKATCHEWAN, TO THE GRAND RAPID AND LAKE WINNIPEG.

Departure from Fort à la Corne.—Object of the Expedition.—Equipment.
—"Bull-Boats."—Birch-bark Canoes.—General Direction, Current, and
Breadth of the Saskatchewan. — Character of its Valley. — Country
through which the River flows well adapted for Settlement.—Pem-
mican Portage.—Cumberland House.—Description of Cumberland.—
The Saskatchewan and surrounding Country between Cumberland and
the Pas. — Indian Hunter.— Sturgeon. — The Pas.— Christ Church.—
Gradual Depression of the Country bordering the River.—Alluvial Flats.
—Marshes.—Delta.—Muddy Lake.—Rock Exposure.—Marshes and Mud
Flats.—Cedar Lake : its Situation and Dimensions.—Surrounding Coun-
try.—The Saskatchewan between Cedar Lake and Lake Winnipeg.—Cross
Lake Rapid : its Dimensions.— Enter Cross Lake.— Meet a Brigade of
Boats.—Cross Lake : its Dimensions and Altitude.—Surrounding Country.
—The Saskatchewan east of Cross Lake.—Rapids : their Dimensions.—
Smooth Reach. — Drift Clay Banks. — The Grand Rapid : Portage ;
Running the Rapid ; its Dimensions ; Character of its Excavated Bed ;
Magnificence of the Upper Portion of the Cataract ; Mode of Ascending
it ; Remarks in Relation to surmounting this Barrier and making the
Saskatchewan available for Steam Navigation.—Indian Encampment.—
Lake Winnipeg.

MR. FLEMING'S NARRATIVE.

DEAR SIR,—On the 9th of August, 1858, we set out from
Fort à la Corne to continue the canoe voyage down the
main Saskatchewan to Lake Winnipeg, thence to coast
along the western shore of that lake to the Red River. The
principal object of this journey was to complete the track-
survey and reconnaissance of the Saskatchewan, which had
been began at the Elbow of the South Branch, and to
survey and examine the west coast of Lake Winnipeg

with a view to describe the topography and general character of the region bordering the line of exploration.

The birch-rind canoe in which we embarked was eighteen feet long, two feet six inches wide, with a round bottom, and drew about a foot of water, a depth sometimes too great in places where the river abounds in mud-shallows and sand-bars. This small and light craft was one of those which we had hauled overland from Selkirk settlement, and although new at starting, it had become battered and worn during our summer's campaign on the prairies. The crew consisted of two half-breeds, one of Ojibway and the other of Blackfoot origin. Their experience and skill in canoeing, "woodcraft," and hunting rendered them well adapted for the service to be performed.

A canoe-voyage on these north-western waters is generally monotonous and not often accompanied by that excitement and adventure which is still not unfrequently met with in journeying across the great prairie-plains of the North American Continent. A relation of incidents, and a description of objects to which observation was directed, on this voyage of exploration, may therefore possess but little interest compared with a portrayal of prairie life and travel, buffalo hunts, and scenes among the savage and often hostile tribes distributed over the great plains of the far West.

Bark canoes are not often seen so high upon the Saskatchewan, there being a scarcity of birch-bark in the region through which the north and south forks flow. These great prairie-rivers are generally crossed and often descended in "bull-boats" or "parchment canoes" by the Indians, for great distances. These bull-boats are made of one or two buffalo skins, stretched on a light frame, stitched together, and the seams covered with tallow and ashes. Hunters and trappers frequently set out from Fort à la Corne, on horseback or on foot, to the

Moose Woods or the great prairies on the south Saskatchewan, and return in bull-boats laden with dried-meat and skins, both craft and cargo being the proceeds of their hunt. Bark canoes, although more durable than bullboats, are nevertheless very fragile and require to be handled with great care; the seams and cracks in the bark require constant "gumming" and attention. Our canoe being leaky, owing to injuries it had sustained in crossing the plains from Red River to the Elbow of the South Branch, was the source of much trouble until we reached Cumberland House, where, through the courtesy of the gentleman in charge, we were enabled to procure a new canoe and some other necessaries.

The general direction of the Saskatchewan from Fort à la Corne to Cumberland House is north-easterly. The current continues strong for a considerable distance below Fort à la Corne, where the average rate was found to be three miles an hour. In some places the mean velocity of the current exceeds this, as ascertained by repeated trials; and at the points or alluvial promontories at the great bends a small rapid is frequently seen, generally caused by a submerged spit or reef of boulders and gravel protruding into the river; but the water is turbulent in its passage over these shoals only, which are always on one side of the river: in the bay opposite them it is quite smooth and deep, averaging in the channel nineteen feet.

At Fort à la Corne the breadth of the Saskatchewan according to trigonometrical measurement is 965 feet, and its immediate banks are high; the sides of the valley, which are much higher, being no great distance from the river. The breadth of the river continues very uniform, but its banks become gradually lower, the hill sides of the valley at the same time diverging. About twenty miles below Fort à la Corne the banks of the river are low, and the general character of the adjacent country considerably changed. The high cliffs before seen at the great bends give place to rich alluvial flats, supporting a forest of fair-sized balsam-spruce and poplar, and the valley becomes so broad that the high banks are nowhere observed.

Having been occupied some time near Fort à la Corne in making a transverse section of the river, ascertaining its fall by leveling, and measuring its rate of current by log-line (adopting the mean of a series of observations), it was at a late hour when we got fairly under-weigh, and we did not accomplish more than twenty-three miles the first day.

As the day began to wane we drew up our canoe on a low boulder-promontory of this fast-flowing river, and were soon reclining upon the polished and rounded pavement, beside the ruddy and cheerful blaze of a fire of driftwood. The stillness of night gradually crept on, until nothing was heard but the rippling and surging of the water over the smooth boulder-stones at our feet. The Saskatchewan or " the river that runs swift" is truly well named, for even upon the smoothest and deepest parts of the river, long lines of bubbles and foam, ever speeding swiftly but noiselessly by, serve to indicate the velocity with which this mighty artery courses unceasingly onward, swelling as it goes, with the gatherings of its many wide-spreading tributaries, to mingle its restless and muddy waters in the Arctic seas.

The second day of our journey (August 10th) we embarked at 6 A.M., and passed during the day the "Big Birch Islands," and many others; they are all alluvial deposits, and some of them are overflowed in spring. The banks of the river are now quite low, and the country on either side is very flat; but it still continues well adapted for agricultural purposes and settlement, the soil being a rich alluvial loam of a considerable depth, well watered and drained by many fine creeks, and clothed with abundance of timber for fuel, fencing, and building. In some places stony points, projecting into the river, contract it to a width of five or six chains; stretching out from these points there are shoals over which, as already observed, the current is very strong and rough. Among the islands the river attains a width of a quarter to half a mile, but where it is broad its depth is diminished in many places by mud flats. We stopped to camp for the night about half-past six P.M., nearly fifty-three miles from where we started in the morning.

August 11*th.*—We left our last night's resting-place at daybreak this morning, and passed through an excellent tract of country all day; the soil on both sides of the river consisting of a very rich alluvial deposit, ten feet in thickness, above the surface of the water, well wooded with large poplar, balsam-spruce, and birch, some of the poplars measuring two and a half feet in diameter; and, as far as I was enabled to ascertain, the land continues good for a great distance on either side, but more especially on the south side of the river. In many places the river is studded with large alluvial islands supporting a most luxuriant growth of poplar and willows. Among these islands the channel is sometimes intricate, being occasionally interrupted by sand-bars and snags. We encamped about 6 P.M., having attained a distance of about forty-seven miles to-day. We passed a sleepless night; a terrific thunder-storm coming on after dark, and having no tent to protect ourselves from the driving rain, we were drenched to the skin. Our constant tormentors, the mosquitoes, were also excessively annoying.

The general character of the country we passed during the next day is excellent, the soil being rich and the timber of fair quality. The depth and breadth of the river is variable; in one or two places it is impeded by mud-flats and shoals, sometimes holding snags and saw-yers. About noon we came to the mouth of a tributary stream 100 feet broad, flowing into the Saskatchewan from the north, which we supposed to lead to Cumberland, as it corresponded to the description given to us at the Nepo-wewin, but being desirous of keeping the main river, we went on until reaching an old carrying place, called "Pem-mican Portage," leading to the fort, where we discharged and hauled up the canoe. I despatched the Blackfoot half-breed to the fort, and he returned in the evening reporting the road very wet and marshy. We came to-day nearly twenty-nine miles, so that the distance between Fort à la Corne and Cumberland House, by the windings of the river, is upwards of 150 miles.

August 13*th.*—Owing to the thick growth of rushes and the shallowness of the water in many parts of the marsh between the Saskatchewan and Pine Island Lake, we had to go over to Cumberland this morning in the empty canoe, pushing it through the marsh until we reached a strip of dry ground, about half a mile wide behind the fort. Mr. Edward M'Gillivray, the gentleman in charge *pro tem.*, received us very hospitably. I obtained from him some pemmican and flour, and got him to procure for me a new canoe, for which I had to wait, as it was not quite finished. In the forenoon a brigade of boats from the McKenzie River arrived and departed *en route* to York Factory. One of the boats contained Mr. Anderson, chief factor, who was going direct to Red River and Canada. Although Mr. Anderson left Cumberland three days before us, in a boat of four or five tons burthen, well manned and equipped, and infinitely better adapted for encountering the boisterous gales of Lake Winnipeg, than our little canoe, we reached the mouth of Red River only twenty-four hours after him.

On Saturday the 14th August we were aroused at daybreak, by the singing of the voyageurs of another brigade of boats just arriving. It proved to be a detachment from York Factory, bringing Mr. J. G. Stewart, chief trader, in charge of Cumberland. Our canoe was not finished till late in the afternoon, when I should have started had I been supplied with a guide for Cedar Lake and the Grand Rapid; but the only man that was competent and willing to go, being one of Mr. Stewart's boatmen, and they having received their usual holiday allowance of rum on reaching their destination, no arrangement could be made with him. I was consequently compelled to remain till Monday. During the day, Mr. Stewart, from whom I received the most kind and hospitable attention, opened some packs and enabled me to get one or two articles of clothing, of which I stood greatly in need.

Cumberland House, the chief depôt or fort of the Cumberland District of the Hon. Hudson's Bay Company, is situated on the south shore of Cumberland or Pine-Island Lake; in latitude 53° 57′ N., and in longitude 102° 20′ W. of Greenwich (according to Sir John Richardson). It is about two miles in an air line north of the Saskatchewan, on the north side of what is called "Pine Island," a tract of land of considerable extent, between the Saskatchewan and Pine-Island Lake, isolated by two branch rivers connecting the lake with the Saskatchewan. The stream we passed before reaching Pemmican Portage is the western connection, and bears the name of Big-Stone River; it is about six miles long by its windings, and about two chains wide. When the water of the Saskatchewan is high, it passes through this channel or canal into Pine-Island Lake, and when low, the water from the lake flows into the Saskatchewan. At the time my survey was made (16th August, 1858), Big-Stone River was flowing into the Saskatchewan, at the rate of one and a half miles an hour. The eastern connection is about the same size as Big-Stone River, and joins the Saskatchewan some distance below Pemmican Portage; it is called " Tearing River," and is the route followed by the McKenzie River boats. The Saskatchewan boats go by these rivers when they require to call at Cumberland.

The country around Cumberland is low and flat; the soil in some places is a stiff clay, but in general it consists of a gravelly loam a few feet in thickness, covering a horizontal bed of white limestone, and supporting a light growth of poplar and birch. Occasional groves of spruce (the so-called pine of Rupert's Land, from which Pine Island derives its name) are seen here and there. The land being so little raised above the lake and river, a great deal of it is submerged during the spring floods, and some portions upon which the water remains become marshes and swamps, but many of them could be drained and improved without much difficulty.

There are about ten acres of ground enclosed and under cultivation at Cumberland. I observed a field of barley, and another of potatoes, both looking well, within the fort palings; and there is an excellent garden adjoining the chief factor's house; the soil appeared rich and fertile, bearing an exuberant growth of rhubarb, cabbage, peas, carrots, and other vegetables.

Cumberland House being at the junction of two great lines of water communication, one leading from the Pacific, and the other from the Arctic Seas to the Winnipeg basin, is a place of importance, and was formerly one of the Company's principal depôts. Within the fort there are a number of buildings, one of them (the store-house) is a very large edifice, containing extensive machinery and appliances for pressing and packing furs, and making pemmican.   Cumberland has been visited by several cele- brated Arctic explorers.   In the garden there is a sun- dial which was brought from England, and erected by Sir John Richardson, and Sir John Franklin remained here a portion of the winter of 1819, while on his first over- land expedition to the Polar Seas *viâ* the McKenzie River.

Cumberland House, Pine Island Lake.

The weather was very fine during our short sojourn at Cumberland, and although enjoying the hospitality of the gentleman in charge, I regretted the unavoidable delay as the season was already far advanced.   Mr. Anderson very kindly offered me a passage in his boat to Red River, which of course I could not accept as his route lay along the east side of the lake.

The gentlemen at the fort appeared astonished that a bark canoe had come such a distance down the south

Saskatchewan, and that it had been dragged thither over the prairies from Red River; they looked upon it as a novelty in traveling in the north-west.

Messrs. Stewart and Anderson went in search of Sir John Franklin in 1855, under the auspices of the Hon. Hudson's Bay Company. They descended Back's Great Fish River to the Arctic Seas in bark canoes, and obtained some relics near Point Ogle; they also met Esquimaux in the vicinity of Montreal Island who had actually seen the whites, and who confirmed the account which had been previously given to Dr. Rae.

Sunday, August 15th. — A beautiful day. Another brigade from Methy Portage came in and left about noon to-day; bound for York Factory under the pilotage of the veteran guide, L'Espérance. The guide's boat contained a gentleman in the Company's service, from the McKenzie River district, who was proceeding to Montreal by the round-about way of York Factory and England.

August 16th.—We left Cumberland this morning in our new craft, a three-fathom birch-bark canoe. Not being so deep nor of the same beam as the old one, our load of baggage, instruments, and provisions, sank it to within a few inches of the gunwale, rendering it rather unsafe in a heavy sea. I succeeded in getting an Indian guide, through the kindness of Mr. Stewart, but could not prevail upon him to accompany us farther than the Grand Rapid, which ultimately proved fortunate for us, as, had he continued with our party, the pemmican, upon which we had now solely to depend till we reached Red River, would have been exhausted much sooner than it was. We returned to the Saskatchewan *viâ* Big Stone River; and passed the mouth of Tearing River about fourteen miles further down. Between the mouths of these rivers, the Saskatchewan flows occasionally among low alluvial islands, wooded with small poplar and willows, and in many places its depth is lessened by mud-flats and sand-bars; its banks are now low, alluvial flats, only two to three feet above the water, covered with grey willows and sapling poplar. The current in this part of the river is slacker than before, the average rate, as measured by the log, being two miles an hour. We camped about a quarter to seven, P.M., and made a section of the river, which gradually increases in breadth and volume of water; a number of soundings, taken at intervals across the river with the hand-lead, showed a mean depth of twenty feet; the width of the river at this point being 980 feet. I leveled about three quarters of a mile along the bank of the river here, to ascertain its fall.

To-day we met an Indian hunter and his family in a small canoe, the first we had seen since we set out on our journey; indeed the only signs of animal life yet observed in this solitary region were a young black fox, that came down to the brink of the river to slake his thirst, but scampered away as soon as a shot was fired at him, and a beaver in the vicinity of some trees, felled by these industrious animals. The Indian had been catching sturgeon and drying them for future use. This excellent fish abounds in many parts of the Saskatchewan, and it is one of the chief articles of food in the country. The Indians, as well as those in charge of the posts, have frequently nothing else to live upon for months, and the failure of the sturgeon fishing is often the cause of much distress and starvation. The sturgeon sounds are collected at the forts, and form an important article of export.

August 17th.—We embarked at 4 A. M., and observed no material change in the general character of the river and adjacent country during the day. The banks of the river are similar to those already described, being low alluvial flats not exceeding two feet above the water, and covered with willows and patches of balsam-poplar. The tract of country back from the river is rather low and wet; and the Indians make portages in one or two places from the river to small lakes north of it. The current is now much slacker than before, being only one to one and a half miles an hour.

About thirteen miles below Tearing River, Fishing-Weir Creek falls into the Saskatchewan; by which, during high water, boats sometimes go to Cumberland. About fourteen miles farther down, at what is called the Big Bend, the general direction of the Saskatchewan changes from a north-easterly course, which it has maintained from the Grand Forks, to a south-easterly one. This Big Bend is the most northerly point on the river, being very near the 54th parallel of latitude. The Pas or Cumberland missionary station, where we arrived about sunset, is nearly twenty-two miles below the Big Bend. About three miles above, or west of the Pas, the Saskatchewan makes an abrupt semi-circular curve, (called by the Indians " The Round Turn,") causing eddies and whirlpools, the river being at the same time diminished in width. The depth of the river was here found to be thirty-three feet, and its breadth about ten chains. Near the Round Turn there is a wooded ridge, upwards of fifty feet high, about half a mile from the north bank of the river. About three-quarters of a mile above the Pas, Root River, a long affluent with a width at its mouth of two chains, empties into the Saskatchewan.

The Pas, or Cumberland Station, is a missionary post of the Church of England, situated at the confluence of the Saskatchewan and the Basquia River, a tributary about three chains wide at its mouth.

Christ Church, as will be seen in the sketch I made of the Pas, is a neat and rather imposing edifice ; and it seemed like getting back to civilization again, after all our wayfaring, when, on rounding one of the majestic sweeps of the river, the pretty white church, surrounded by farmhouses and fields of waving grain, burst unexpectedly upon our view. It was on a calm summer's evening, and the spire was mirrored in the gliding river, and gilt by the last rays of the setting sun.

The Pas, or Cumberland Mission.

The church is situated on the right or south bank of the river; near it is the parsonage, a large and commodious building, occupied by the Rev. E. A. Watkins, the present incumbent. Adjoining the church there is a neat school-house with several dwelling houses ; and on the opposite side of the river there are six or seven houses, but they seemed to be uninhabited and in a dilapidated condition ; the Indians, for whom they were erected, disliking a settled life devoted solely to the pursuit of agriculture ;

and preferring the wandering and precarious life of a hunter in their native wilds. The river banks at the Pas are ten to twelve feet high, composed of light-coloured drift clay holding boulders and pebbles of limestone, and the surface soil is a dark gravelly mould, well adapted for cultivation; but the surrounding country is said to be low and swampy with marshy lakes. Barley and other crops growing here looked well and were just ripening. Mr. Watkins' garden also looked well, and he kindly supplied us with some onions to make our pemmican more palatable. Mr. Watkins had just arrived from Red River in a freighter's boat. He had been twenty-five days on the route, having encountered much stormy weather on Lake Winnipeg.

August 18*th.*—Having to make some observations this morning, and Mr. Watkins wishing to send some letters with me, we did not leave the Pas till about 9 A.M. From the Pas, the Saskatchewan flows in a north-easterly direction through a low flat country, wooded with scrub poplar, and balsam-spruce, for about eight miles; when again turning suddenly it resumes its south-easterly course, forming a great bend or elbow. About a mile below the mission, a branch, three chains wide, leaves the Saskatchewan, and cutting across the tongue of land embraced by this elbow, affords a navigable passage about three miles shorter than by the main river; although it is the route generally followed by the boats, had I availed myself of it I must have left a considerable portion of the Saskatchewan proper unsurveyed. It was with the greatest difficulty that our Indian guide could be prevented from taking us by this short cut instead of the main river. When remonstrated with, and requested to return, after proceeding some distance down the smaller branch, he said, shaking his head, "you will find yet that the river forks off in many a branch."

About six miles from where this branch or canal rejoins the Saskatchewan, another branch, leading from Moose Lake and House, falls in; before uniting with the great river it separates into two branches forming a Y, the distance between the mouths being about half a mile. From the Pas to this point the character of the country bordering the river gradually deteriorates, the banks becoming lower and lower, and the timber more scrubby and scanty. The alluvial flats are in many places only one to two feet above the water, and they are at some points covered with drift wood, showing that they are flooded at certain seasons.

We stopped to cook dinner opposite the Moose Lake branch, where, by ascending a tree, I succeeded in getting a view of the surrounding country. The banks are here three feet above the river, supporting a thin strip of grey willows along the water's edge; and about half a chain back from the river there commences an extensive marsh or swamp with rank reeds and rushes, interspersed with ponds of open water, and dotted with clumps or islands of balsam-spruce and willows as far as the eye can reach. From Moose Lake Fork to where we camped, about sixteen miles further down, a slight improvement is observed on the immediate banks of the river; occasional groves of young ash, elm, and ash-leaved sugar maple are seen, but the flats behind are generally very low, and covered only with grey willows and sapling poplar.

We started on Thursday, August 19th, at break of day, with wet baggage and blankets, a thunder-storm with heavy rain having come on during the night. About four miles below our camping place, one or two branches leave the main river and flow to the north into a marshy expanse of water, about one mile broad and two to three miles long, called "Marshy Lake." Between this point and Cedar Lake are seen all the characteristics of a great alluvial delta. The Saskatchewan ramifies into many different channels, some of them return to the parent stream forming large islands, and several flow into Muddy Lake and other expansions of the main river, before finally emptying into Cedar Lake.

The country bordering the Saskatchewan from Marshy Lake towards Muddy Lake and Cedar Lake, consists of low mud-flats not exceeding eighteen inches above water, supporting along the river's edge a belt of willows, alder, dogwood, and long rank grass; in the rear is an extensive marsh with occasional islands of small poplar and spruce. These flats, being so little above water, are flooded every spring after the ice breaks up, and no camping place can then be found for a considerable distance up the river. A very rich mud is deposited during these floods, raising and extending the flats every year.

Muddy Lake, near which we were compelled to remain for some time, owing to a boisterous head wind, is apparently a dilatation of the Saskatchewan in a northerly direction; it is about two miles wide, and extends to the north for about four miles. We effected a landing on a point of the river four to five feet above the level of the water, where we found an exposure of light-coloured limestone in horizontal beds along the water's edge, and several large detached masses adjacent. This was the first outcrop of rock *in situ* we met with on the main Saskatchewan, and I made a very careful search for fossils, but, being unsuccessful, had to content myself with some specimens of the rock. On examining the point it was discovered to be an island eight chains long and four broad, with the river on one side, and on the other a vast reedy marsh interspersed with large ponds. This island is a favourite camping and fishing place of the Swampy

Crees, there being on it a clump of good sized poplar, the only timber fit for fuel for miles around ; and here they hold their great councils, dog feasts, and medicine dances. Its name in Swampy is *Kash-ke-bu-jes-pu-qua-ne-shing*, signifying, "Tying the mouth of a drum."

Between Muddy Lake and Cedar Lake the Saskatchewan meanders through an immense marsh with tall reeds and rushes. It is now no longer an integral stream, but is divided into a maze of reticulating branches. According to our Indian guide, land is being formed here very fast ; and what is now marsh and mud-flats was, within his recollection, open navigable water for a considerable distance back from where the Saskatchewan at present debouches into Cedar Lake through its numerous mouths. In one or two places we saw the trunks and branches of stranded trees sticking above water, where alluvial flats or shoals of mud and drift timber are in course of formation.

The Indians informed me that beyond these extensive alluvial flats and shallow marshes there is not to their knowledge anything but " muskeg " or boggy swamps for a very great distance on either side. I could see no high ground of any kind, and the character of the country bordering the Saskatchewan, as above described, may be said to continue back from the river for many miles.

We entered Cedar Lake on the morning of the 20th August, and coasted along the north shore till about noon, when we ran into a fine little harbour to eat dinner, after making a long traverse. In the afternoon, while crossing a wide and deep bay or sound stretching far to the north (the extremity being below the horizon), a stiff breeze sprang up, soon raising a very heavy sea, in which our canoe became almost unmanageable, pitching tremendously and shipping a great deal of water. On the 21st we breakfasted at the Rabbit Point, and entered the portion of the Saskatchewan issuing from the east end of the lake about noon.

Cedar Lake (so called from the occasional groves of cedar — a tree rarely seen in Rupert's Land — growing on its shores, particularly at its western extremity) is an expanse of water of considerable extent, in which the turbid waters of the Saskatchewan are allowed to disseminate and settle before re-uniting into one great river and rushing down the Grand Rapid into Lake Winnipeg. It is situated in about 53° 15′ N. latitude, and 100° W. longitude; and is nearly thirty miles long, with a breadth at its widest part of about twenty-five miles; its coast line embracing an area of water of about 312 square miles. Cedar Lake being more than sixty feet higher than Lake Winnipeg, is consequently upwards of 688 feet above the sea level. The only tributary it has of any size, besides its principal feeder the Saskatchewan, is a branch leading from Moose Lake and House, which enters it from the north. I was unable to obtain soundings of the lake in consequence of the high winds and stormy weather that prevailed during our voyage through it, but so far as I could learn it has sufficient depth of water for the largest craft, except at the west end, where the Saskatchewan is rapidly filling it up.

The northern coast of Cedar Lake is deeply indented and very low, and the country continues flat for a long distance back. At some of the points and on many of the islands along the coast, there are exposures of limestone in horizontal beds, the top of the strata being a few feet above the surface of the lake. It is to be regretted that, owing to the stormy weather, and the rate at which we were obliged to travel, no opportunity was afforded for collecting specimens. The mainland and islands being well wooded with balsam-spruce, birch, poplar, tamarack, cedar, and Banksian pine, could furnish an abundant supply of fuel ; thus offering, like the Saskatchewan, facilities to steam navigation ; but a considerable portion of the land is reported to be swampy and unavailable for agricultural purposes.

Cedar Lake is separated from Lake Winnipego-sis (the Little Winnipeg) on the south, by a low isthmus called the Mossy Portage, by which the Hudson's Bay Company formerly sent their supplies to the Swan River district. (They are now sent *viâ* the Little Saskatchewan.) The distance between the lakes at this point is about four miles, and Lake Winnipego-sis is about four feet higher than Cedar Lake, according to Mr. Dawson's measurements.

The portion of the Saskatchewan between Cedar Lake and Lake Winnipeg is nearly twenty miles in length, and its general direction is easterly. Through this channel, the great volume of water brought down for many hundred miles by the main river, and its north and south branches, together with that collected by many tributaries through a wide extent of country, is disembogued by one grand mouth into Lake Winnipeg.

Where the Saskatchewan issues from Cedar Lake the bed of the river is divided for a short distance into two channels by an island. We entered the smaller or south channel, and found it only two or three chains wide, for a distance of about a quarter of a mile. At its narrowest part, near the beginning, the Indians have a fishing station, and white fish and sturgeon are caught there in abundance. Along the side of this watercourse there is an outcrop of horizontal limestone, three to four feet in thickness, above the water, covered with a thin coating of vegetable mould, supporting small poplar, willow, and dogwood. I brought away some specimens of the rock, but could find no fossils. The current in this channel, as in most places where the river is narrower than usual, is strong, measuring two and a half to three miles an hour.

About half a mile below Cedar Lake, on the right or west bank of the river, which is now more than half a mile in width, is situated Cedar Lake House, a winter trading-post of the Hon. Hudson's Bay Company, lately established with a view to compete with the "Freemen," who come annually from Red River to trade with the Indians in this locality.

Between Cedar Lake and Cross-Lake Rapid, a little below which the Saskatchewan expands into Cross Lake, the river is very broad, and widens here and there into deep bays and funnel-shaped indentations. It grows narrower again a little above the rapid, where a projecting point of limestone, obstructing the current, causes a small smooth rapid on the south side, with a fall of about eight inches. The Cross-Lake Rapid is occasioned by a band of limestone intersecting the bed of the Saskatchewan nearly at right angles ; and this is the first interruption of any magnitude to the even flow of the river. The Saskatchewan is let down by this rapid about five and a half feet in a short distance. There is a large island near the south side of the river, extending the length of the rapid, and dividing it into two channels. The broadest or northern channel is that which came under my observation. It is about thirty chains wide, and is the route followed by the Hon. Hudson's Bay Company's boats. In order to ascend the rapid, the Company's boats of four to five tons burden have to be "tracked" or dragged up with half cargo, and the other half of their load has to be carried over the portage, a distance of 230 yards. The fall from the west to the east end of the portage (obtained by leveling) is 4·08 feet, and from the east end of the portage to the quiet water below, about one and a half feet, making a total fall of 5·58 feet. Loaded boats run the rapid without difficulty, and if the channel were cleared of boulders and improved, it might be ascended by a powerful steamer.

Having spent some time in making observations at Cross-Lake Rapid, it was late in the afternoon when we entered Cross Lake; where our Indian guide left us, although he had agreed to pilot us down the Grand Rapid. He expressed himself anxious to return to his family at Moose Lake, and could not be induced to go farther. During the return journey, upon which he set out in a little canoe that he picked up coming down the river, he would have several days' hard paddling against a swift current. His departure was not objectionable, however, inasmuch as, had he continued longer with us, our slender stock of pemmican must soon have given out. We could spare him but a handful to carry him to his destination.

At the east end of Cross Lake we met Mr. Christie (a gentleman in the service of the Hon. Hudson's Bay Company, who had recently been appointed to the charge of Edmonton House,) in command of a brigade of boats, *en route* from York Factory to Edmonton and the Rocky Mountain district. Mr. Christie's heavily laden boats (fourteen in number) were manned by a motley group of Indians, half-breeds, Orkney-men, Norwegians, and negroes; they had just made the laborious ascent of the Grand Rapid, and thus far their progress had been very slow. Mr. Christie represented the many difficulties which had to be contended with in a boat voyage; the detentions on the lakes by contrary winds; the strong currents and rapids that had to be encountered in ascending the rivers; and the difficulty of procuring men suitable for the work; (each boat requiring six to eight experienced voyageurs;) and he expressed a hope that the long talked of steamers would soon make their appearance on Lake Winnipeg, to replace the present tedious, toilsome, and expensive mode of conveyance.

In reply as to whether there would be sufficient business to warrant the placing of steam vessels on these north-western waters, (irrespective of the establishment of a continental route to the Pacific, through British territory;) I was informed that there would be plenty of freight to carry for the present requirements and traffic of Rupert's Land; as during the year (1858) no fewer than 167 freight boats of the largest class, (four to five tons each) belonging to private traders and merchants, as well as the Hon. Hudson's Bay Company, (many of them loaded with valuable furs,) had passed Norway House, at the northern outlet of Lake Winnipeg, *en route* to York Factory; and returned with heavy cargoes of merchandise brought by sea to York, consisting chiefly of the usual supplies for Selkirk settlement, ammunition, and a variety of goods for the prosecution of the Indian trade both by the Company and "Freemen." The aggregate quantity of freight transported by this fleet of boats from the seaboard to Lake Winnipeg, and thence distributed along its principal feeders would be upwards of 800 tons. It is well known that there are large quantities of goods imported by other lines of communication — chiefly through the United States territory at present; and as the York Factory route is to be partially abandoned, a large portion of the importations of Rupert's Land will have henceforth to enter the Winnipeg Basin from the south, so that there will doubtless be sufficient commerce in view of the great water facilities afforded by the country, to encourage the initiation of steam navigation.

After remaining at Mr. Christie's encampment about

an hour, we set off again in the hope of reaching the Grand Rapid before dark. We soon entered a rapid by which we were lowered about two and a half feet in a distance of ten chains, followed, after an interval of smooth water, by another about a mile long, but with an easy inclination, the descent in that distance not being above seven and a half feet; it being nearly dark when the foot of the latter was reached, we camped for the night (August 21).

Cross Lake doubtless derives its name from its shape and the peculiar position it bears in relation to the Saskatchewan, of which it is evidently a dilatation. It is an oblong sheet of water, upwards of eight miles in length, having its longitudinal diameter at right angles to the general trend of the river; three miles is its greatest transverse diameter, and this breadth is about the distance between the termination and beginning of the bed of the river on either side of the lake. The altitude of Cross Lake in relation to Cedar Lake and Lake Winnipeg, acquired by leveling the rapids and measuring the currents in the river, would make its approximate elevation above the sea about 680 feet. It is reported to be deeper than Cedar Lake; and its banks on the east and west side are more abrupt and rocky, but its northern and southern shores are very low. Along the coast there are some fine groves of balsam-spruce, and aspen, but the land back from the lake is very flat and poorly wooded, a great portion of the original forest having been destroyed by fire; large tracts of burnt and dead timber are seen here and there; the blackened trunks of poplar and spruce indicating the ridges or dry areas over which the conflagration extended, and the lifeless tamaracks revealing the swamps or flooded land. The lake extends so far to the north, its extremity in that direction is not seen from the traverse line, being below the horizon of the spectator. In the northern arm of the lake there are several wooded islands, but as they were some distance from our track I was unable to ascertain the nature of their formation.

There being two rapids between Cross Lake and the Grand Rapid, the Saskatchewan may be said to descend by four distinct steps from Cedar Lake to Lake Winnipeg ; the first one east of Cross Lake, having a length of about ten chains with an estimated fall of two and a half feet, occurs half a mile below the recommencement of the channel of the river, and appears to be attributable to a low and nearly level belt of limestone, through which the river has gradually excavated its way by three separate channels. The middle channel, by which we descended the rapid is only three to four chains wide, and could apparently be ascended by a steamer without difficulty, as it is deep and appears to be free from boulders. The other channels might even be more favourable for steam navigation, being broader as far as could be observed, and containing a greater volume of water ; they are, however, a little out of the direct course, and for this reason are not followed by the boats. The smooth portions of the river are broad here ; the width above the two islands formed by these three channels being more than half a mile, and below them upwards of three-quarters of a mile. About a mile below the foot of the first rapid the second one begins. Its length is fully a mile, and its approximate fall is not more than seven and a half feet. It has a long gradual slope, with a deep channel of rolling, but comparatively unbroken water in the middle ; the water is more turbulent at the sides, where the current is interrupted by points of limestone rock, boulders, and débris. The exposures of limestone on the points, are four to six feet in thickness above the water, with a horizontal stratification. The loaded boats of the Hon. Hudson's Bay Company descend this rapid easily, and as they are generally " tracked" up with the whole of their lading, a lightened steamer, with powerful engines, might surmount it by taking the best channels and other precautions.

It is about four miles from the foot of this last rapid to the beginning or summit of the Grand Rapid. In that distance the river is smooth and deep, but has a very swift current, especially where its bed is contracted. The width of the river in this interval is much diminished, varying from nine chains to a quarter of a mile, and the rate of current is from three to three and a half miles an hour. There are one or two large boulders in the bed of the river here, over and around which the water boils and bubbles like a cauldron; and now and then shoals on the north side of the channel are indicated by the rippling water and ground-swell occasioned by the current in passing over them. The land between Cross Lake and the Grand Rapid is generally low and flat, but thickly timbered with balsam-spruce, poplar, tamarack, and birch. At the second rapid, east of Cross Lake, the banks on the north side of the river are eight to ten feet above the surface of the water, and are composed of a light-coloured drift clay. These clay banks gradually increase in height towards the Grand Rapid, where they attain an elevation of upwards of twenty feet; but it is probable that the surface of the country is nearly level, and that it is the descent in the river which causes the apparent rise in its banks.

August 22nd.—This being Sunday we did not proceed on our journey till after breakfast (about 8 A.M.). However desirable it might have been, under other circumstances, to have remained inactive on this day, in the position in which we were placed, like a ship at sea, with a limited supply of provisions, and a long and hazardous voyage before us, it would have been altogether out of the question; indeed, the loss of a day or even an hour might have compromised the safety of the whole party.

In about an hour we reached the beginning or west end of the portage at the head of the Grand Rapid, where my various instrumental observations and measurements in relation to the rapid began. In order to commence operations we disembarked and made the portage, which of course is never done by boats in descending the river. Yet, notwithstanding that boats invariably "run" the whole of the rapid, it would be extremely perilous to descend the upper portion of it in a small heavily laden canoe without a guide.

So much having to be done with so few hands, our little party exhibited a scene of unusual activity and exertion from the time we landed at the top of the rapid until we camped in the twilight on the coast of Lake Winnipeg. The first thing to be accomplished was the transportation of the canoe and the heavier articles of luggage to the east end of the portage; to effect this the united energies of the party were required, and owing to the length of the portage it occupied some time. Whilst the Ojibway was carrying the remainder of the lading, I was engaged with the Blackfoot in making a survey of the portage and rapid, chaining across in one direction and leveling back in another, and so forth.

About 4 P.M. the various observations were completed, and everything had arrived at the east end of the portage. The different operations involved the crossing of the portage (more than a mile in length) many times during the day. While dinner was preparing I occupied myself in making a sketch of the cataract and examining the character of the perpendicular limestone cliffs at its side.

After eating a hasty meal we re-embarked to run the lower portion of the rapid.* The voyageurs wished me to walk through the woods to the foot of the rapid (probably to lighten the canoe), but as the day was already far advanced, and being anxious to reach Lake Winnipeg, as well as for other reasons, I deemed it expedient to go down " in canoe."

In running the rapid we followed as closely as possible the instructions given to us by our old guide on the Plains (John Spence), who had often piloted the old North-West Company's *North* canoes down its entire length. In attempting, according to his directions, to cross from the north to the south side of the rapid in order to get into what was reported to be the best channel for a small canoe, such was the fierceness of the current, and the turbulence of the great surges and breakers in the middle, that we were nearly engulfed ; and although every nerve was strained we were swept down with impetuous velocity, and did not get near the other side till we were about three quarters of a mile below our starting point. We were then impelled with astonishing swiftness along the south side of the torrent, often in dangerous proximity to the rugged wall of rocks bounding the channel, and now and then whizzing past—almost grazing—sharp rocky points jutting out into the river, against which the thundering waters seethed and foamed in their fury. During the descent the voyageurs exerted themselves to the utmost of their strength, and evinced an admirable degree of coolness and dexterity.

The Grand Rapid is acknowledged by those who have witnessed it, and who have had opportunities of traversing

---

* The part here designated as the *lower portion*, although the Grand Rapid is one continuous torrent from beginning to end, is that below the east end of the portage, and is more than one mile and a half in length.

the great river systems of the continent, to be unsurpassed (as a rapid) in magnificence and extent, as well as in volume of water. It is certainly a formidable barrier to the navigation of the Saskatchewan.

The following are the dimensions of the leading features of the Grand Rapid :—

1. *Its Length.*—The portage path is nearly straight, with a magnetic course from the upper to the lower end, of S. 60° E. ; it is 87 chains 40 links in length ; the

The Grand Rapid of Saskatchewan.

distance between its extremes by the river is a little more than this, as the river describes an arc of which the portage is the chord, but as the head of the rapid is a little below the west end of the portage, this distance may be adopted as the length of the upper or most precipitous portion of the rapid. The distance from the east end of the portage to the foot of the rapid by our track is 129 chains. This would make the whole length of the rapid 216 chains 40 links, or nearly 2¾ miles.

2. *Its Descent.*—By leveling carefully along the por-

tage path, I ascertained the fall between the smooth
water at the head of the rapid to the general level of the
water at the east end of the portage to be 28·58 feet;
and after observing instrumentally the descent in the
lower portion of the rapid as far as the nature of the
country would allow, I closed my levels on a bench-
mark at the surface of a pond of still water, fed by an
eddy at the lower end of the portage. The fall in the
lower portion of the rapid, acquired by leveling and by
careful estimation, is about 15 feet; this would give 43½
feet as the total descent of the rapid.

3. *Its Breadth and Depth.*—The width of the river, at
the upper end of the portage, is about 20 chains; at
the head of the rapid, about 7 chains further down,
where there is an island in the bed of the river, it is
about 30 chains; and at the lower end of the portage,
where the rapid emerges from the highest limestone
plateau, its width is about 10 chains. From thence it
gradually widens towards the foot of the rapid, where it
attains a width of 25 chains. I was unable to obtain
soundings of the rapid, but, from the depth and volume
of water above and below it, where the river is much
broader, it is undoubtedly deep.

The Grand Rapid, throughout almost its entire length,
washes the bases of perpendicular escarpments of rock.
It passes through two plateaux of brittle buff-coloured
limestone, with a horizontal stratification; the top of the
first, or upper plateau, being nearly on a level with the
surface of the water at the head of the rapid, and under-
lying a stratum of light-coloured clay, twenty-three feet
in thickness, in which are embedded boulders and pebbles
of limestone; the whole overlaid by about eight inches of
vegetable mould, and clothed by a forest of balsam-
spruce, tamarack, and poplar. The surface of this plateau

continues nearly level as far as the lower end of the portage, where the top of the rock is 25·36 feet above the surface of the water, and about the same height above the lower plateau.   The lower plateau continues some distance further down, but is soon hidden by drift clay-banks, which, at the foot of the rapid, have an altitude of twenty to thirty feet above the water.

It is not improbable that the Grand Rapid is the result of the eroding influence of the great body of water in the river, upon the rock through which it flows—the lime-stone being of a crumbling and yielding nature.   At a remote period, the water of the Saskatchewan was per-haps lowered from the top of this rock formation, by a perpendicular cataract; the precipitous leap most pro-bably began at the lower end of the portage, or at the eastern limit of the highest limestone plateau, from whence the river gradually wore away the rock, at the same time diminishing the height of the fall, until it became a foaming rapid from beginning to end.

The upper portion of the Grand Rapid—of which I succeeded in getting a sketch—presents a scene that strikes the beholder with wonder and admiration.   The great body of water that has been stealing along, swiftly but silently, for many miles, appears to be suddenly imbued with life—the rippling of the river becoming gradually more turbulent, until the tumultuous surges grow into huge, rolling billows, crested with foam, like waves in a tempestuous sea.   The great rollers and breakers seem, to the spectator, to be continually changing in shape and appearance, owing to the lines of surf and the peculiar colour of the water; but although the mighty cataract thus appears to be for ever changing, it really rolls on for ever the same.

The ascent of the Grand Rapid is one of the most

laborious duties that has to be performed on a boat voyage from Lake Winnipeg to the Saskatchewan district. The Hon. Hudson's Bay Company's brigades surmount this fearful interruption to the upward navigation of the Saskatchewan in the following way: On arriving at the foot of the rapid, every boat discharges one-half of its cargo of four to five tons. Thus lightened, they are then " tracked" (towed) up to the beginning of the portage— the whole of the crew of six or eight voyageurs, with the exception of the bowsman and steersman who remain in the boat, being engaged in the labour of tracking. Each man is attached to the tracking-line by a leather belt, or portage-strap, passing round his body; and harnessed in this manner they drag the boat along, running and scrambling barefooted over the slippery and jagged rocks at the sides of the cataract. When the lower end of the portage is reached, the boat is emptied, and " run" back again to the foot of the rapid, and from thence hauled up as before, with the remainder of its load. The whole of the lading is then carried over the portage, exclusive of fifteen pieces, or about 1,350 lbs., which is left in the boat. With this ballast, the boat is pulled across to the south side of the rapid, to be tracked up, as the towing-path is better there than on the north side. In consequence of the rapidity and violence with which the upper portion of the rapid flows, in ascending it, it is necessary to employ the " main line,"—a much thicker and stronger rope than is generally used for tracking To this line the crews of one or two boats are lashed, and thus they run along the top of the cliffs of limestone,— there being no footing at the bottom of these walls of rock—hauling the heavy craft up the surging cascades. The utmost strength of the bowsman, with his pole, and the steersman, with his long sweep oar, is required, to prevent the boat from being dashed to pieces among the rocks.

Small brigades, feebly manned, often haul their boats over the portage. The portage road bears evidence of this, as it is deeply scored and furrowed by the keels of boats from beginning to end.

The boats used by the Hudson's Bay Company are built at the principal posts of the various districts. They are heavy, and strongly built, but last only about three years, having to undergo such severe strains and hard usage in the rapids and on the portages on the routes followed. They are of the whale-boat build, about thirty feet long and six broad, sharp fore and aft, with flat floors, which make them very leewardly.

Although the Grand Rapid is the most serious obstacle that the Company's boats have to encounter, it is not the only difficulty they meet with on the Saskatchewan. The whole ascent of the river is one of labour and fatigue. The current is so swift—as the name of the river is well known to imply—that the voyageurs would track nearly all the way to the Rocky Mountains, if the banks of the river would allow ; but where the river passes through marshes and swamps, they have no alternative but to pull against the current, however strong it may happen to be.

Before finally determining upon any works or measures for overcoming the Grand Rapid, in order to render the whole of the Saskatchewan navigable for steam vessels from Lake Winnipeg, without interruption ; it would be necessary to make a more extensive and elaborate survey ; but probably sufficient information and data have been acquired during this reconnaissance from which schemes might be devised, and suggestions offered, for surmounting the difficulty. To navigate the Saskatchewan at present, a steamer would evidently have either to be built above the rapid, hauled over the portage, or "warped" up the rapid itself. Seeing that the Company's large bateaux are hauled up the rapid by manual labour, it does not seem impracticable for an empty steamboat, with engines of great power, to ascend it, by the aid of hawsers and guy-ropes stretched from the steamer to the land, using, along with capstans, the motive power of the steamer as far as available. But in any case, unless a canal were constructed, a transhipment of cargo bound upwards would have to take place, whether there were steamers plying above and below the rapid, or whether steamers were forced up the rapid; so that it would be necessary to construct a good road or tramway on the present line of portage. The features of the country in the vicinity of the Grand Rapid are very favourable for a road, and even for a settlement, as the banks of the river are high, with a considerable depth of good soil, from the second rapid east of Cross Lake to near Lake Winnipeg. There is also abundance of timber for fuel and building.

From the foot of the Grand Rapid, the Saskatchewan flows with a pretty strong current, in a northerly direction till it enters Lake Winnipeg. Its mouth has a width of about twenty-eight chains, and is a little over two miles below the lower end of the rapid. On the coast of Lake Winnipeg, immediately east of the mouth of the Saskatchewan, there are several deep and narrow bays, or estuaries, marshy at their inner extremities, and separated by narrow points or spits of gravel, by which it seems not improbable the Saskatchewan entered the lake at some period of its existence, and that north-easterly gales and shoves of ice have driven up these barriers, and caused the river to excavate new outlets.

We visited an Indian encampment on the north bank of the river, a little below the foot of the rapid, in the

expectation of procuring some sturgeon, but were unsuccessful—the fishery carried on here by the Indians
having failed this year. This encampment of two
lodges was the only one we saw on the main Saskatchewan. It had·been a larger camp, but eight families had
just left it, previous to our arrival, for their winter
quarters at the Little Saskatchewan. They are Swampy
Indians, and generally winter at Fairford, whence they
proceed in summer to the Grand Rapid; where, by
assisting in dragging the boats and *portaging*, they get a
small recompence in the shape of tea, tobacco, or
pemmican. They occupy the time between the arrivals of
the different brigades of boats, in catching and drying
fish, and generally leave after the last fleet has passed up
in the autumn.

We entered Lake Winnipeg at sunset, and camped not
far from the mouth of the Saskatchewan, upon a narrow
spit of gravel, separated from the wooded shores by a
marsh. The night was clear and beautiful, and the lake
wonderfully calm. From our bivouac, where we lay
with cramped limbs outstretched on the shingle-beach,
could be seen the great headland "Kitchi-nashi," vanishing away to the south-east in the far distant horizon. A
view very extensive and beautiful, but which betokened
many hours of paddling and tracking out of the direct
course to Red River. To the east and north the only
limit to our gaze was the dim horizon of the great lake
which lay tranquilly outspread before us like an unruffled
sea.

CHAP. XXII.

FROM THE GRAND RAPID OF THE SASKATCHEWAN TO THE
RED RIVER SETTLEMENTS *viâ* THE WEST COAST OF LAKE
WINNIPEG.

Enter Lake Winnipeg.—Cape Kitchi-nashi.— Storms.— Detained on an
Island.—Windbound on the Mainland.—Tempest.—Repulsed by the Wind.
—Character of the Coast: the Sand Beaches and Swamps.—War Path
River.— Verifying Rate of Canoe. — Indians. — Tracking. — Limestone
Point.—Encountering a head Wind and Storm.—Lightening Canoe.—
Starving Indians.—The Little Saskatchewan.—The Prominent Features
of the Coast.—Formation of Cape Kitchi-nashi.—Limestone Exposures.
—Tributary Streams.—General Character of the Country.—Indians and
Fishery at Little Saskatchewan.—Indian Chart.—Inaccuracy of the Maps
of the Lake.—Depart from the Little Saskatchewan.—Windbound again
for three Days.—Provisions exhausted.—Contrary Winds.—Driven back
and stopped.—The Cat Head.—Windbound again by a Hurricane.—Barrier
of Boulders. — Eagle. — Stopped by foul Winds again at the Wicked
Point.—Pike Head and River.—Indian Fishing-Weir.—Opportune Supply
of Fish.—Wide Traverse to Grindstone Point.—Grassy Narrows.—Sandy
Bar.—Arrive at the Settlements.—Conclusion.

MR. FLEMING'S NARRATIVE — (*continued*).

IT was on the evening of Sunday the 22nd August,
1858, that we glided from the mouth of the Saskatchewan
into Lake Winnipeg, but our voyage through this great
inland sea was not fairly begun until the following morning, when we embarked at an early hour (4.20 A.M.) in
our little canoe. Being favoured with a light breeze for
a few hours we reached the neck of the great promontory,
Cape Kitchi-nashi, about noon. From the mouth of the
Saskatchewan to this point the coast trends to the south-
east, and is indented in a remarkable manner by a series
of deep bays of every shape and size. As it would re-

quire unlimited time and resources to penetrate into every
sinuosity of the coast, we generally steered straight from
point to point, although in doing so some long traverses
had to be made.

The northern coast-line of the promontory being nearly
straight with fine sand-beaches, affording tolerably good
footing, we tracked along the shore for the remainder of
the day; although this was hard enough work, the men
were glad to avail themselves of it, as a change or relief
from paddling.   By working fifteen hours to-day we were
enabled to camp at the extreme point of the headland,
where, the night being favourable, the magnetic variation
of 15° E. was observed.   The Ojibways call this cape
" Kitchi-nashi," and the Swampys " Missineo," both names
signifying " Big Point."   By some it is called " The
Détour."

August 24*th*.—A fine morning, the lake quite calm.
After doubling the cape we overtook eight small canoes
containing the band of Indians (Swampy Crees) who left
the Grand Rapid on Sunday, 22nd.   In a short time
a light breeze sprang up, and by hoisting a blanket we
sailed at a pretty good rate for some hours.   About
2 p. m. the wind began to increase in strength and
turned suddenly against us, so that we had to run in
behind a low point of sand and gravel for shelter.
Although the wind still continued high we started again
and made a traverse to a small sand island on which we
were obliged to remain, being then over two miles from
the main land, and the storm having increased in violence.
A storm of wind soon raises a very heavy sea on Lake
Winnipeg on account of its little depth of water.

The island on which we were detained is one of the
Gull-egg group, which, with a point of sand protruding
from the main land, forms a pretty good harbour on the

south side of the neck of the great promontory. The Indians were nearly destitute of provisions, and followed us to the island, where they fortunately got a plentiful supply of eggs and young gulls ; but having little ammunition, they brought down only a few old ones, although they hovered in countless numbers over the island, screaming at the wholesale destruction of their young brood.

August 25th.—The storm raged all night, and this morning we found ourselves surrounded by a foaming sea on a low island of sand about 100 yards in length, and so narrow that the spray from the breakers dashed completely over it. The gale blew hard from the east till about noon, when it began to subside ; I then determined upon starting on our course, but, seeing a thunder-storm approaching, decided upon taking dinner before making the attempt. It was well that we did so, because just as we were hastily swallowing our meal of pemmican, the thunder-storm, accompanied by strong wind and heavy rain, burst upon us with great violence. Some of the Indians were endeavouring to reach the next island in the line of traverse, but had to abandon the attempt and drive before the gale to the main-land, three miles off.

The storm soon abating again, we crossed to the next island and thence to the main-shore ; and after coasting along for some miles, encamped on a sandy point, where we found a small clump of poplar and spruce.

August 26th. —Last night the Northern Lights or Aurora Borealis were unusually brilliant, darting and playing about with extraordinary rapidity in all directions, sometimes extending to the zenith and sometimes to the south of it. The voyageurs said they portended a coming storm, and their prognostications proved correct. The night was clear with a bright moon till about midnight, when a cold north-westerly wind arose, followed in a very short time by a stormy sea. The gale soon veered round to the north increasing to a perfect hurricane, and during the day the lake was white in all directions with breakers and foam. A heavy surf breaking along the coast and tearing away large portions of the bank on which we were camped, warned us to move our canoe and lading back from the shore ; yet, notwithstanding every precaution, some of our paddles and poles were swept away during the night. A large marsh being in our rear we could retire but a few yards from the raging lake to wait for the abatement of the storm.

August 27*th.*—After midnight the wind began to decrease gradually, and by daybreak it had so far subsided as to permit us to continue our voyage. By breakfasting at a point where we found an outcrop of limestone I was enabled to procure some fossils. This, the first rock exposure observed since leaving the Saskatchewan, is apparently the termination of a ridge running at right angles to the coast-line, and bounded on either side by marsh and swamp. The top of the rock is ten feet above the surface of the lake, and is covered by a stratum of boulders and drift two feet in thickness, supporting small poplar, tamarack, spruce, birch, and Banksian pine ; there are only six feet of the limestone exhibited, the remaining four feet being concealed by a talus of boulders and débris. The high water-mark of the lake reaches to the top of the talus.

A contrary wind arising about noon detained us four hours at the mouth of a creek, which we ascended a short distance. The entrance, or where the creek cuts through the sand-beach enclosing a marsh, is one chain wide ; within the sand-beach the creek expands into a deep pond thirty chains in diameter, surrounded by a marsh ; this pond is fed by the inner portion of the creek, a broad and sluggish stream five feet deep, meandering through a tamarack swamp. It is reported by the Indians to have its source a long distance inland. As there is but one and a half to two feet of water over the bar this could only be used as a harbour for bóats. It is about half-way between the Gull Islands and War-Path River.

We set off again after the wind had moderated a little, but were compelled to camp in an hour and a half in the lee of a point, on the weather side of which an adverse wind was blowing hard, driving before it a heavy sea. Being thus repulsed again by the wind, I directed my attention to the character of the coast in the vicinity. of our bivouac. Along the shore there extends a long straight sand-beach, sixty feet wide and arched like a roadway; on the inner side of this beach there is a tamarack and black spruce swamp, with a bottom of black muck and moss two feet in thickness, covered with water. This "muskeg," is said to continue for a great distance back. By leveling I found the surface of the water in the swamp to be only eight inches higher than the lake; and as the crown of the sand beach is only four and a half feet above the level of the water, and is covered with driftwood, it is evident that the lake washes into the marsh during high water.

Leaving camp at 4.30 A.M., August 28th, we reached the mouth of War-Path River at 1 P.M. The Indians say this river rises in lakes, and draining a great extent of swampy country, is very large in spring. There are three feet of water over the bar at its mouth; the channel at the entrance is contracted in summer by the sand to a width of forty feet, with an average depth of four feet; within the entrance there is a basin thirty chains broad, forming a boat harbour of easy access.

After tracking for several hours along straight sand-beaches, which separate marshes from the lake, we camped nearly opposite Caribou Island, on a coast similar to that which we left in the morning. The Indians came up with us, and erected their lodges in our neighbourhood. They were now quite friendly, but at first their chief and principal men looked upon us with suspicion, especially while noting or making any observations. They seemed to be under the impression that there was something going on which would ultimately deprive them of their country. The failure of their summer fishery at the Grand Rapid had rendered them badly off for provisions, and the only food they could get on this inhospitable coast was an occasional gull. Whoever had the good fortune to obtain one of these sea-birds was soon surrounded by a group of Indian children, watching with anxious eyes for the intestines of the animal, which they threw upon a fire for a few minutes, and then ate apparently with great relish.

August 29*th*.—Embarking this morning at daylight, we reached Limestone Point about eleven o'clock, after making a traverse of three miles against a strong head wind. On this point there is a very fine exposure of light-coloured limestone, containing numerous fossils, some of which I succeeded in procuring. The outcrop on the point is fourteen feet in thickness above the lake, in massive horizontal layers, overlaid by two and a half feet of drift and fragments of-limestone that have evidently been broken up by ice. This headland is the abrupt termination of a narrow ridge of limestone clothed with aspen, spruce, and birch; it is about two miles long, running nearly north and south. On the west side of it is Pòrtage Bay, so called by the Indians, as they sometimes make a portage from the foot of it, across the neck of the point.

After remaining here about two hours we proceeded on our journey. On rounding the point we found the wind on the east side of it blowing directly in our teeth, and it required the utmost exertions for two hours to force the canoe against a high gale and stormy sea, until we got into the lee of a small island, it being impossible to land on the main shore. The canoe leaked and shipped so much water during this traverse, that in order to lighten her we were compelled to throw overboard some of the heavier of our geological specimens. It was with great regret I saw one of them, a very large and fine *orthoceratite*, consigned to the deep.

On the island we found part of the Indian band, but the greater portion were hurrying on to the Little Saskatchewan to get fish, as they had nothing to eat. We saw them in the distance, battling against the wind and sea, their little canoes like specks tossing among the swells and breakers. The occupants of some of the canoes were only women and children; one canoe was paddled by two Indian women and two little boys eight or nine years of age; they, especially, would have hard work in reaching the mouth of the river. The Indians remaining on the island were chiefly the more feeble of the party, and being ravenously hungry they were all in the marshes busily engaged in pulling up and eating the roots of bulrushes. The storm increased towards evening, and we were obliged to camp on the island ourselves.

August 30*th.*—Although the unfavourable wind had diminished but little this morning, we plied our paddles so well, and made such good headway against it, that we entered the mouth of the Little Saskatchewan or Dauphin River about 11 A.M. We tracked up the river to the Indian encampment, about four miles from its mouth, for the purpose of procuring fish, and found the Indians at the rapids scooping large numbers of excellent white-fish from the eddies.

It would perhaps be as well to give here a short re-capitulation of the character and general topography of the west coast of the lake between the main Saskatchewan and the Little Saskatchewan.

The distance from the mouth of the Main to the mouth of the Little Saskatchewan, by our track along the coast, or by the course that canoes or row-boats would be likely to pursue is about 140 miles; but the distance by the coast-line embracing every sinuosity of the shore is much greater.

The most prominent feature in the line of coast is the great headland, Cape Kitchi-nashi. This immense promontory begins to stretch out into the lake in a direction a few degrees north of east, about fifteen miles south of the Saskatchewan. Its extreme point is about twenty-four miles in an air line from the general line of the coast, and its width varies from three to six miles and upwards; its neck is indented by several deep bays, some of which might be available as harbours or roadsteads. The formation of the cape is peculiar; it is very low and flat on the north side, while on its southern boundary the coast is comparatively high and abrupt. Its northern side consists of a series of marshes separated from the lake by a narrow sand-beach, and these marshes gradually blend into a tamarack and spruce swamp. Along the south side of the cape there is a continuous escarpment of light-coloured clay, twenty-five to forty feet high, yet even on the top of these high banks the character of the land is of the poorest description, being nothing but a " muskeg " or trembling swamp, containing a thin growth of very scrubby tamarack and spruce, covered with drooping moss.

The extremity or apex of the promontory is a very low and broad sand-beach covered with water-worn boulders; the lake is also dotted with boulders a long way out from the shore, there being a sand-bar or continuation of the point under water, on which they rest. From the size and position of the cape, and the dangerous shoals extending out from it, if beacons or lighthouses are ever required on the lake for the safety and convenience of shipping, no more suitable place could be selected for the erection of one than here.

The coast north-west of the cape, as already stated, is very low, and much broken by deep and narrow bays.

From Cape Kitchi-nashi to the Little Saskatchewan the coast trends generally to the south-east. Between these points limestone is exposed in six places. The exposures are the precipitous extremities of ridges, forming points at intervals along the coast. The stratification in every instance is horizontal, but the escarpments vary in height above the lake; they increase in altitude from four to fourteen feet towards the south. These ridges are generally wooded with aspen and other deciduous trees; and the swamps intervening are timbered with tamarack and spruce; some of the spruce near the coast are pretty large. Between the ridges, low sand-beaches extend along the coast. These beaches separate ponds and open marshes, averaging from a quarter to one mile wide, from the lake; in the rear of the marshes is the great tamarack and spruce swamp, or " muskeg."

The tributary streams in this part of the coast are not numerous, and they are generally of no great size. The chief are the Gull-Egg Rivers or the Two Rivers, the War-Path River, Jumping River, and one or two others without name; they are not in themselves large, but their estuaries might be available as harbours for boats.

The character of the country exhibited on the coast extends almost an unlimited distance back; indeed the Indians report the whole of the country between Lake Winnipeg and Lake Winnipego-sis as one vast "*muskeg*" —the great moose hunting-grounds of the Swampys.

Although the country here described is quite unfit for agricultural purposes, it is not altogether valueless; there are large areas of good timber along the coast, available for fuel, and the limestone cropping out at the various points is well adapted for building.

A person ascending or descending the Little Saskatchewan would be greatly deceived in relation to the character of the country through which this rapid river flows, if he were to judge by the appearance of its banks without penetrating beyond. I made a section of the river and its excavated bed at the Indian encampment, about three and a half miles from its mouth. At this place the river was 360 feet broad, four and a half feet deep, and ran over a stony bottom. Its banks were from twenty-five to thirty-five feet high, composed of light-coloured clay, holding boulders and pebbles of limestone. Along the margin of these banks there was a zone of tall poplars, extending about 300 feet back from the face of the bank. This narrow strip of good timber and land soon blended into a great muskeg or tamarack and spruce swamp. This swamp consisted of dead and dwarfish tamaracks covered with pendant moss, and struggling through several feet of wet and trembling moss and black mould, through which a pole could be shoved for several feet.

The Indians were catching very fine white-fish in large quantities at this point with little trouble. The river seemed to be crowded with them. At various places along the brink of the stream, enclosures of stones were constructed, beside which an Indian stood with a large scoop-net attached to a pole, filling the stone enclosure from time to time as he scooped them from the eddies in his vicinity. The excellence and abundance of these white-fish, and the proximity of this locality to the salt region, together with the facilities for water communication, may lead to the establishment of a fishery here on an extensive scale. The Indians were curing the fish without salt, by splitting them very thin and drying them in the sun.

We were invited into one of the tents to partake of some fish which they had cooked for us. A large birch-bark dish of fish was set before us, the whole of which we had to eat, according to Indian custom, not to give offence to our entertainers. As an extra mark of favour, an old fellow in a very dirty blanket strewed a few grains of black-looking salt over my portion. This salt he carried beneath his belt in a piece of birch-bark.

Being without a guide I got one of the Little Saskatchewan Indians to draw me a map of the lake between Bushkega Islands and Grassy Narrows, showing the traverses and route to be taken between the islands in order to cross the great arms of the lake, Fisher Bay and Washow Bay. This Indian chart was of great service to us; the best and most recent maps of the lake to which I had access being so incorrect: on them the general contour of the coast north of the Little Saskatchewan is tolerably well delineated, but to the coast north and south of the Dog's Head Straits they bear very little resemblance; the large islands are omitted altogether, and the Great Black Island is represented as forming the extremity of a promontory on the mainland between two bays.

From the beginning, our canoe was very weak, the bark being of the poorest description and badly put together, and having now become quite frail I tried to barter with one of the Indians for a new and stronger one; but, taking advantage of our situation, he placed upon it a much greater value than I felt inclined to give.

Having made sections of the river, and examined the country bordering the Little Saskatchewan, we left it on the 31st of August, but were detained the greater part of the day on a point only a few miles from the mouth of the river, by unfavourable wind and in consequence of the sickness of Louis, our steersman, who, being a pretty old man, was disabled from over exertion in the storm on Sunday.

On the 1st of September, while sailing with a side wind across the mouth of a deep bay, in which there was rather a heavy sea rolling, a large swell broke over us, throwing in a great deal of water, and from the evening of the 1st of September until the morning of the 5th we were windbound on a low marshy point on the north-east side of the great bay into which the Little Saskatchewan empties.   The spot on which we were imprisoned is very much circumscribed, being a narrow sand-beach, about a chain in length, and bounded on three sides by an extensive marsh.   During the three days that the storm lasted the wind blew a hurricane from the N.N.W., raising a tremendous sea on the lake, and the surf beating along the shore, washed away several yards of the sand-beach on which we were encamped.   The weather was clear the first day, but on the second and third days it rained almost incessantly, and it was then for the first time on our voyage that we really felt the want of a sufficiency of food, as our stock of provisions was reduced to a few pounds of rather mouldy pemmican, which I determined to eke out as long as possible, being still a great distance from Red River (upwards of 170 miles by the canoe route), and with that object in view we made it a rule to eat only one meal a day while we were windbound, unless we were fortunate enough to procure some additional food, in the shape of wild fowl or other animals. We succeeded in getting a grey gull on the second day, on which we made an excellent repast.

On the morning of the 5th, just before we started, an Indian and family from the Dog's Head came to us, they had been windbound seven days on an island not far from where we were, they said they never saw such a continu ous succession of winds and storms on the lake before, and informed us that a freeman's boat which passed during the night had been thirty days between Red River and the Saskatchewan, a distance that has been accomplished by a boat, with a favourable wind, in three days. After bartering with this Indian for a small "rogan" of fish pemmican (dried fish pounded and mixed with sturgeon oil), we proceeded on our journey, glad to get away from the dreary spot. Although there was still a heavy retarding ground-swell on the lake, we paddled many miles before halting. On stopping to cook breakfast we were greatly disappointed to find that the fish pemmican which we were so thankful to get, was nearly all rotten, there being only a small portion on the top that could be eaten, the remainder had to be thrown away.

A contrary wind freshened up again about noon, but we continued struggling against it, until in attempting to round a point we were completely driven back, and narrowly escaped foundering among the huge swells and breakers that dashed high over the boulders extending out from the beach; we saved the canoe by jumping into the surf and throwing the lading rapidly ashore. As soon as we got everything out of the reach of the waves that were dashing their spray over the dripping shingle beach into the swamp behind, I sent the Ojibway off into the marshes to try to procure us some food. Not

making his appearance at night-fall I despatched the
Blackfoot in search of him; they both returned very late,
having wandered many miles along the coast, but brought
nothing with them. The Blackfoot half-breed (who had
received missionary instruction at Red River) attributed
the Ojibway's want of success to the fact of his hunting
on Sunday.

Embarking at daylight on the 6th we reached the Cat
Head at 2 P.M., after a hard paddle against an adverse
wind and rough sea. This bold headland consists of a
perpendicular escarpment of buff-coloured limestone in

The Cat Head, Lake Winnipeg.

massive horizontal layers, the top strata overhanging the
base; the summit of the rock is thirty to thirty-five feet
above the lake, and is covered with drift and boulders to
the depth of three feet, on which grow scrubby poplar,
spruce, and tamarack. The water is quite deep up to the
foot of the cliff, and as no landing can therefore be
effected, I was unable to make a minute examination of
the rock. There is a series of low, arched caverns in the
base of the cliff, in which the waves and swells washing
to and fro make a singular hollow noise, and for this
reason the Indians think it is the abode of a *manitou*.

Some of the Swampys say Cat Head is so named because an Indian hunter was killed there by falling over the precipice while chasing a wild-cat or lynx. The profile of the upper, or overhanging portion of the cliff, bears a singular resemblance to the "cat-head" of a ship.

The wind becoming more foul we were compelled to camp on a point about a mile and half south-east of the Cat Head, at the extremity of the north-western side of Kinwow (Long) Bay.

During the next day (7th September) the wind blew hard from the east, and the waves on the lake rolled mountains high, so that we could not venture out, having a long traverse before us. The narrow point or peninsula upon which we were detained, is of a peculiar character, consisting of a straight barrier or ridge of boulders about three-quarters of a mile long, running at right angles to the coast, and connecting it with a small area or island of limestone a few feet high; this barrier resembles very much a railway embankment, or a rip-rap breakwater; although it is twenty to twenty-five feet high, the waves wash over it during the great storms on the lake in the fall of the year. A spruce tree growing on this peninsula has been trimmed into a "lopstick," by Angus Macbeth, from which the locality has derived the name of Macbeth's point. There are also two high cairns of stones on the point, but whether they were erected by the Indians, or by half-breeds I did not ascertain.

The morning of the 8th dawned, but there still seemed to be little chance of our getting off, and our prospects now began to look cheerless enough; we had but a handful of pemmican and one charge of ammunition left; while deliberating whether to eat the last remnant of our food, a bald-headed eagle came wheeling in great circles over us; he poised himself for an instant as if about to descend upon his prey, when he was fortunately brought down with our last charge of shot. He proved to be a large bird with magnificent plumage; a Cree or Blackfoot would have given a good horse for his wings or tail. By eating nearly every portion of the animal, except his feathers, we managed to make him serve for two or three meals.

The wind moderated sufficiently at last to permit us to resume our journey, but we had a fatiguing paddle for two hours in crossing Kinwow Bay. The extremity of this long arm of the lake was below our horizon, and the wind came sweeping out of it in great squalls. The wind veered round to the east and stopped us again about noon at the Wicked Point, where we spent the afternoon in drying our clothes and blankets, and gathering sand cherries.

September 10*th.*—The wind fell and allowed us to reach Pike Head yesterday morning. We at once ascended the Pike or Jack-Fish River to the " basket " or weir erected across it by the Indians, about half a mile from its mouth, for the purpose of procuring fish. The basket was much broken, and when we arrived, was covered with turkey-buzzards waiting to pounce on any fish that might get entangled in its meshes. By repairing the basket and watching it all night we caught an abundance of fish of four species, viz. gold-eyes, wall-eyed pike, suckers, and pike. It rained without intermission during the day, and as the wind continued unfavourable we remained at ·the basket, gutting fish to take with us.

We generally boiled our fish, making use of the liquor in which they were cooked as a substitute for tea; and having succeeded in capturing a small badger, by pouring water into his burrow, we got sufficient fat or oil to enable us to have fried fish occasionally.

The Indians catch fish by means of these weirs or traps in large numbers, and this mode of capture is practised by most of the North American tribes; the traps varying in construction according to the locality or the ingenuity of those by whom they are erected. The one by which we were fortunately enabled to procure a supply of fish at the Pike River consisted of a fence of poles, stretching from one side of the river to the other; they were sloping in the direction of the current, like the inside of a mill-dam, and allowed the water to pass through but not the fish. Near the river bank, on one side, there was an opening in the weir about a yard in width to allow the fish descending the river to pass into a rectangular box, with a grated bottom sloping upwards, through which the water flowed and left the fish dry. The fish very seldom entered this pound in daylight, but during the night they poured in in great numbers. In order to secure all that come into the trap when it is prepared for catching fish, an Indian sits beside it all night with a wooden mallet in his hand, with which he strikes the larger fish on the head to prevent them jumping out. He is kept busily employed pitching them out on the bank, and in the morning there is a large heap for the women to clean and cut up. The fish came into the trap almost as fast as we could pitch them out, and we caught in a short time 111 gold eyes, 44 wall-eyed pike (called perch by the half-breeds), 16 sucking carp (or suckers), and 11 pike, making 182 altogether.

The average width of the Pike River is about a chain, and its depth about five feet, with a moderate current; its banks half a mile from its mouth are of light-coloured clay five to ten feet high, and covered with a rich dark mould, supporting a thick growth of aspen, spruce, tamarack, birch, and balsam. Near the basket there is an old log-house, formerly a missionary station, but now abandoned. When the Indians come to fish here, they cut up the flooring and timber of this house for fuel, instead of availing themselves of its shelter.

September 11*th.*—Having stowed away as many fish as we could find room for in the canoe, we left the Jack-Fish River in the morning, and being favoured with a fair wind, sailed without stopping till dark, when we camped on a small island in the entrance to Fisher Bay. On Sunday the 12th we had to encounter a brisk contrary wind from the south, but, by working sixteen hours against it, and making some wide traverses between the islands, we succeeded in reaching the point opposite Dog's Head, at the beginning of the narrows, before night set in.

By starting at daylight and sailing along the east coast of the lake on the 13th we got in sight of the Grindstone Point about 2 P.M., when we set out on a longer and more dangerous traverse than any we had yet accomplished. We had to cross from the east coast of the lake to the Grindstone Point on the west coast, a distance of about twelve miles. From the shape of the lake with its many deep and broad bays this great traverse is unavoidable. When we started from the east side of the lake, the high escarpment of rock forming the point seemed quite low and blue in the distance. By spreading a blanket we were assisted for a while by a side wind ; but the wind soon changed and freshened, so that we had to lower sail and ply our paddles with all our strength until reaching the point, nearly four hours from the time we left the east shore. Taking advantage of a little moonlight, which enabled us to coast along a straight shore after dark, we did not stop to camp till arriving at the Little Grindstone Point. The east coast of the lake from the Dog's Head to where we left it to cross to Grindstone Point, consists of a succession of knolls or low domes of granite and gneiss rising generally eight to ten feet above the water, and clothed on their flanks with a scrubby growth of timber, chiefly Banksian pine, spruce, and a few aspen; there are, generally, ponds and swamps between the granite knolls, and the coast line is much broken by deep inlets and small well-sheltered bays, forming excellent harbours and coves for boats. The east coast, north and south of the straits, is described as being similar to this; abounding in harbours, and for this reason it is the route by which boats invariably go to York Factory, and generally to the Saskatchewan. Opposite the mouth of Great Washow (Deep) Bay there is an inlet or passage called Loon's Straits, formerly a canoe route of the old North-West Company. There is always more or less current through the narrows of the lake at the Dog's Head. This current is sometimes flowing north and sometimes south, the direction depending upon the prevailing direction of the wind.

By making an early start on the 14th, and creeping along in the shelter of the land, we were enabled to dine at Grassy Narrows. Although our fish had not improved any since leaving Pike River, we always possessed keen appetites, and were now by no means fastidious. Sailing from Grassy Narrows across a bay into which White-Mud River empties, we arrived at the Sandy Bar a little after dark and camped.

September 15*th.*—The wind and weather being favourable to-day, by working fifteen and a half hours, we reached the marsh near the mouth of Red River about dark. We found an Indian encamped on the sand-beach hunting the ducks which are in countless numbers in these marshes at this season. He had killed 100 " stock " ducks during the day, and generously gave us a liberal supply.

We reached the Stone Fort about dark on the 16th September, where I succeeded in procuring a horse, and left the men to track the canoe up the river. A ride of twelve miles brought me to the middle of Selkirk Settlement, and by 11 P.M. I was in our old quarters, after a canoe voyage of forty-eight days in all; nine of which were occupied in descending from the Elbow of the South Branch of the Saskatchewan to Fort à la Corne, fourteen from thence to the mouth of the Saskatchewan, and twenty-five days in traversing Lake Winnipeg.

The whole distance traveled and explored in canoe is over 940 miles; 600 of which being down the Saskatchewan, and 340 miles open lake navigation. In performing this latter part of the journey with a little frail canoe, heavily laden, we were completely windbound for twelve days, and had to contend nearly all the time we were moving with boisterous head winds, foul weather, and a hand to mouth sustenance. This will, in some measure, account for the slow rate of progress we unwillingly made through Lake Winnipeg. I must take this opportunity of bearing testimony to the unwearied labour, patient endurance, and unflinching devotion of my two voyageurs. Their conduct while they were my companions, for nearly two months was beyond all praise; and they sustained hardships and risks of no ordinary description without a murmur.

Very truly yours,
JOHN FLEMING.

H. Y. Hind, Esq. &c. &c. &c.

END OF THE FIRST VOLUME.

LONDON
PRINTED BY SPOTTISWOODE AND CO.
NEW-STREET SQUARE.